PERGAMON INTERNATIONAL LIBRARY
of Science, Technology, Engineering and Social Studies
The 1000-volume original paperback library in aid of education,
industrial training and the enjoyment of leisure
Publisher: Robert Maxwell, M.C.

Time in Animal Behaviour

THE PERGAMON TEXTBOOK
INSPECTION COPY SERVICE

An inspection copy of any book published in the Pergamon International Library
will gladly be sent to academic staff without obligation for their consideration for
course adoption or recommendation. Copies may be retained for a period of 60
days from receipt and returned if not suitable. When a particular title is adopted or
recommended for adoption for class use and the recommendation results in a sale
of 12 or more copies, the inspection copy may be retained with our compliments.
The Publishers will be pleased to receive suggestions for revised editions and new
titles to be published in this important International Library

Other titles of interest

CORSON, S. A.
Ethology and Nonverbal Communication in Mental Health

HANIN, I., and USDIN, E.
Animal Models in Psychiatry and Neurology

KEEHN, J. D.
Origins of Madness
The Psychopathology of Animal Life

REINBERG, A., and HALBERG, F.
Chronopharmacology

SAUNDERS, D. S.
Insect Clocks

UPPER, D., and CAUTELA, J. R.
Covert Conditioning

Time in Animal Behaviour

BY

MARC RICHELLE

AND

HELGA LEJEUNE

with contributions by

Daniel Defays, Pamela Greenwood, Françoise Macar,
and Huguette Mantanus

PERGAMON PRESS

OXFORD · NEW YORK · TORONTO · SYDNEY · PARIS · FRANKFURT

U.K.	Pergamon Press Ltd., Headington Hill Hall, Oxford OX3 0BW, England
U.S.A.	Pergamon Press Inc., Maxwell House, Fairview Park, Elmsford, New York 10523, U.S.A.
CANADA	Pergamon of Canada, Suite 104, 150 Consumers Road, Willowdale, Ontario M2 J1P9, Canada
AUSTRALIA	Pergamon Press (Aust.) Pty. Ltd., P.O. Box 544, Potts Point, N.S.W. 2011, Australia
FRANCE	Pergamon Press SARL, 24 rue des Ecoles, 75240 Paris, Cedex 05, France
FEDERAL REPUBLIC OF GERMANY	Pergamon Press GmbH, 6242 Kronberg/Taunus, Hammerweg 6, Federal Republic of Germany

First edition 1980

British Library Cataloguing in Publication Data

Richelle, Marc
Time in animal behaviour.—(Pergamon international library).
1. Animals, habits and behaviour of
2. Biological rhythms
3. Time perception in animals
I. Title II. Lejeune, Helga
591.5   QL753   79-40953

ISBN 0-08-023754-1 (Hardcover)
ISBN 0-08-025489-6 (Flexicover)

Printed in Great Britain by A. Wheaton & Co. Ltd, Exeter

TO JEAN PAULUS, WHOSE TEACHING WAS SEMINAL TO ALL
CONTRIBUTORS OF THIS BOOK, AND WHO CONVEYED TO US
PIERRE JANET'S INTUITIONS ABOUT TIME

Contents

List of Contributors   xi

Foreword   xiii

Part I: Scope and Methods

I.   INTRODUCTION (M. R. and H. L.)   3

   1.1 Behavioural study of time and chronobiology   3
   1.2 Animal clocks and the psychology of time   6
   1.3 Historical survey of research on temporal regulation and estimation in animals   7

II.   METHODS (M. R., H. L., H. M. and D. D.)   9

   2.1 Instrumental procedures   9
   2.2 Pavlovian conditioning procedures   10
     2.2.1 Delayed reflexes   11
     2.2.2 Trace conditioning   11
     2.2.3 Temporal conditioning   11
     2.2.4 Discrimination   12
     2.2.5 Temporal conditioning and disinhibition   12
   2.3 Operant conditioning procedures   12
     2.3.1 Temporal regulation of the subject's own behaviour   14
       A. Spontaneous temporal regulation   14
         a. Fixed-Interval schedule (FI)   14
         b. Modified FI schedules and other contingencies inducing
           spontaneous temporal regulations   19
       B. Required temporal regulation   22
         a. Differential Reinforcement of Low Rates (DRL)   22
         b. Modified DRL schedules and other procedures requiring a
           temporal regulation   27
       C. Negative contingencies   30
         a. Non-discriminated Sidman-type avoidance   31
         b. Fixed-cycle avoidance   32
     2.3.2 Temporal discrimination   33
       A. Discrimination of duration of external events   33
         a. Stimuli used   36
         b. Methods of presentation of the temporal stimuli   36
         c. Types of responding   37
         d. Data analysis in duration discrimination   38
       B. Discrimination of complex temporal properties of external events   43
         a. Rhythm   44
         b. Movement velocity   44

Part II: Basic Results

III. TEMPORAL PARAMETERS AND BEHAVIOUR
    (H. L., M. R., H. M. and D. D.)   49

3.1 Temporal regulation of the subject's own behaviour   49
    3.1.1 Properties of spontaneous regulation   49
        A. Absolute delays and relative timing behaviour   49
        B. The determinants of spontaneous regulation   52
        C. Suprasegmental properties of FI behaviour   56
    3.1.2 Properties of required regulation   57
        A. Rate of responding, frequency of reinforcement, and temporal
            adjustment   58
        B. Absolute or relative time   64
        C. Suprasegmental traits in required regulation   70
3.2 Control of behaviour by temporal properties of external events   71
    3.2.1 Generalization   71
    3.2.2 Thresholds   73
    3.2.3 Scaling   74
3.3 Summary and discussion   78
    3.3.1 One clock or several   78
    3.3.2 The relativeness of behavioural time: Weber's law   82

IV. COMPARATIVE STUDIES (M. R. and H. L.)   85

4.1 Studies of various species   85
    4.1.1 Mammals   87
    4.1.2 Birds   97
    4.1.3 Fish   98
    4.1.4 Insects   99
4.2 Problems of comparison   99
4.3 Comparative studies   101
4.4 Temporal regulation and evolution: some hypotheses   105

V. FACTORS INFLUENCING TEMPORAL REGULATIONS
    (H. L., M. R. and H. M.)   108

5.1 Contingencies of reinforcement   108
5.2 The reinforcer   111
    5.2.1 The nature of the reinforcer   111
    5.2.2 The magnitude of the reinforcer   112
5.3 The response   120
5.4 Developmental and historical factors   124
    5.4.1 Developmental factors   124
    5.4.2 Historical factors   125
5.5 Interindividual differences   130
5.6 Chronobiological factors   135

Part III: Mechanisms

VI. PHYSIOLOGICAL MECHANISMS
    (F. M., P. G., H. L. and M. R.)   143

6.1 The search in the central nervous system   143
    6.1.1 Lesional studies   143
        A. Role of the septal and hippocampal structures   143
        B. Role of other cerebral structures   151

6.1.2 Cerebral stimulation   153
6.1.3 Electrophysiological correlates   155
6.1.4 Single unit studies   160
6.2 Peripheral mechanisms   161
6.2.1 Visceral pacemakers   161
6.2.2 Proprioceptive feedback and timing   165
    A. Anticipatory timing   169
    B. Related animal studies   171
    C. Proprioceptive feedback and collateral behaviour   174
    D. Feedback and voluntary movement   175
6.3 Drugs and timing   176
6.4 Miscellaneous: body temperature   181

VII.   COLLATERAL BEHAVIOUR (P. G. and M. R.)   188

7.1 Experimental data   188
7.2 Origin and function of collateral behaviour   193

VIII.   TEMPORAL INFORMATION: TEMPORAL REGULA-
TION AND EXTERNAL CUES (H. L. and M. R.)   200

8.1 External clocks   200
8.1.1 Types of external clocks   200
8.1.2 Control exerted by external clocks   201
8.1.3 Limits of the external clock control   203
8.1.4 Discriminative or reinforcing function of external clocks   206
8.2 The structuring power of contingencies   207
8.2.1 Definitions and hypotheses   207
8.2.2 Control by the periodic structure of contingencies   210
8.2.3 The periodic reinforcing event: a further analysis   212
8.3 Reinforcement and response related events   214
8.3.1 Stimuli associated with the reinforcement   214
8.3.2 Feedback from the operant response   217

IX.   TEMPORAL REGULATION AND INHIBITION
(M. R. and H. L.)   222

9.1 Disinhibition   224
9.2 Generalized inhibition   226
9.3 Aversiveness of schedules involving temporal regulations   229
9.4 Behavioural contrast   230
9.5 Omission effect   231
9.6 The inhibition-discharge balance   232

References   237

Index of names   263

Subject index

List of Contributors

Marc Richelle (M.R.), Professor of Experimental Psychology, Director Psychological Laboratory, University of Liège, Belgium.

Helga Lejeune (H.L.), Dr. Psychol., Chargée de Recherches F.N.R.S., Psychological Laboratory, University of Liège, Belgium.

Daniel Defays (D.D.), Dr. Sc. (Math.), Assistant and Research Fellow, Institute of Psychology and Education, University of Liège, Belgium.

Pamela Greenwood (P.G.), Lic. Psychol., Fellow Fondation Hela, University of Liège, Belgium.

Françoise Macar (F.M.), Chargée de Recherches C.N.R.S., Institut de Neurophysiologie et de Psychophysiologie, Département de Psychobiologie expérimentale, Marseille, France

Huguette Mantanus (H.M.), Lic. Psychol., Psychological Laboratory, University of Liège, Belgium.

Foreword

THE PRESENT book aims at filling a gap in the scientific literature on biological and psychological time. There are dozens of books on biological rhythms, reflecting the tremendous development of chronobiology and, after years of pioneering work, the wide acceptance of this field by biologists at large. There are also a few excellent books on the psychology of time mainly or wholly devoted to the human level. There is a missing link between these two aspects of the study of time in living organisms, namely the experimental analysis of behavioural adjustments to time in animals. Numerous facts and hypotheses accumulated during the last 25 years or so have not yet been put together in a concise synthesis. This is what has been attempted here, for the benefit, it is hoped, not only of psychologists, but of all those fascinated by the problem of time in living matter. If it helps chronobiologists in becoming familiar with and in integrating behavioural data in their endeavour to build a general theory of biological time, and if it induces psychologists to take into account chronobiological data and thinking in their future research, this book will have fulfilled its goal.

This book stems from a long-lasting interest in the problem of time in our laboratory, an interest that materialized in a number of researches, many of which remained unpublished. This has been the case, especially, for final dissertation or even doctoral theses, done by psychology students. Their work has been abundantly quoted and used to illustrate some of the problems to be discussed here. Their contribution is fully acknowledged. The nine chapters have been written by a group of authors who have been or are currently working at our laboratory. Except for one of them (Pamela Greenwood), the authors do not use English as a native language. Despite careful reading and correcting by our British colleague Derek Blackman and by Pamela Greenwood herself—both of whom deserve gratitude from the team—and after expert final revision by the publisher, the text certainly still suffers from many stylistic and lexical defects. These are due to the linguistic habits of the authors, and are the price to pay for the very success of the English language as a tool for scientific communication.

The help of the members of the technical staff of our laboratory has been crucial at one stage or another in the preparation of the manuscript or in the research on which it is backed. We express our gratitude to all of them: S. Lénaerts, N. Fayen, R. Lénaerts, C. Vanderbeeken, M-A. Thunissen and F. Letihon.

MARC RICHELLE
HELGA LEJEUNE

PART I
SCOPE AND METHODS

I. *Introduction*

1.1. BEHAVIOURAL STUDY OF TIME AND CHRONOBIOLOGY

Time is an omnipresent dimension of living systems. On one hand, structures and processes change in an irreversible way through time. Individual organisms go through a number of developmental and involutive stages from conception to death. At another level, new, generally more complex forms emerge and eventually disappear in the course of biological evolution. In a sense, all biological phenomena are marked by the arrow of time. This notion has been central in modern biological thinking for more than a century. On the other hand, living systems also exhibit recurrent, periodic phenomena, which are basic to their normal functioning. Cardiac, respiratory, neuronal rhythms, and the like, have long been familiar to physiologists. More subtle periodicities, now described as *biological rhythms*, have only recently (if we exclude a few forerunners) received systematic attention from biologists (Bunning, 1973; Palmer, 1976; Reinberg, 1977). These cyclic phenomena have become, in the last 25 years or so, the realm of *chronobiologists*. Their main concern is with rhythms which appear in plants and animals in close correspondence with natural periodicities such as day-night cycles, tidal rhythms, yearly seasons and so on. So-called *circadian rhythms* (with a period of about 24 h), or, in marine organisms, *circatidal rhythms* (with a period of about 12 h) are the most extensively studied among biological cycles. It has been shown, for example, at many different levels of the organism, from the cellular level to general activity, and in a wide range of organisms, from unicellulars to man, that circadian rhythmicities persist even when all external cues (synchronizers) associated with the day cycle are eliminated. Though in the presence of such cues rhythms are strictly synchronized with the 24 h natural cycles, they have a slightly different period, a little longer or shorter than 24 h, in *free-running* conditions (hence the name *circadian*, "about one day"). While a few specialists argue that unidentified or very subtle cues (such as very low frequency electromagnetic fields) might be at work even in supposedly free-running experiments (see Brown, 1976), it is now widely admitted that some endogenous and autonomous timing mechanism accounts for the persistence of rhythms. This mechanism is metaphorically referred to as the *biological clock*. It is, to some extent at least, built into the genetic endowment of the organism. In fact, it is not clear whether one should talk of biological clock or clocks; in other words, whether all manifestations of biological cycles in a given organism are derived from one single timing source, or whether there exist several sources and possibly some co-ordinating systems between them.

While chronobiologists are still at work trying to elucidate problems as to the nature, the origin, the locus of biological rhythms, they currently treat them as independent variables in the study of other biological events. One typical example, of great practical consequence, is the variation in the effect of pharmacological agents according to the moment of the circadian cycle at which they are administered. In experimentally leukemic mice, arabinosyl cytosine, an antimitotic agent, can be fatal if

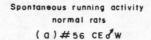

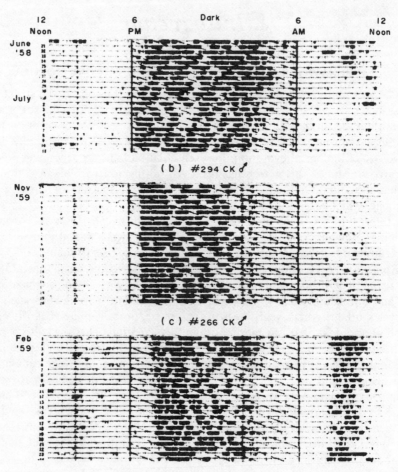

Fig. 1.1. Activity distribution records from three normal rats. The laboratory was in total darkness from 6 p.m. to 6 a.m. The period of the activity rhythm is synchronized with the illumination cycle. (From Richter, 1965. Copyright 1965 by Charles C. Thomas, Publisher.)

administered irrespective of the cyclic variations in susceptibility, while it has therapeutic value, if doses are adjusted to these variations. Chronobiology has also moved to applications to human situations where rhythms are disrupted, as is the case in shift-work, in rapid crossing of several meridians by jet flight, in prolonged exposure to monotonous environments (atomic submarines, speleologic experiments, etc.) or in space flights.

One should note that many observations, now constituting the basic material of chronobiology, were made using animal activity as a dependent variable (Richter, 1965). Spontaneous locomotor behaviour recommends itself as a convenient expression of biological rhythms, and is easily recorded by means of various kinds of actometers (such as the running wheel). (Fig. 1.1 and 1.2).

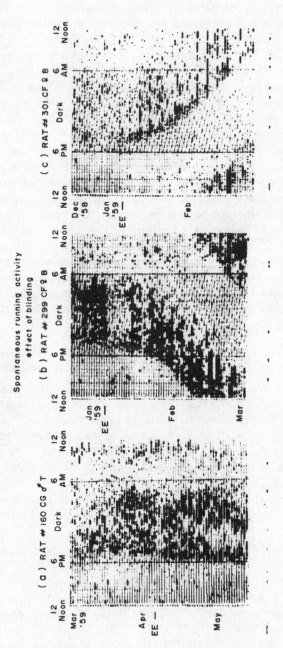

FIG. 1.2. Activity distribution records for three rats after blinding. Note the lengthening (rat 301) or shortening (rat 299) of the period of the activity rhythm after removal of the synchronizer (light). (From Richter, 1965. Copyright 1965 by Charles C. Thomas, Publisher.)

Animals not only adjust to basic periodicities typical of the terrestrial or marine environment in which their species have evolved, they also exhibit the capacity to adjust to durations which, compared to natural periodicities reflected in biological rhythms, seem arbitrary. This has been shown and extensively analysed in behavioural laboratory studies which are the subject matter of this book. If we take the word *chronobiology* in its full etymological meaning, these studies are clearly within the scope of this recent multidisciplinary field of biology. They are, however, rarely considered in this perspective in the technical literature, apart from a few exceptional attempts at integration (see, for example, Ajuriaguerra, 1968; Rusak and Zucker, 1975). Questions concerning the relations between the biological clocks on the one hand and the performances of the behavioural clock(s) on the other are plausible. Is some common mechanism at work in both cases? Or are there completely different mechanisms accounting for these two different kinds of temporal adaptation, and if so, do they work in perfect independence, or are there interferences? Are behavioural adjustments to time easier, or of better quality, when the temporal parameters approach or coincide with some natural periodicity? Do performances in temporal regulations or estimation change as a function of the moment in the circadian (or other) cycle? How shall we explain in an evolutionary perspective the emergence of behavioural clocks; or, in other words, what environmental selective pressure could have been at work to account for timing behaviour—if it is not merely a by-product of biological timing devices bound to biological rhythms? These and similar questions will eventually be answered if we integrate facts and problems from the behavioural study of time in animals and from chronobiological studies.

1.2. ANIMAL CLOCKS AND THE PSYCHOLOGY OF TIME

The study of behavioural clocks in animals, rooted, as suggested, in chronobiology, is obviously part of the psychology of time. This area of psychology looks, at first glance, like a rather heterogeneous field, including classical psychophysical experiments on time estimation in humans, studies on rhythms, whether at the primitive level of motor stereotypies or at the sophisticated level of musical or poetic productions, as well as investigations on the concept of time or on the time span of human consciousness. One might be tempted to discard the idea of putting together such disparate aspects as far-fetched speculation. Though we shall not indulge here in such possibly premature theorizing, one should, however, keep in mind interesting connections that might prove to be fruitful for future research.

It will appear quite evident from the description of techniques in Chapter 2 that part of the study of behavioural adjustments to time in animals is very close to the study of time estimation in humans using psychophysical methods (such as production, reproduction, estimation, bisection of time intervals). As has been the case in many fields of psychophysics, animal research brings an important complement to human data, and occasionally, with the help of highly sophisticated techniques of modern laboratories, allows for an elegant solution to difficulties that seem unavoidable when dealing with human subjects (for instance, it is almost impossible to prevent human subjects from counting in a time estimation task). It is also obvious that hypotheses concerning the mechanism of time estimation in humans suggest explanations suitable for animal data and vice versa.

It is by no means foolish to look for continuities from basic biological processes up to the highest forms of adaptation emerging in the human species as conceptual thinking and knowledge. Piaget has accustomed us to this line of reasoning. If the concept of time is rooted, as he claims, in the actions of the sensorimotor stage of development, it might derive, more primitively, from the elementary capacities of the behavioural clock already present in animals (see Richelle, 1968). A similar proposition was phrased by Skinner as far back as 1938. Outlining the scope of a chapter on temporal discrimination, he wrote: "stated more generally the problem is how time as a dimension of nature enters into discriminative behaviour and hence in human knowledge" (Skinner, 1938). The most elaborate ways of dealing with time, of integrating past and future admittedly involve symbolic and cultural processes not present in infra-human animals. This does not mean that they are fully freed from the elementary processes by which an organism experiences duration. Half a century ago, the French psychologist Janet (1928), theorizing about the psychology of time in humans, suggested deriving the most complex forms of conduct relating to time from the basic properties of *waiting behaviour*. What Janet had in mind has certainly more than superficial analogy with the kind of behaviour we shall analyse below. We shall later come back to his interesting intuitions.

1.3. HISTORICAL SURVEY OF RESEARCH ON TEMPORAL REGULATION AND ESTIMATION IN ANIMALS

Important as it is, time has not been a favoured topic for psychologists. The reasons for this may be partly extrascientific. As far as the study of time in animals is concerned, the main reason is undoubtedly a technical one. Until the middle of the twentieth century, experimenters did not have at their disposal the appropriate techniques to explore in any precise manner an animal capacity to estimate time. Two essential requirements were, in fact, lacking. First, phenomena of interest, for the most part, show up only through prolonged study of individual behaviour. Prolonged studies are only feasible with the help of highly automatized technologies. Second, for the control of temporal parameters and of their relations to other variables, we cannot rely upon a human experimenter operating a stop-watch. What we need is high performance and flexible chronometric devices. Such devices have become available to experimenters in the last quarter of a century or so.

Decisive technical progress has taken place especially in the field of operant conditioning, where automatized procedures have facilitated long-term experiments on individual behaviour with an equivalent degree of precision in the control of variables and in data recording. Though there is no claim to exclusivity, it turns out that most of the techniques and most of the data we shall review here pertain to that field.

This is not to say that earlier studies are negligible. In fact, Pavlov and his students deserve credit for having designed the first experimental procedures for the study of time in animal behaviour, and for having described a number of solidly established data. We also owe Pavlov several important theoretical proposals concerning temporal conditioning, especially their interpretation in terms of active inhibition. Until now, Pavlov's theory has not been, in this respect, clearly proved, but as we shall see it still remains the most convenient working hypothesis in face of the accumulated experimental facts in the field of operant research as well as in Pavlovian conditioning studies. Work on temporal conditioning was continued in Pavlovian laboratories (Dmitriev and Kochigina, 1959) and taken as a major research theme by some followers outside the U.S.S.R., such as Popov in Paris (Popov, 1948, 1950).

There have also been a few studies on time among the innumerable studies on animal learning in American behaviouristic laboratories before the modern development of operant techniques. Some illustrations of the procedures used will be given in the next chapter.

II. *Methods*

2.1. INSTRUMENTAL PROCEDURES

Instrumental learning, as a general label, applies to a variety of experimental situations in which the subject learns to perform a given action that is instrumental to obtaining a reward. Familiar examples are mazes, jumping stands, shuttle-boxes, puzzle-boxes and the like. Food, water, or the avoidance of an electric shock are the most common types of reward. Operant techniques, to be discussed below, derive from instrumental procedures, but they present specific features that amply justify separate treatment for our purpose.

Among the hundreds of studies of instrumental learning, mainly performed in behaviourist laboratories in the first half of the century, very few were devoted to time estimation. Sams and Tolman (1925) and Anderson (1932) used a simple maze in which the rat had to select one out of two or four alleys; all alleys were identical except for the fact that the subject was confined in each of them for a different duration. Choosing the alley with the shortest period of confinement was taken as reflecting time estimation. A correct choice, however, as Hull objected (Hull, 1943), might have depended upon the delay to reinforcement rather than upon estimation of the duration of confinement.

In other experiments, the confinement took place in a starting-box before the choice of one of the alleys of a Y or T maze. Either the right or the left alley was rewarded according to the short or long duration of the confinement (Cowles and Finan, 1941; Heron, 1949; Yagi, 1962). Here a correct choice clearly depended upon time estimation, irrespective of the delay to reinforcement. Performances in this situation are poor. Though results give evidence of time estimation, they do not provide for the measurement of a valid differential threshold. There are large interindividual differences, and most subjects fail to learn, when the confinement durations are not extremely contrasted: 5 s versus 25 or 45 s, 10 versus 50 s, that is a 1:5 or even 1:10 ratio. A similar technique was recently used with cats by Rosenkilde and Divac (1976a) in a psychophysiological study. Reinforcement was contingent upon the time spent running from the start box to the goal box in Logan's experiments (Logan and others, 1955; Logan, 1960). Bower (1961), using a somewhat similar technique, delivered immediate reward if the running time had been longer than a specified *cut-off time*, or he delayed the reward if the rat had reached the goal box before the cut-off time. Kelsey's study (1976) is a mixture of a *duration of locomotion* requirement and a *two-lever DRL* schedule (to be reviewed under 2.3.1).

Another procedure, involving aversive stimulation, was designed by Ruch (1931). The experimental space is divided into three compartments A, B, C. Both A and B are equipped with electrifiable grid floors. C is a safe compartment, but separated from B by a door that can be locked. When the rat is put in A, B is electrified and the door from B to C is locked, so that if the rat leaves, it gets shocked and blocked at the door. Compartment A, however, is safe for a time *t1*. After a time *t2* shorter than *t1* has passed, B becomes safe and the door is unlocked. As soon as *t1* has elapsed, both A and B are electrified again. The subject will avoid the shock in A and B if it crosses B during the *safe interval* between the end of *t2* and the end of *t1*. This technique was explicitly designed for the purpose of measuring differential threshold for duration. But Ruch, as well as other experimenters that followed him (Blancheteau, 1965, 1967a, 1967b; Buytendijk and others, 1935; Stott and Ruch, 1939) did not succeed in obtaining more than 50 per cent correct responses, despite the long duration of the safe interval (i.e. within the range of 1 to 5 min).

These procedures, though involving time estimation, were obviously not very adequate to test the subject's real possibilities in discriminating or estimating duration. They required the continuous presence of the experimenter, but what appeared to him as long and fastidious experimental sessions might have been far too short to expect any consistent temporal adjustment from the subject. More important, a number of variables were likely to interact with time estimation proper: all three kinds of situations—choice maze, runway and double-avoidance—maintained confusion between the spatial configuration of the response and the time discrimination. Choice maze and double avoidance both involved some sort of conflict or stress: in the first case, estimating the delay of confinement had to take place while the required response was physically prevented; in Ruch's rather ambiguous procedure, the two compartments were equally aversive and the same behaviour could be a safe or a punished response depending on its position in time. Some subjects might actually never experience the temporal arrangement set by the experimenter to make avoidance possible. These shortcomings most probably suffice to explain the poor performances of animals, compared with what has been obtained since then, using operant techniques. Therefore, early instrumental procedures of the type described here have but historical interest.

An earlier study by Woodrow (1928) allowed for a less contaminated appraisal of time discrimination in monkeys. The duration of an *empty interval* between two brief acoustic signals was the stimulus. A long duration—initially 4.5 s—was associated with the presence of food in the food-box; when the duration was short, that is 1.5 s, the subject would find no food. He eventually learned to open the box only after long stimuli. The two monkeys were still able to perform with 75 per cent correct responses when the comparison positive stimulus was progressively shortened to 3.75, 3.2 and 2.25 s. The technique, however, was still time-consuming for the experimenter, requiring thousands of trials—what is by no means unusual in that kind of training but makes automatized techniques highly desirable.

2.2. PAVLOVIAN CONDITIONING PROCEDURES

In Pavlovian conditioning, a neutral stimulus, for instance a tone, is paired with an unconditioned stimulus (US). An unconditioned stimulus is one that elicits a given response because the physiological machinery is wired that way. For example, food elicits salivation, electric shock elicits leg flexion, etc. After pairing, the formerly neutral stimulus becomes a conditioned (or conditional) stimulus (CS): it elicits a response quite similar, if not strictly identical, to the unconditioned response. Generally, in a Pavlovian experiment, the CS and the US are in close temporal contiguity, the former occurring approximately 0.5 s to 1 s or so before the latter (with or without overlapping) as shown in Fig. 2.1 A. Pavlov explored several slightly different arrangements which revealed interesting properties of conditioned responses with respect to time.

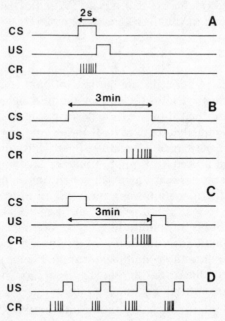

FIG. 2.1. Schematic representation of Pavlovian conditioned reflexes (CR). A: usual procedures, with close temporal contiguity between conditioned stimulus (CS) and unconditioned stimulus (US); B: delayed reflex; C: trace reflex; D: conditioned reflex to time.

2.2.1. Delayed Reflexes

In *delayed conditioning*, the CS duration is extended from the usual short value of simultaneous conditioning (up to 5 s or so) to a much longer value (Pavlov and his co-workers generally used a 3 min interval). The lengthening of the CS may be progressive or abrupt. The conditioned response that, in simultaneous conditioning, occurs immediately after the CS is presented, is now delayed, appearing toward the end of the CS, close to the presentation of the US. Figure 2.1B shows the absence of conditioned responses during the first minutes of the delay, and a maximum of responses in the last minute.

2.2.2. Trace Conditioning

In a variant of delayed conditioning, the CS is not continued to the US presentation. It is separated from the US by an *empty interval*, as shown in Fig. 2.1C. Here again, conditioned responses occur towards the end of the interval.

2.2.3. Temporal Conditioning

The control of conditioned behaviour by time is still more clearly demonstrated in so-called *temporal conditioning*, as illustrated in the experiment performed by

Feokritova (1912) in Pavlov's laboratory. The paradigm of temporal conditioning involves no CS. The US is presented periodically (every half-hour in the princeps experiment). Conditioned responses occur toward the end of the intervals between US, reflecting the periodicity of the reinforcing event (Fig. 2.1D).

2.2.4. Discrimination

Consistently pairing a conditioned stimulus and not pairing another stimulus of the same physical category with a US provides for the development of discrimination: conditioned responses are elicited only by the positive stimulus after the extinction of generalized responses eventually produced by the negative stimulus early in discriminative conditioning. If the positive and negative stimuli differ only with respect to the temporal dimension, the procedure is appropriate for studying time discrimination. This was the case in many classical Pavlovian studies involving the metronome as a source of stimulation. Discrimination was easily established to different beat frequencies. The technique, however, is not appropriate for the study of discrimination between different durations of a stimulus: in order to discriminate a stimulus duration, the subject must wait until the stimulus is over. The Pavlovian procedures make no provision for withholding the conditioned response, and, therefore, they are not convenient for measuring differential threshold for duration. This problem has been elegantly solved with operant techniques.

2.2.5. Temporal Conditioning and Disinhibition

Without anticipating on later discussions, we shall conclude this section by mentioning three important consequences drawn by Pavlov from the situations described above (and especially the first three of them). First he felt it was necessary to modify and extend the definition of the conditioned stimulus to include not only any event that could trigger a sensory receptor but also the temporal dimension of changes occurring in the environment of the organism. Second, he was confronted with the problem of the mechanisms underlying these time-adjusted conditioned responses, and as we shall see (Chapter 6), he suggested several possible time-bases. Thirdly, whatever the timing device proper, the behavioural data led him to assume an inhibitory mechanism to account for the distribution of conditioned responses in time. In Pavlov's theory, inhibition is not just an absence of excitation; it is an active process counteracting excitation, as evidenced by the phenomenon of *disinhibition*: a novel stimulus presented during an excitatory phase provokes a reduction of the conditioned response. If presented during the no-response phase of a delayed or trace conditioning, it has the opposite effect of reinstating the conditioned response, as seen from Fig. 2.2. We shall come back to the concept of inhibition in the theoretical discussion concerning the mechanisms involved in behavioural adjustments to time (Chapter 9).

2.3. OPERANT CONDITIONING PROCEDURES

In the operant conditioning paradigm, a reinforcement is delivered to the subject if a given response has been emitted. For instance, food is delivered to a rat when it presses a lever. Food is thus contingent upon the emission of the response (this is a

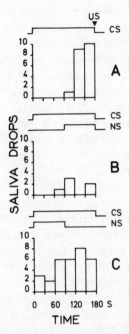

FIG. 2.2. Disinhibition of a Pavlovian delayed reflex. A: normal control reflex; B: inhibition produced by a novel stimulus during the final (excitatory) phase; C: disinhibition produced by the same stimulus during the initial (inhibitory) phase. (Drawn after numerical data from Pavlov, 1927.)

main procedural difference distinguishing operant from Pavlovian conditioning). Reinforcing a bit of behaviour in this way results in an increased frequency of that behaviour. The relation *Response → Reinforcement*, basic to all operant experiments, is generally complicated by resorting to *schedules of reinforcement* in which not every response is reinforced. Instead, reinforcement is intermittent and contingent upon, for instance, the emission of a fixed number of responses (*Fixed-Ratio schedule, FR*) or of an average number of responses (*Variable-Ratio schedule, VR*). Temporal specifications are involved in several schedules that will be described in detail below. Preceding stimuli may be introduced into the situation. They do not, however, actually trigger the response but specify the occasion on which it is more or less likely to be reinforced. For example, a lever press will be rewarded by food in the presence of a 1000 Hz tone, and will not be rewarded in the presence of a 3000 Hz tone, or in the absence of a tone. Such stimuli in operant conditioning experiments are referred to as *discriminative stimuli*, the word making for a clear distinction with eliciting stimuli. Whatever the role of discriminative stimuli, the main controlling variable is the reinforcing stimulus that exerts a selective action upon the behaviours produced by the organism.

The modalities of the relation between response, reinforcer, and discriminative stimuli, technically called *contingencies of reinforcement*, may be varied at will, sometimes reproducing in the laboratory typical features of the natural environment, sometimes with the aim of exploring the potentialities of an organism (as in determining psychophysical thresholds). Contingencies of reinforcement can be classified according to the nature of the reinforcing event. *Positive contingencies* involve a positive reinforcer such as food for a hungry animal, or intracerebral stimulation in so-called pleasure centres of the mid-brain. *Negative contingencies* involve aversive stimuli, that is stimuli from which an organism would normally withdraw. They subdivide further into the three categories of escape, avoidance and punishment. In the first case, the response suppresses an aversive stimulus already present; in the second, it anticipatively prevents the stimulus from occurring; in the third, the aversive stimulus is contingent upon a response otherwise positively reinforced.

The favoured measure of operant behaviour is the rate of responding, and the most widespread way of recording it is the cumulative curve. Of course, in some cases, other measures are preferred or used in addition to the rate of responding. We shall meet a few of these in temporal schedules.

Whatever one likes to think about the theories of behaviour derived from or associated with operant conditioning, it has become one of the most fruitful techniques for the experimental analysis of animal behaviour. It offers the student of time a number of possibilities unmatched by previous methods. One reason for this is the degree of automatization of experimental operations: refined contingencies can be realized and precise data can be recorded only with the help of fully automatized equipment.

A number of schedules have been designed which are of interest in the study of time in animal behaviour. To put some order in this diversity, we shall first dichotomize them into two large categories: in one category we shall describe those schedules typically inducing some sort of distribution of responses through time that exhibits *temporal regulation* of the subject's own behaviour; in a second category we shall consider those schedules in which the subject has to make a *temporal discrimination*, that is to estimate some time-related aspect of an external event (duration, rhythm, velocity). Temporal regulation of behaviour, in its turn, may develop as a by-product of the schedule, without being required as a condition for reinforcement, or it may be the very condition for reinforcement. Consequently, we shall subdivide our first category into subcategories. One we shall label *spontaneous temporal regulation* and the other *required temporal regulation*. *Spontaneous* does not mean, of course, that the regulation has no causes; nor does the term *required* imply that the subject always satisfies the specification of the schedule. These terms are convenient labels. We shall separately treat as a third subcategory schedules involving aversive stimulations, that is pertaining to what has been defined above as negative contingencies.

Admittedly, some border-line cases are not easy to classify in clear-cut categories. The decisions concerning them might appear somewhat arbitrary. The adopted classification is not a purely formal one. As will also appear from the discussion, it is based on hypotheses drawn from the body of data now available.

2.3.1. Temporal Regulation of the Subject's Own Behaviour

A. Spontaneous Temporal Regulation

a. Fixed-Interval Schedule (FI)

Description of the FI Schedule. This schedule was first studied by Skinner (1938) under the name *periodic reinforcement* and extensively explored by Ferster and Skinner (1957). The contingencies of the FI schedule are schematically represented in Fig. 2.3. The reinforcement is contingent upon the first response that is emitted after a

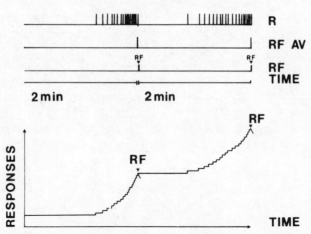

FIG. 2.3. Schematic representation of main events in a Fixed-Interval schedule. From top to bottom: operant responses, availability of reinforcement, obtained reinforcement, time, and projection on a cumulative record.

specified delay has elapsed since the previous reinforcement. Responses emitted during the delay have no consequences. Typically the schedule generates a pattern of behaviour exemplified in the cumulative curve of Fig. 2.3. A period during which operant responses are not emitted follows reinforced response. After this pause the subject resumes responding and continues to respond at an increasingly accelerated rate until the delay has elapsed, so that there is usually no time gap between the end of the delay making the reinforcement available and the response by which the subject obtains it. Reinforced responses are thus as periodic as the contingencies permit. It will be noted that the first response emitted after the delay is completed is signalled to the animal by the delivery of the reinforcement, and the visual, auditory or other stimuli associated with it.

The pause after each reinforcement evidences a temporal regulation of the subject's own responding. This regulation is all the more striking since it is by no means required by the contingencies. The duration of the pause frequently extends over 50 to 80 per cent of the total cycle duration. The phase of responding that follows gives rise to various patterns of responding. The one described above was first identified by Skinner (1938) and has been observed by many experimenters since then. It is metaphorically called *scalloping*, after the shape of the cumulative curve. The progressive acceleration in rate, however, is not a universal phenomenon. Another pattern has been described under the name *break-and-run* (Schneider, 1969). After the pause, the subject resumes responding immediately at a sustained rate, that it will maintain throughout the rest of the interval (Fig. 2.4). It is not clear which variables cause one pattern or the other, both having been observed in the same species (namely pigeons and rats). The duration of the delay (Schneider, 1969) and the history of conditioning (Dews, 1978) might be critical. The question need not detain us in this methodological section. In either case, the pause in responding is present. It can occur that, under the influence of certain variables (motivational, pharmacological, and the like) the pause extends over the entire cycle and beyond. Conversely, pausing can be disrupted by various factors, and responding can appear throughout the interval.

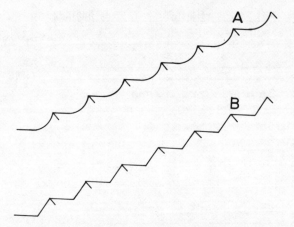

FIG. 2.4 *Scalloping* (A) and *break-and-run* (B) patterns in Fixed-Interval behaviour.

The FI schedule in its classical version specifies after what delay the reinforcement is available. A new cycle will not begin until a response is emitted, and reinforced. If, for some reason, the subject does not respond after the delay has elapsed, there will be no new cycle. Dews has called these contingencies *fixed-minimum-interval schedule* (Dews, 1970). The reinforcement can be made available only during a limited period, that is a limit is specified beyond which a response will not be reinforced. This limited period of availability is called a *limited hold*, and the schedule providing for that restriction has been called *fixed-minimax-interval schedule* (Dews, 1970). This procedure was used by Gollub (1964) and, implicitly, for some values of the temporal parameters, by Cumming and Schoenfeld (1960) in their T–t schedule.

Statistical Expression of Temporal Regulation in the FI Schedule. Methods of data analysis applied to FI behaviour usually retain one main character of operant responses, that is their position on the temporal continuum defining the interval. This is best expressed as the time elapsed between the reinforcement, taken as the origin of the interval, and the response. For practical reasons, the interval is subdivided into successive classes and each response assigned to the appropriate class of what could be called *Reinforcement-Response-Time* (Rf–R–T). Distributions can be drawn interval by interval or from averaged data for a whole session or whatever experimental unit one might wish. Figure 2.5 shows an example of such distributions, in histogram form. This, of course, is just a more accurate representation of the data recorded by the cumulative recorder. It does not give a mathematical estimate of the temporal regulation. Several methods have been proposed for that purpose. In some of them, the performance is characterized by one or several unidimensional indices which are intended to summarize some aspects of the distribution of responses through time. In others, the distributions themselves are studied by means of various adjustments.

Unidimensional Indexes. Figure 2.5 provides an example for which the values of the various indexes are given.

Characterizing the Pause. $t(1)$ is the time from reinforcement to the first response, eventually averaged over a whole session. It has been objected (Schneider, 1969) that this simple measure is based on the assumption that pause and phase of activity are opposed as all-or-none processes, and that it does not account for the *scalloping* typical of FI behaviour. In that respect, the following index, almost as simple, appears more appropriate. $t(4)$ is the (average) time from reinforcement to the fourth response. This index gives less weight than $t(1)$ to an isolated response emitted in the beginning of the interval. It cannot be computed, of course, if the subject has not produced at least four responses (or an average of four responses). The *Quarter life*, Q, is the time elapsed from reinforcement until one fourth of the responses has been emitted. (Herrnstein and Morse, 1957). It corresponds to the first quartile of the distribution of responses. A linear interpolation is required to calculate this index. To do so is to assume that the rate of responding is fairly constant throughout the part of the interval being considered (i.e. the class of the distribution). Strictly speaking, this assumption is of course never correct when we deal with scallops. For all practical purposes, however, these refinements can be ignored provided that we have a reasonably large number of classes (let us say ten rather than four). It must be noted that the Quarter Life (as well as $t(1)$ and $t(4)$) computed from averaged data from a whole session will not necessarily be similar to the average Q (or $t(1)$, $t(4)$) computed by averaging the Qs obtained for successive individual intervals. In our example, the Q

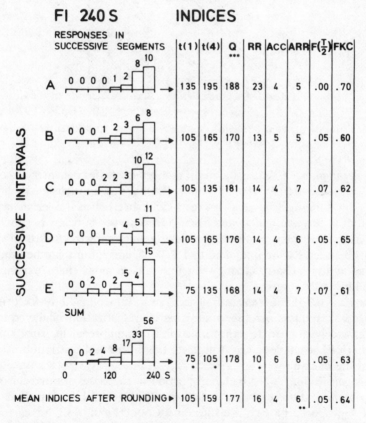

FIG. 2.5. *Fixed-Interval schedule.* Distributions of responses in successive segments of the interval (left to right) for successive intervals (top to bottom). Data in the bottom row are averaged from data given in rows A to E above. Columns on the right give the various indices discussed in text. (*Those indices are uninteresting. **The two methods for the computation of the ARR have the same outcome. ***Q may also be expressed as relative value.)

value is 178 in the first case, 177 in the second. The difference can generally be considered negligible (Schneider, 1969). Different authors have shown that these three indexes intercorrelate, so that the experimenter might use the one most readily available, given his technical setup.

Characterizing the Activity Phase. The *Running Rate*, RR, is the mean rate of responses per minute during the period of activity. The definition of the period of activity is complementary to the definition of the pause. If the latter is defined by $t(1)$, the running rate index will suffer similar objections.

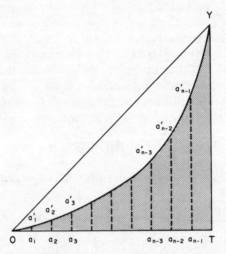

FIG. 2.6. Cumulative distributions of responses in a Fixed-Interval illustrating the formula of the Fry, Kelleher, and Cook's index (see text).

The *Acceleration index*, Acc, is the mean number of classes per interval containing a number of responses larger than the immediately preceding class (the first class is neglected) (Dews, 1978). This index makes a difference between constant and accelerated rate, that is between *scallop* and *break-and-run* types of behaviour.

Studies correlating the two indexes are lacking. According to Dukich and Lee (1973) correlations between RR on one hand and $t(1)$, $t(4)$ and Q on the other are low. This suggests that an index characterizing the pause and an index characterizing the active phase should be used jointly.

Characterizing the Whole Interval. The *Average Response Rate*, ARR, is the average number of responses per time unit. This characterization of the overall rate of responding, widely applied to other schedules of reinforcement, is not particularly appropriate to FI schedules since it neglects the differential distribution of responses throughout the interval.

The *First-half-Ratio*, $F(T/2)$ is the proportion of responses emitted during the first half of the interval; it corresponds to the value of the cumulative function in $T/2$.

The index proposed by Fry, Kelleher and Cook (1960), *FKC* or *curvature index*, characterizes the area located under the cumulative distribution function of the responses. The formula is easily understood from the illustration in Fig. 2.6. $[O, a_1]$ is the first temporal class, $[a_1, a_2]$ is the second, ... $[a_{n-1}, T]$ is the last class. The curve $(O a_1' a_2' \ldots a_{n-1}' Y)$ corresponds to the cumulative-distribution polygon. A is the area of the triangle OTY. A_1 is the area of the figure OTY $a'_{n-1}, a'_{n-2}, a'_{n-3} \ldots a'_3 a'_2 a'_1$ O. This area is *shaded* in the illustration. The index is equal to $(A - A')/A$. It varies between $-(n-1)/n$ (in the case where all responses would occur in the first class) and $+(n-1)/n$ (all responses in the last class). The closer the index is to the positive maximal value, the better the temporal regulation. As for other indexes, the FKC may be computed from the distribution of the whole session (in our example it amounts to 0.63) or it may be averaged over the indexes computed from each successive interval (which gives 0.64 in our example). As shown by Gollub (1964) the two methods may produce somewhat different outcomes. This is related to the variations in the total number of responses from interval to interval. In the first method (computing the index over the whole session), the more responses in an interval the more weight this interval is given in the index. In the second method, individual intervals are given equal weight irrespective of the absolute number of responses they contain. Also, it seems that the interinterval variations of the FKC index are larger than those of Q.

The FKC index has a number of properties that recommend it. It reflects the characteristics of the whole cycle. It is sensitive, easily computed, and applies to deceleration as well as acceleration of rate. It has some limitations, however: the maximal value depends upon the number of classes (therefore, one should be careful not to compare FKC indexes computed from distributions with different numbers of classes); in the rather unusual case where acceleration and deceleration would alternate, they would compensate each other giving the index a low value not correctly reflecting the distribution; arbitrarily including the reinforced response in the last class, as is usually done, introduces a bias that is especially important when the reinforced response is delayed after the end of the interval and when the total number of responses is low. The FKC index and Q are highly correlated. As expected, the ARR is little correlated with any of these indexes (Gollub, 1964).

Mathematical Models. Methods of the second type aim at adjusting mathematical models to the distribution of responses rather than simply characterizing it by one or two indexes. This has been done, for example, by Schneider (1969). He simultaneously adjusts two straight lines on the distribution; the intersection of the two lines—one with a small slope, the other with a steep slope—or *break-point*, marks the passage from the pause to the activity phase. The model provides for separate analysis of each phase. It assumes constant rates in the pause and in the activity phase. Other authors have represented the distribution of responses by the left tail of a normal curve (Killeen, 1975).

b. Modified FI Schedules and Other Contingencies
Inducing Spontaneous Temporal Regulations

Fixed-Mean-Interval Schedule (*Dews, 1970*).   It differs from the classical FI in that the beginning of a new interval coincides with the end of the preceding interval, irrespective of the moment at which the reinforced response is emitted. If the response is delayed, it will occur somewhere during the interval; the interreinforcement interval will be defined as an average value. No external event, such as the reinforcement and the stimuli associated with it, signals the beginning of the interval, as is the case in the classical FI (Fig. 2.7). This schedule, called *clock dependent FI* by Snapper and co-workers (1971), has been studied by Cumming and Schoenfeld (1958) in pigeons and

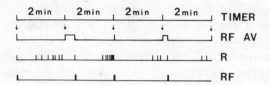

FIG. 2.7. Schematic representation of the Fixed-Mean-Interval schedule (see text).

by Snapper and co-workers (1971) in rats. In the latter study, comparison with classical FI shows a close similarity of performances as analysed by several mathematical indexes.

FI Schedule with no Exteroceptive Signals Associated with the Reinforcement. In the classical FI schedule, cues associated with the delivery of the reinforcement tell the subject that the food is in the tray, so that it has to make no estimate about the time at which it will be delivered. The temporal regulation is evidenced by the regularity of the pause, but is by no means involved in the phase of responding. The question arises: what would be left of the patterns of behaviour generated by this classical FI schedule, were the cues eliminated? This has been done by Deliege (1975) in a schedule identical to the classical FI except for the fact that all auditory and visual cues associated with the reinforcement were eliminated. Technically, this was achieved by putting the pellet dispenser at a distance from the experimental cage in a soundproof compartment, and by similarly soundproofing the tube to the tray and the tray itself. The inside of the tray was hidden by a flap valve. Rats initially trained on a classical FI 2 min schedule, when transferred to this *No-Cue FI schedule*, showed a reduction in overall rate and a marked disruption of temporal patterning (scalloping, pause-activity alternance). Prolonged exposure to these contingencies eventually resulted in long phases of extinction. Rats could be trained from the beginning under that schedule, with performances markedly different from those obtained with classical FI. When cues were reinstated, the classical pattern rapidly appeared.

Discrete Trial FI Schedules. In the classical FI, successive intervals are contiguous to each other: a new cycle begins when the previous cycle is terminated. In the variant considered here, successive intervals are separated by a *time-out*, or *intertrial interval*, during which the FI contingencies are interrupted. The FI cycle is signalled by a discriminative stimulus (Fig. 2.8A). The duration of the intertrial interval is fixed or variable. In the case of a time-out fixed duration equal to the FI cycle duration, the regular periodicity of the classical FI schedule is preserved to some extent. In technical terms, a discrete trial FI schedule could be named a *multiple schedule. Mult. FI5 Ext. 5*, for instance, would designate a schedule in which FI cycles of 5 min and Extinction periods of 5 min alternate, with a stimulus signalling which component of the schedule is currently prevailing.

Performances of pigeons or squirrel monkeys under such contingencies were explored for example by Caplan and co-workers (1973), De Weeze (1977), Dews (1965b), Gollub (1964) and Marr and Zeiler (1974). Rather than being signalled by a discriminative stimulus, the FI cycle may be initiated by the subject itself. For example, the experimental cage is equipped with two response levers, A and B. A response on lever A initiates the FI cycle, that will be reinforced for a response on B after the delay has elapsed. Intertrial intervals are, of course, determined by the subject's behaviour (Fig. 2.8B). This procedure has been studied by Mechner and co-workers (1963) in rats with FI values of 30 and 60 s, and in monkeys with a FI value of 60 s. The response initiating the cycle also switched on a light, clearly signalling the FI condition. Typical FI scallops appear under these contingencies. Response-initiated schedules using one single lever are very similar in form to the one just described (Fig. 2.8C), but they produce quite different patterns of behaviour. They are known as *Tandem FR1FI* schedules (which means that Fixed-Ratio components, with

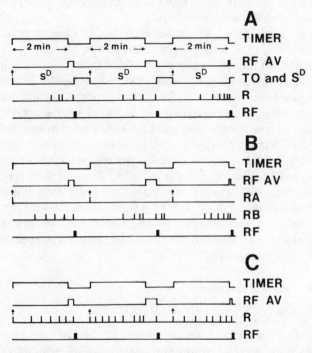

FIG. 2.8. Schematic representation of discrete trial FI schedules, (A) using discriminative stimuli (S^D) that signal FI and Time-Out (TO) periods; (B) with a subject's response on lever A (RA) initiating the interval and using a lever B for the eventually reinforced operant (RB); (C) with the interval being initiated by the subject using one single response lever.

one response requirement, alternate with Fixed Interval) (Chung and Neuringer, 1967; Shull, 1970a) or as *Chain FR1FI* when external discriminative stimuli are associated with each component (Shull and Guilkey, 1976). A pause is observed after the reinforcement, but not in the FI component proper. In other words, the subject makes no distinction between the FR and the FI components: the first response (fulfilling the FR requirement) is immediately followed by sustained responding until reinforcement.

The expression *discrete-trial FI* has been used by some authors (Hienz and Eckerman, 1974; Schneider and Neuringer, 1972) to designate an experimental manœuvre initially used by Dews (1962) to explore some properties of the behaviour generated by FI schedules. Dews superimposed alternating positive and negative discriminative stimuli of fixed duration on the FI cycle. The results of these studies will be summarized and discussed in Chapter 3. The expression *discrete-trial FI* is improperly used in this case and it leads to confusion with the kinds of procedures described in the

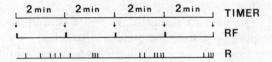

FIG. 2.9. Schematic representation of Fixed-Time (FT) schedule.

present section (to which it straightforwardly applies, by analogy with similar procedures in other conditioning situations).

Fixed-Time Schedule (FT). This schedule is not an operant conditioning schedule *sensu stricto* for the reinforcement is not contingent upon the response emitted by the subject: it is delivered automatically after a fixed interval of time irrespective of the subject's behaviour. It is analogous to the Pavlovian temporal conditioning procedure (Fig. 2.9). One would not expect a naïve animal exposed to these kinds of contingencies to produce systematic responding not required by the schedule unless the response has been adventitiously reinforced (so-called *superstitious* behaviour) or has a respondent rather than an operant status. However, if the subject has previously been exposed to a classical FI schedule, something is retained from the FI pattern, though the rate of responding after the pause is lower in FT than in FI (Appel and Hiss, 1962; Shull, 1971b; Zeiler, 1968). The activity phase does not always appear with the typical scallop or break-and-run (Shull, 1971b). Details of the results obtained under FT schedules will be reviewed in Chapter 5 and Chapter 8. They offer interesting data for the discussion of the role played by the periodicity of reinforcement as such in spontaneous temporal regulations.

B. Required Temporal Regulation

a. Differential Reinforcement of Low Rates (DRL)

Description of the Classical DRL Schedule. In the DRL schedule (Skinner, 1938; Ferster and Skinner, 1957) a response is reinforced if, and only if, it follows the preceding response by a specified temporal interval. Thus the temporal regulation of the subject's own behaviour is the condition for reinforcement. In a DRL 60 s, for instance, the subject must space its responses by at least 60 s in order to be reinforced. Responses emitted before the critical delay reset the timer and, of course, are not reinforced (Fig. 2.10). As a consequence of these contingencies, the response rate is reduced rather than increased, as expected from the classical definition of the reinforcement effect in operant conditioning. This is only seemingly a paradox as the *Inter-Response Time* (IRT) rather than the motor response proper, is the reinforced event in DRL. The requirements of the schedule can be further refined by setting an upper limit, or *limited hold*, LH, for reinforced IRTs. Thus, in a DRL 60 LH 6, responses must be spaced by at least 60 s and at most 66 s (see for example Kelleher and others, 1959; Powell, 1974; Staddon, 1969a).

The terms *Differential Reinforcement of Low Rates* have also been applied, in fact more properly, to other kinds of contingencies in which the reinforcement is delivered if the number of responses emitted during a defined period of time is equal or inferior to *n* (Ferster and Skinner, 1957; Morse, 1966). This acceptation, however, has not become common usage and we shall neglect it here.

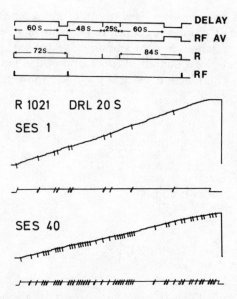

FIG. 2.10. Schematic representation of the schedule of reinforcement of low rates of responding (DRL 60 s), and samples of cumulative records obtained early in training and after stabilization (DRL 20 s).

Classical DRL schedules, as defined above, usually generate low and regular rates of responding, inversely related to the requirements of the programme (Wilson and Keller, 1953; Zimmerman and Schuster, 1962). The control of rate by the contingencies is easily evidenced in situations where a DRL schedule alternates with schedules imposing no such temporal requirements or inducing high rates (Ellen and others, 1978; Richardson, 1973, 1976; Ross and others, 1962; Skinner, 1938). IRTs show a characteristic evolution as learning progresses. IRTs close to the reinforced value become more and more frequent, up to an asymptotic level, but very short IRTs are also often observed. These are generally produced by response bursts, the origin and function of which are still unclear (for a discussion, see Ferraro *et al.*, 1965 and Kramer and Rilling, 1970; and *infra* Chapter 3 and Chapter 9). Such response bursts, which typically do not fall under the control of the temporal contingencies, might, at least, be a species-specific phenomenon, or an artifactual by-product of technical factors (Blough, 1963; Harzem, 1969; Kelleher and others, 1959; Kramer and Rilling, 1970; Ray and McGill, 1964).

Methods of Data Analysis. The most relevant data, in DRL performance, are IRTS. Maximal information is available when the recording equipment provides the experimenter with the complete sequence of IRTs throughout the session. The smaller the time unit used, the finer the possible treatment of data. A convenient recording method distributes IRTs into a number of classes. Lacking the sequential information, this method precludes part of the analysis to be discussed.

We shall use an example of the maximal information type (Fig. 2.11). The raw data are in the form of a sequence of IRTs that we shall note $(x_1, x_2, x_3 \ldots x_N)$. x_i is the duration of the ith IRT. The sequence is limited to twenty IRTs for the sake of simplicity. In an actual experiment, the sequence for a single session would usually be much longer. The analysis can bear on the duration of IRTs or on their sequence.

Analysis of IRT Duration.   Efficiency indices. It follows from the definition of DRL contingencies that a large number of short IRTs (relative to the reinforced value) shows poor temporal regulation. Several authors have proposed unidimensional efficiency indices that would give a quantified expression of the performance.

If we denote by f any real function, efficiency indices are of the kind

$$\frac{\sum_{i=1}^{N} f(x_i)}{NT} \tag{1}$$

where T is the critical delay.

Thus if $f(x_i) = x_i$, the index becomes

$$\frac{\sum_{i=1}^{N} x_i}{NT}. \tag{2}$$

If

$$f(x_i) = 0 \text{ for } x_i < T$$
$$= T \text{ for } x_i \geqslant T,$$

the index becomes N'/N, where N' is the number of reinforced IRTs.

This simple efficiency ratio has an important defect in that all non-reinforced IRTs are taken as equivalent. On the other hand, index (2) weights reinforced IRTs according to their duration, which one might consider as irrelevant as far as sheer efficiency is concerned. Both defects could be eliminated by stating that

$$f(x_i) = x_i \text{ if } x_i < T$$
$$= T \text{ if } x_i \geqslant T$$

we then derive

$$\frac{N' + \sum_{x_i < T} \frac{x_i}{T}}{N} \text{ (Heuchenne and others, 1979).} \tag{3}$$

IRT Distribution. The most classical way to present DRL data is to divide the distribution of IRTs into a number of classes, the class width being a submultiple of the reinforced delay. This can be presented in histogram form as shown in our example (Fig. 2.11).

The quality of the temporal regulation is easily assessed by visual inspection. With a well-conditioned subject, the mode of the distribution coincides with the class of the reinforced IRT. Observed frequencies increase as one approaches the reinforced class.

IRTs per Opportunity (IRT/OP). However, the fact that the mode is not in the class of the reinforced IRT does not necessarily mean a lack of temporal regulation. Anger (1956) has suggested that the number of responses in a given class depends upon the number of opportunities to respond that the animal is offered in that class; in other words, it is conditioned (in the mathematical sense of the word) by the IRTs

DRL 10 S

SUCCESSIVE IRTs (in seconds):

0-0-10-11-9-0-10-9-0-11-10-6-0-9-0-10-13-11-10-13

EFFICIENCY INDICES :   .71 (INDEX 2)
                        .62 (HEUCHENNE 1979-INDEX 3)
EFFICIENCY RATIO   :   .60

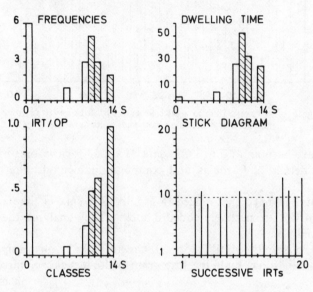

FIG. 2.11. An example of IRT distributions obtained using various methods of data analysis (see text).

produced in the preceding classes. The probability associated with each class is estimated by dividing the number of IRTs in that class by the sum of IRTs in that class
plus IRTs in all subsequent classes, that is by the number of opportunities offered.
Hence the name *IRT per opportunity*.

These estimates make sense only if we can assume a certain stationarity in behaviour. Were this not the case, it would be difficult to figure out how, at any time, the
probability of responding could possibly be estimated by resorting to information
from previous or subsequent intervals.

Dwelling Time. The expression has been forged by Weiss (1970) to designate a third
type of IRT analysis. It is based on the simple idea that the longer an IRT, the better
the performance. The Dwelling Time is computed by multiplying the observed frequencies of each class by the central value of the class. The distribution obtained
reflects the total time covered by IRTs of each class in the whole session (or whatever
experimental period). Here again, a good temporal regulation will be shown by the
location of the mode at the class of the minimal reinforced IRT.

Sequential Analysis.   In none of the preceding methods is the order of successive
IRTs taken into account. However, simple visual inspection of a stick diagram (Weiss
and others, 1966), as shown in Fig. 2.12, is often sufficient to reveal sequential dependencies: IRTs of similar size, for instance, appear in a row. Analysing these dependencies is a classical problem of time series analysis. Several of the techniques used in that

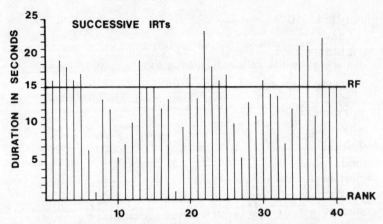

FIG. 2.12. Stick diagram of successive IRTs in a DRL schedule. Each IRT is represented by a vertical bar, the head of which is proportional to its duration.

kind of study have been applied to DRL data. We shall review some of them without going into the details of formulas and computation methods, that can be found elsewhere.

Weiss and Laties (1965) dichotomized IRTs into IRTs smaller than and IRTs larger than the median IRT (respectively coded 0 and 1). They analyzed the sequence of 0 and 1 so obtained.

The influence of reinforced IRTs on subsequent IRTs is obviously an important question. Weiss (1970) has built the histogram of frequencies of post-reinforcement IRTs.

Another question is whether, and to what extent, a given IRT depends upon the immediately preceding, upon the second, ... upon the nth preceding IRT. *Autocorrelation* techniques (Weiss and others, 1966), provide a means to answer that kind of question, provided that the stationarity of the series can be assumed, that the statistical properties of IRTs and their dependencies are not affected by chronological position. Data do not always permit such an assumption.

Some authors resort to "*smoothing*" *procedures* in order to reveal general trends in IRT series. For instance, moving averages have been used to that purpose. These tools must be used with caution since an unfortunate choice of parameters might mask interesting periodic components.

Another procedure has been proposed under the name *expectation density* (Weiss and Laties, 1964; Weiss, 1970). It gives an estimate of which IRT is the most likely to be produced during the *n* seconds following a given response. Each response being successively taken as time zero, the times at which following responses occur are computed within an interval of *n* seconds. The computation is done with the first response as origin, then with the second response, and so on. The histogram of the times obtained has modes corresponding to the dominant periodicities.

The sequence of IRTs is also amenable to *spectral analysis* (Weiss and others, 1966). This basically consists in decomposing x_i into a linear combination of trigonometric functions having decreasing periods. This type of analysis, which is also used in chronobiology, aims at analysing and quantifying the respective importance of the various periodic components of the series.

Such a variety of procedures confront the experimenter with a problem of choice. In the absence of any systematic comparison between them on the same set of experimental data, it is difficult to give preference to one on purely theoretical or practical grounds. From the above summary, it will be clear that efficiency indexes are useful but very crude expressions of the behaviour observed, leaving out important properties of the temporal regulation. These are better assessed by means of IRT analysis, such as IRTs/OP or Dwelling Time. The latter recommends itself as particularly revealing, while requiring no hypothesis as to the IRT sequence. The most complete and less sophisticated among sequential analysis procedures is doubtless the autocorrelation method. It leads to highly refined mathematical models (Anderson, 1976). A consistent effort is still needed to test the various mathematical procedures on a variety of experimental situations (using various types of responses, of delays, of reinforcers) and on a variety of species, and also to explore other possible models. Only then can we hope to develop a fully satisfying tool for measuring temporal regulation of the type considered here.

The techniques reviewed under the present section also apply, with some adaptations and restrictions, to other experimental paradigms presented below, that, as DRL, involve a required temporal regulation of behaviour.

b. Modified DRL Schedules and Other Procedures Requiring a Temporal Regulation

Discrete Trial DRL: Two-Response Procedures. In classical DRL schedules, IRTs that follow a reinforced response are not strictly comparable to IRTs following a non-reinforced response, since the subject spends part of them consuming the reinforcer. This is a source of ambiguity when one comes to problems concerning the mechanisms involved in temporal regulation. Two-response procedures offer a simple way to avoid the difficulty. The experimental cage is equipped with two levers (or whatever response device). A response on lever A (R_A) starts the delay. When the delay has elapsed, a response on lever B (R_B) is reinforced. A R_B during the delay ends the trial. A new trial will begin with another R_A. Responses on A following R_A simply reset the delay (i.e. they cancel the ongoing trial and immediately start another one). R_Bs following R_B have no consequences. The critical delay may be defined further by a maximum, as in the DRL limited hold (Fig. 2.13). The intervals between R_As and R_Bs $(R_A - R_B T)$ can be assimilated to IRTs in classical DRL and treated statistically in

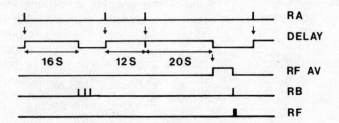

FIG. 2.13. Schematic representation of a discrete trial two Response DRL procedure, using Response A (RA) to initiate the delay and Response B (RB) as the reinforced response. The critical delay for reinforcement availability (Rf AV) is 20 s.

similar ways, except that sequential analyses are irrelevant. A two-lever DRL procedure was first conceived by Mechner and Guevrekian (1962) but it differed from the contingencies considered here in that responses on A following a response on A did not end the trial, they were without any consequence. The critical datum for the study of temporal properties of behaviour was not $R_A\ R_B\ T$ but the time elapsed between the first R_A (following an R_B) and R_B. Actually, extra R_As were rare in Mechner and Guevrekian (1962) and Mechner and Latranyi (1963) studies with rats, contrary to Nevin and Berryman's study with pigeons (1963). Mechner called this schedule *fixed minimum interval* (FMI) schedule (not to be confounded with the *fixed-minimum-interval* in FI contingencies).

The schedule where each R_A starts the delay provides for a more rigorous definition of the temporal requirement. It has been used with cats by Macar (1969) and by Greenwood (1978); with rats by Lince (1976) with a critical delay of 8 s LH 5 s, and by Blackman (1970) with delays of 5, 10 and 15 s. Blackman has added an auditory signal, started by R_A and interrupted by R_B. This signal gives no temporal information but it makes a clear contrast between trials and intertrial intervals (generated by the subject itself) where R_Bs have no chance of being reinforced. Blackman's procedure also makes the schedule in some respect similar to the schedule of reinforcement of response latencies (see below).

In all these studies, the high frequencies of very short IRTs (bursts) obtained in classical DRL are not observed. The distribution of IRTs is unimodal, with the mode around the class corresponding to the critical delay.

Differential Reinforcement of Response Duration (DRRD). What an animal does between operant responses on a DRL schedule has been a matter of concern for many experimenters. According to one hypothesis that will be discussed later (see Chapter 7), the so-called *collateral behaviour* would provide the timing mechanism needed in temporal regulations. Experimental manipulations of these collateral behaviours are relevant to this issue. Reinforcing responses of a given duration is a way to restrict the range of movements, that is of possible collateral behaviours, in which an animal can engage while estimating time. In a typical schedule of this category, the subject presses a lever, thus starting a delay, and holds the lever down until the delay has elapsed. Releasing it after the delay produces the reinforcement. An upper limit to the delay may be added, making for a LH contingency (Fig. 2.14). Response duration is amenable to the same statistical treatment as IRTs.

Skinner (1938) had already conditioned rats to press a lever for as long as 30 s, but the subjects did not really have to estimate time: when the lever had been pressed for the specified delay, an auditory stimulus was presented; releasing the lever in the presence of the discriminative stimulus produced the reinforcer. This type of con-

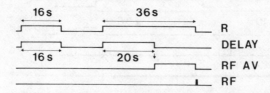

FIG. 2.14. Schematic representation of a schedule of reinforcement of response duration. The critical delay for reinforcement availability (RF AV) in this example is 20 s.

tingencies was also used by Stevenson and Clayton (1970) with rats, and earlier by Blough (1958) with pigeons. In the latter study, the response was defined as the relative immobility of the head cutting two perpendicular light beams (the traditional pecking response does not lend itself, of course, to the reinforcement of long durations). This kind of schedule could be called a *waiting schedule*, since the subject has only to wait until the reinforcement or a discriminative stimulus occurs; it does not have to estimate how long it is waiting. Waiting schedules provide a useful and simple way for demonstrating that a motor response can be maintained for a specified time, and that learning to estimate the response duration is not limited by purely motor problems.

Differential reinforcement of responses of short duration has been used by Molliver (1963), who reinforced cats for vocal responses lasting at least 0.4 s or not exceeding

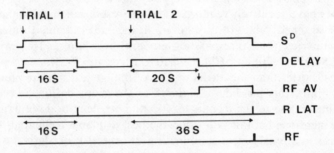

FIG. 2.15. Schematic representation of a schedule of reinforcement of response latencies. The critical response latency (R Lat) for reinforcement availability (RF AV) in this example is 20 s.

0.5 s; by McMillan and Patton (1965) on rats, monkeys (and humans), with critical durations between 1 and 1.27 s; by Ferraro and Grilly (1970) on rats, with the minimal value of the critical duration ranging from 0.1 to 1.6 s and a LH of 0.2 s. Longer durations have been successfully reinforced in rats by Platt and co-workers (1973) (0.4 to 6.4 s), and by Kuch (1974) (2, 4 and 8 s, with a LH equal to 25, 50 or 100 per cent of the minimal value). These authors used a retractable lever that was withdrawn from the cage during intertrial intervals. They obtained bimodal distributions of response durations, with one mode at very short values, and another mode around the reinforced value. Reducing the limited hold, as Kuch did, results in a reduced dispersion around the second mode, without affecting the first.

Response duration contingencies can be included in a two-lever situation. A minimal duration is specified for R_A, but releasing lever A is not sufficient: in order to receive the reinforcer, a response on B must be produced, eventually within a limited hold (Greenwood, 1977).

Differential Reinforcement of Response Latency (DRRL). In classical DRL, as in the two-lever procedure and in DRRD schedules, the subject initiates the delay that it has to estimate. The origin of the delay may be given from the environment, by an exteroceptive stimulus, that remains present (in a manner analogous to Pavlovian delayed conditioning) or that is presented only briefly (in a manner analogous to trace conditioning). The response is reinforced if it is emitted after the delay has elapsed and, if a limited hold is being used, before the limited hold is over (Fig. 2.15). Response latencies are amenable to the same statistical treatments as IRTs, R_A R_B T and response durations (with the same restrictions as to the relevance of sequential analysis in situations involving intertrial intervals).

Though there is an external stimulus, DRRL schedules clearly require correct positioning of the operant response in time, that is a temporal regulation of the subject's own behaviour. Therefore they are properly reviewed under this heading.

They have been explored in the rat by Zimmerman (1961) for values of 18 s LH 12 s; by Logan (1961) for 5 s; in the pigeon by Schwartz and Williams (1971) for 6 s, and by Catania (1970) for values ranging from 0.6 to 48 s; in monkeys by Saslow (1968) whose requirements were within the range of 200 to 600 ms. This provides interesting data on temporal regulations in the lower part of the time scale. Quite generally, the distributions of response latencies are unimodal, with the mode close to the reinforced value and few very short latencies. Moreover, the mean latency exceeds the specified value when the latter is small, and is inferior to it beyond 10 s or so.

Some authors have selectively reinforced "short" response latencies, that is latencies shorter than a specified value. In this situation, the subject tends to produce shorter and shorter latencies, which reveal no temporal regulation. This was done, for example, by Catania (1970) and Church and Carnathan (1963) who selectively reinforced latencies shorter than the median value of the latencies of the previous session. Pushed to its limits, this procedure is closer to the methods for collecting reaction times from animals than to the techniques for studying temporal regulation.

It should be mentioned at this point that the best available techniques for the study of reaction times in animals are a combination of several procedures reviewed in the present section. In a typical situation, a cat presses a lever at a warning signal, and holds the lever down until the imperative stimulus is presented (after an unpredictable delay). It has then to release the lever within the shortest delay in order to be reinforced. The first part of this sequence is similar to what has been called a *waiting schedule*. The last part involves differential reinforcement of short latencies. Valid reaction times are obtained by setting an upper limit for reinforcement very close to the onset of the imperative stimulus. The criterion for setting the limit may be adjusted to the subject's behaviour so that the reaction time is actually tracked (Macar and others, 1973).

C. Negative Contingencies

Temporal regulations induced by negative contingencies deserve separate treatment, not only for the sake of clarity, but because there might be some hesitation as to the category—*spontaneous* or *required*—into which they should be placed. Some authors have argued that temporal patterning observed under some avoidance schedules are not just by-products of specific temporal parameters of the contingencies, but an actual part of the avoidance-learning process (Anger, 1963; Dinsmoor and Sears, 1973; Rescorla, 1968). A full account of this theoretical problem would be outside the scope of this book. We could summarize the literature on the subject by saying that, on the whole, the arguments are rather favourable to the hypothesis of the spontaneous character of the temporal regulations observed (Sidman, 1966).

We shall omit in this methodological chapter the numerous versions of negative contingencies with added cues, that is schedules in which an aversive stimulus (to be avoided or escaped from) is signalled by a warning stimulus. This is not to say that temporal patterning never emerges under such contingencies—actually it does when certain temporal parameters are selected—for the main control of behaviour is not by time, but by the discriminative stimulus. In addition, few studies have been devoted to the temporal aspects of behaviour in this context. We shall similarly omit a number of intriguing and paradoxical effects obtained with shock presentation or shock termination in FI schedules. These will be discussed later when we consider the reinforcement variable and the role of the individual's history (see Chapter 5).

a. Non-discriminated Sidman-Type Avoidance

This schedule first investigated by Sidman (1953) is a *shock-delay procedure* (Hineline, 1977). In the absence of responses, brief shocks are delivered at regular intervals (S–S interval). A response interrupts this sequence and starts a response-shock (R–S) interval. The shock occurring at the end of the R-S interval reinstates the shock-shock interval sequence if no new response has been emitted. Another response in the response-shock interval resets the R–S delay and starts another R–S interval. Responses emitted at any interval shorter than the R–S delay are shock avoiding. S–S and R–S intervals may be chosen equal or different. The most commonly explored values range from 1 to 20 s or so. Fig. 2.16 represents the contingencies of the Sidman

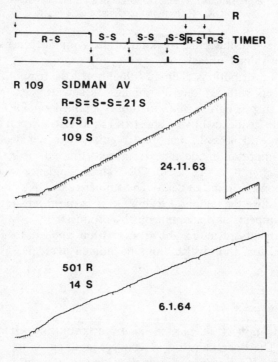

FIG. 2.16. Schematic representation of Sidman avoidance schedule (with R-S different from S-S), and samples of cumulative records obtained early in conditioning and after avoidance behaviour has been acquired in a rat using a R-S interval equal to S-S interval. Deflections of the cumulative pen indicate shocks (S). In the example shown, the number of shocks received decreases with training, though the number of responses (R) remains fairly unchanged. Efficient avoidance behaviour is attained by better distribution of responses through time.

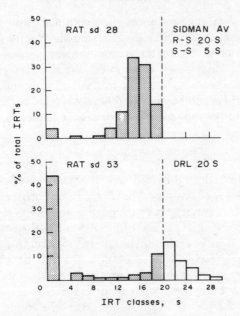

F<small>IG</small>. 2.17. IRT distributions in Sidman avoidance schedule in a rat that developed temporal regulation and in DRL. (Drawn and modified after Sidman.)

avoidance schedule and illustrates the kind of behaviour it generates, early in acquisition, and in well trained animals.

Analysing IRT's distributions, Sidman (1966) noted the temporal regulation emerging in some subjects with prolonged training. As can be seen from Fig. 2.17, IRT distributions are surprisingly similar to those obtained on DRL schedules. A first mode evidences the frequency of very short IRTs (bursts); a second mode towards the critical R-S delay evidences the temporal regulation of behaviour. Of course, this second and more important mode appears, in well-adjusted subjects, *before* the critical value rather than *after*, as is the case in DRL, since avoidance is contingent upon the emission of a response within the delay. The similarity to DRL IRT distribution is all the more striking since the temporal regulation is by no means required in the avoidance schedule. It appears as a spontaneous by-product.

Statistical treatments reviewed above have been applied to (and some of them initially developed for) IRT distributions in Sidman avoidance (see Anger, 1963, or Wertheim, 1975).

b. Fixed-Cycle Avoidance

This schedule is a *shock-deletion procedure*, according to Hineline's classification (Hineline, 1977). The operant response changes nothing in the scheduled cycles of shock presentation. If no response is emitted during a cycle, a brief shock will be delivered at the end of the cycle. One response during the cycle suffices to avoid the shock. Other responses bring no additional benefit to the subject. Sidman (1966) investigated this kind of schedule. He delivered a brief shock to rats every 15 s. As a rule, the subjects produced more than the unique necessary response in each cycle. In

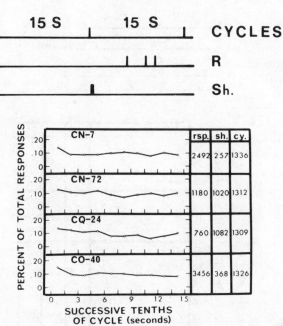

FIG. 2.18. Schematic representation of Fixed-cycle avoidance schedule, and distributions of responses throughout the cycle for four individual rats. Numbers in the columns at the right give the total responses (rsp.), shocks (sh.), and cycles per session (cy.). (Bottom, from Sidman, 1966. Copyright 1966 by Appleton Century Crofts, Inc., and Prentice Hall, Inc. By permission.)

most cases responses were equally distributed throughout the interval (Fig. 2.18). However, under certain conditions of shock frequency, a temporal pattern eventually developed. Other experiments have used a slightly different method derived from Schoenfeld and co-workers (1956) in which each cycle T is divided into two periods t^D and t^Δ. Responses emitted in t^Δ have no consequence; only responses emitted in t^D are instrumental in avoiding the shock that otherwise terminates the total cycle T (Fig. 2.19). t^D and t^Δ can be arranged in both orders ($t^D t^\Delta$, or $t^\Delta t^D$). A temporal pattern emerges depending upon shock frequency, t^D position and duration (Dunn and others, 1971; Hurwitz and Millenson, 1961; Kadden and others, 1974; Sidman, 1962, 1966).

Other variants of shock delay and shock deletion procedures and detailed theoretical discussions can be found in Sidman (1966) and Hineline (1977).

2.3.2. Temporal Discrimination

A. Discrimination of Duration of External Events

The duration of external events can function as a discriminative stimulus as do other dimensions of the environment, e.g. wavelength, frequency, intensity. We shall call the situations involving this particular type of temporal control *temporal discrimination* procedures, after Catania (1970). Here, temporal aspects of the *environment*, rather than temporal aspects of behaviour are correlated with the reinforcement of a given response. Schedule control is not assessed by temporal measures of behaviour, but by discrimination indices.

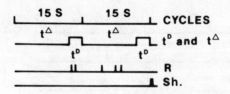

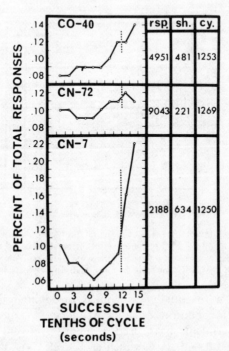

FIG. 2.19. Schematic representation of the limited interval avoidance schedule and samples of results from three individual rats. Shocks can be avoided if a response is emitted during t^D period only. For the columns at the right of the figure, see Fig. 2.18. (Bottom, from Sidman, 1966. Copyright 1966 by Appleton Century Crofts, Inc., and Prentice Hall, Inc. By permission.)

In a typical time discrimination experiment, a pigeon is trained to discriminate between two durations, for example 5 s versus 10 s. The subject faces a panel with a lamp, two response keys and a food aperture. After 5 s illuminations of the lamp, pecking the left key is reinforced, and after 10 s illuminations of the same lamp, pecking the right key is reinforced. The stimulus lamp is turned on at irregular intervals (step 1 in Fig. 2.20). Both response keys are transilluminated as soon as the lamp stimulus is terminated, and provided that no response has occurred during stimulus presentation (step 2). The task of the pigeon is to match the location of its response with the duration of the stimulus that has just been presented. In the case of correct matching, i.e. a right response after a long stimulus, or a left response after a short stimulus, both key-lights are turned off and reinforcement follows (step 3). If a wrong response is emitted, both key-lights are extinguished and no reinforcement is delivered. An intertrial interval then begins (step 4).

This example, which is only one case of the various procedures to be detailed later, contains the main features involved in discriminative control by duration. First, two durations or classes of durations are presented to the subject, each of them being associated with the reinforcement of a given behaviour. It should be noted that stimuli presented are identical in all respects but duration. Teaching an animal to discriminate along such a subtle dimension is by no means an easy task. Several shaping sessions are usually required. In some cases, fading procedures are used to transfer control from purely exteroceptive to temporal dimension. For instance, in the experiment described above one may first train the pigeon to respond on either illuminated key. Then very different long and short durations (that is 10 s and 1 s) are introduced, each value being associated with one of the keys. Only the correct key is illuminated after stimulus presentation, short stimuli are followed by an illuminated left key, the right one being dark and conversely; long stimuli are followed by an illuminated right key. Temporal control is introduced by progressively increasing the intensity of the wrong key from trial to trial until both keys are equally lit after each stimulus.

The example above also illustrates a methodological problem of particular importance in the use of temporal stimuli, namely the problem of *attention*. Sustained attending to the stimuli is especially required since any presented duration must elapse completely before its value is defined. With other dimensions, virtually no more than a reaction time is needed after onset of the stimulus for the subject to react adequately. The attentional constraint on duration discrimination can be a limiting factor as to

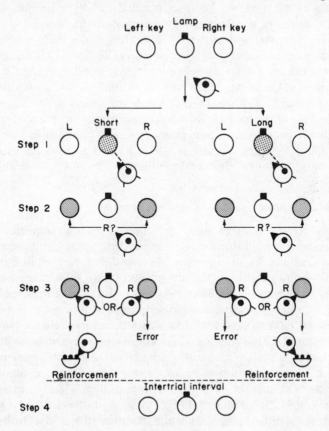

FIG. 2.20. Successive steps in a duration discrimination trial (see text).

the absolute value of the stimuli presented since the longer the stimulus, the more probable the occurrence of extraneous phenomena that may interfere with stimulus control. One of the methodological provisions for the control of attention is, as in our example, the discrete-trial, single-stimulus presentation, with irregular intervals between stimuli to maintain some "uncertainty" about the moment of occurrence of the next trial and thus force the subject to pay attention at every moment of the interval. In some other procedures the stimulus presentations are made contingent upon the emission of an "observing response" which offers the advantage of placing the subject in a given position to attend the beginning of the stimulus. Reinforcement contingencies also prevent the occurrence of responding during the stimulus, a behaviour that is incompatible with that of watching the lamp. This is achieved by punishing key-pecks during the stimulus with an early ending of the trial. Responding may also be controlled by the non-availability of the manipulanda during the stimuli.

Besides these common features, duration discrimination procedures can differ as to the stimuli used, the way the stimuli are presented, and the kind of responding required.

a. Stimuli Used

Visual stimuli: durations to be discriminated are those of the illuminations or illumination changes of a coloured lamp in most studies using pigeons as subjects (Mantanus, 1979; Perikel and others, 1974; Stubbs, 1968, 1976; Stubbs and Thomas, 1974). Similar stimuli are also used with monkeys (Elsmore, 1972). Some authors have simply used the duration of house-light extinction (Church and Deluty, 1977; Elsmore, 1971a; Reynolds and Catania, 1962). Auditory stimuli: white noise duration was used with monkeys (Catania, 1970) and rats (Church and others, 1976; Snapper and others, 1969). Kinchla (1970) trained pigeons to discriminate durations of a 1000 Hz tone. Temporal stimuli can consist of time intervals simply bounded by two external events which may differ in nature. Mantanus (1979) trained pigeons to discriminate empty intervals bounded by a light flash and the illumination of response keys.

b. Methods of Presentation of the Temporal Stimuli

Unless they are designed to study the effect of an independent variable on the discrimination of two fixed durations (Snapper and others, 1969), time discrimination studies always include several values of short and/or long stimuli in order to assess variations in accuracy with increasing or decreasing differences in duration. The various methods by which the different values of stimuli are introduced are analogous to those used in other psychophysical experiments.

"DIFFERENTIAL-THRESHOLD-TYPE" EXPERIMENTS. Several values of the stimulus durations are presented during training. Distinctions can be drawn according to the order of presentation of stimuli. The various possibilities are best described by using the terminology of human psychophysics. In the *method of limits* the durations are introduced in a stepwise manner, with differences in stimuli decreasing over blocks of sessions (Kinchla, 1970; Mantanus, 1973; Perikel and others, 1974). One of the durations is fixed (the standard stimulus), while the other (the comparison stimulus) takes values closer and closer to the standard.

When stepping is submitted to a performance criterion, the procedure can be referred to as an *adjusting method* or even a *titration schedule* where the defined criterion allows up-and-down changes of the comparison stimulus. Such a procedure was used by Church and co-workers (1976) who checked their rats' behaviour every eighth trial. Depending on the proportion of correct responses found in a given block of trials, the difference between long and short stimuli was reduced, increased by a constant duration, or remained unchanged.

With the *method of constant stimuli* all values of the comparison stimulus are presented randomly in each session (Elsmore, 1972; Reynolds and Catania, 1962). With this procedure the discrimination of classes of stimuli is possible; in the experiments designed by Stubbs (1968, 1976) and Stubbs and Thomas (1974) long and short stimuli each take several values.

"GENERALIZATION-TYPE" EXPERIMENTS. The introduction of several, non-reinforced test values distributed around two fixed training values has led to the establishment of temporal gradients of generalization (Elsmore, 1971a). Special interest has been given to the part of the continuum situated between the training durations. When selected interpolated durations are presented in an extinction test, measures of behavioural changes not only indicate the degree of similarity of each new value to the former ones (Catania, 1970), but also locate the "subjective midpoint" between the two durations under study (Church and Deluty, 1977).

c. Types of Responding

The types of responding used in time discrimination experiments are all variants of the usual discrimination procedures.

"SINGLE-RESPONSE" METHODS. These methods allow for one or a few responses after each stimulus presentation. In the so-called "go-no go" procedures, one response manipulandum is available. One duration is correlated with reinforcement and constitutes the positive stimulus, the other duration(s) being correlated with non-reinforcement and thus acting as the negative stimulus. The response alternative for the subject is to respond or not respond according to the duration of the stimulus (Perikel and others, 1974; Richelle, 1972).

Other experiments provide the subject with two manipulanda, each being linked with a given type of duration, long or short. After each stimulus presentation, the subject chooses one of the manipulanda on the basis of a matching relation. In most experiments this matching is based on location, that is after durations of a certain class, reinforcement follows one response on the left manipulandum, and after other durations, one response on the right manipulandum is reinforced (Catania, 1970; Church and others, 1976; Church and Deluty, 1977; Kinchla, 1970; Mantanus, 1979; Perikel and others, 1974; Snapper and others, 1969). Matching can also be based on colour cues. For instance, responding on a red key is reinforced after short durations, responding on a green key after long durations (Stubbs, 1968, 1976; Stubbs and Thomas, 1974). The colours alternate across keys randomly from trial to trial to prevent the development of position habits. Situations with two manipulanda offer the advantage of equalizing the outcomes of responses after long and after short stimuli.

"MULTIPLE-RESPONSE" METHODS. The experiments that allow the occurrence of many responses after each stimulus are designed for the establishment of generalization gradients. The subject is given access to a variable-interval (VI) schedule for some period of time after presentation of the positive stimulus, and to an extinction schedule after presentation of the negative stimulus. Rate of responding is measured after each of the durations presented in the extinction test which follows training. It should be noted that generalization gradients are also obtained when several values of the stimulus correlated with extinction are presented during training (Catania, 1970; Reynolds and Catania, 1962).

d. Data Analysis in Duration Discrimination

In duration discrimination the responses recorded do not possess temporal properties relevant to the duration of stimuli; they are recorded as all-or-none data. The relevant information consists, for instance, in the location of a press being on the right or left lever, or in the number of key-pecks emitted during some fixed period after stimulus presentation. The accuracy of performance is evaluated by relative measures of behaviour like proportions of choice of a manipulandum or relative rates of responding. Such measures can be applied to any discrimination task, whatever the diminsion of the environment under study. With duration discrimination, however, such measures refer to the quality of temporal adjustment displayed by the organism.

The discrimination of temporal stimuli is often analysed as a choice situation, in a manner similar to that used with other stimulus dimensions. The prerequisites for applying choice theory are easily found in the procedures with one or two response manipulanda as presented above.

There is a set of stimuli (S) assumed to differ in only one continuous physical dimension (i.e. duration).

There is a set of responses (R). For instance R can be composed of two responses, like peck on the left key (R_1), and peck on the right key (R_2).

Each stimulus of S is associated with a response of R. The function that maps each response on to the subset of stimuli is called the identification function; in this case it is prescribed by the contingencies of reinforcement and it is assumed to be acquired in a preliminary phase of training,[1] the model dealing only with asymptotic behaviour.

Each experiment is composed of a sequence of trials. On each trial, a given response is linked with a given stimulus.

All the models applied to duration discrimination must take into account the fact that presentations of a same stimulus may produce different responses. This suggests either that the stimulus always produces the same internal effect upon which several decision rules are applied, or that the stimulus as experienced by the subject is intrinsically variable. As noted by Luce and Galanter (1963) a varying decision rule, although actually possible, seems unmanageable with mathematical treatment. The other possibility, that is a varying internal effect with a fixed decision rule, requires the use of probabilities. The problem is, then, to describe the probability of the different responses as a function of the durations taken by the stimulus, and to attempt to develop models for the information received and decision rule applied by the subject.

---

[1] If identification learning is not complete, this will introduce some distortion in the results. As far as we know this aspect has not yet been taken into account in data analysis, though it has been pointed out by Bush and co-workers (1963).

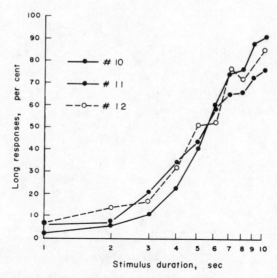

FIG. 2.21. Psychometric function of three pigeons (10, 11 and 12) in a duration discrimination experiment using the method of constant stimuli. The cut-off for choice is between 5 and 6 s. Abscissa: stimulus duration (log scale). Ordinate: percent responses on the key associated with long (6 to 10 s) stimuli. (From Stubbs, 1968. Copyright 1968 by the Society for the Experimental Analysis of Behavior, Inc.)

THE PSYCHOMETRIC FUNCTION. A description of data that is often favoured in duration discrimination is the psychometric function. The context for its use can be defined as follows. Let x denote the duration of the stimulus. Let s be the value of x which defines the cut-off for choice; values longer than s will be linked for instance with R_1 responses, values of x shorter than s will be linked with R_2 responses. If the probability of R_1 given x increases from 0 to 1 with x and is continuous and differentiable, this relation is called the *psychometric function* (Urban, 1907, quoted by Luce and Galanter, 1963). The typical plot of this function is an S-shaped curve, as illustrated in Fig. 2.21. The probability of R_1 given x is estimated by the proportion of R_1 responses given x. In the example shown (Fig. 2.21) the behaviour of pigeons discriminating stimulus duration ranges (1 to 5 and 6 to 10 s) is described by the proportion of responses on the coloured key that is associated with long stimuli. Consequently, responses on this key after 1 to 5 s stimuli are errors, and responses after 5 to 10 s stimuli are correct choices.

The psychometric function may take a different aspect when the experiment consists of reducing the difference between standard (S) and comparison (x) stimuli over blocks of sessions (a variant of the method of limits). When the values of the comparison stimulus are varied in a one-tailed fashion, that is with all differences between standard and comparison stimuli bearing the same sign, then the probability of correct choice is preferred to the probability of R_1 given x. The probability of correct choice is estimated by the proportion of total correct responses. The function that relates probability of correct choice to x values takes the form of the upper half of the psychometric curve, with higher probabilities corresponding to larger differences between standard and comparison stimuli. Figure 2.22 is drawn from an experiment

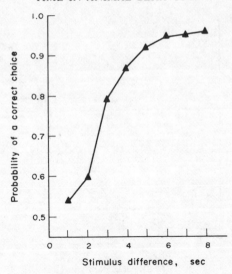

FIG. 2.22. A pigeon's psychometric function in a duration discrimination task with fixed standard stimulus value. Abscissa: difference between the standard stimulus (10 s) and the various comparison stimuli (2 to 9 s). Ordinate: probability of a correct choice, estimated by the proportion of correct responses both to short and long stimuli. (After Mantanus, 1971, unpublished data.)

by Mantanus (1979). Pigeons discriminate two durations; one is long (10 s) and fixed, the other takes increasing values (from 2 to 9 s in 1 s steps over blocks of ten sessions). The proportion of total correct responses is plotted against the difference between standard and comparison values.

Different measures of sensitivity, assumed to be independent of the subject's decision rule, have been proposed and will be reported later in this chapter. The relation between one such measure of sensitivity and stimulus duration is sometimes called a *psychometric function*. As Wright (1974) pointed out, according to this definition, psychometric functions turn into lines. He argued that this simple picture has been obscured in much psychophysical work because of procedures that introduce strong response bias, i.e. they result in decision rules that vary with the value taken by the comparison stimulus.

The plot of a response measure against stimulus values may also give rise to a *generalization gradient*. The context for this type of analysis was defined earlier. Let s_1 denote the stimulus duration associated with the VI schedule, and s_2 the stimulus duration associated with the extinction schedule. When stimulus values other than s_1 or s_2 are presented, responding is maintained to some degree as a function of the duration of the stimulus. The plot of response rate variation as a function of stimulus duration constitutes the generalization gradient. It reflects the spreading of the association between the response and the training stimuli. Generally, there is a peak in the vicinity of the duration associated with reinforcement during training, then response rate decreases as the distance from the positive duration increases, arriving at a near-zero level at the stimulus value correlated with extinction. This is illustrated in Fig. 2.23, reprinted from Elsmore (1971a). Pigeons were trained with a 9 s duration as the positive stimulus and a 21 s duration as the negative stimulus. Stimuli ranging from 3 to 27 s in 3 s steps were presented during testing. The ordinate of the plot is the rate of responding during the 30 s period that followed each stimulus presentation. Gradients may take a variety of shapes, depending on training conditions and stimulus range presented during testing. Little quantification has been developed, in spite of some attempts by Shepard (1965). In many studies the analysis is restricted to visual inspection of the steepness of the gradients.

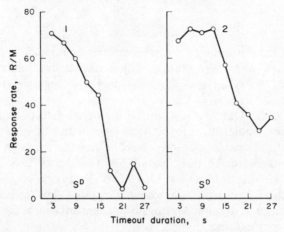

FIG. 2.23. Generalization gradients after training with stimulus (timeout) durations of 9 and 21 s in two pigeons (1 and 2). Abscissa: duration of the different test stimuli. The positive training value is referred to as "S^D". Ordinate: response rate in extinction during 30 s after stimulus presentation. (Modified from Elsmore, 1971a. Copyright 1971 by the Society for the Experimental Analysis of Behavior, Inc.)

MAIN CHARACTERIZATIONS OF THE PSYCHOMETRIC FUNCTION. A psychometric function looks like a cumulative function. Two important features of such a function are the measure of its central tendency—usually the mean or the median, and the measure of its dispersion—the standard deviation or the semi-interquartile range. Median and semi-interquartile range require weaker assumptions about the metric properties of the independent variable—they are usually preferred to mean and standard deviation. The *median*, that is the value of x which gives a probability of R_1 given x (alternatively, a probability of correct choice) equal to 0.5, is called in the discrimination context the *point of subjective equality* (PSE). This point may not always equal s because of bias in the responses. The half of the interquartile range is called the *just noticeable difference* (JND). When the proportion of correct responses is used as the measure of discrimination, the JND is computed as the difference between the 0.75 quartile and the PSE. Other measures of dispersion have been used for example by Luce and Galanter (1963) and Treisman (1963).

WEBER'S LAW. A classical question in discrimination experiments is: how does the JND vary with the value of the standard stimulus? This function was first studied by Weber. He suggested the following form for the function:

$$JND(s) = ks$$

where the constant $k > 0$ is called the Weber fraction. Weber's law states that if discriminability is characterized by the JND value, the discriminability is proportional to the standard stimulus value. Luce and Galanter (1963) have proposed a more general form of Weber's law; the psychometric functions should superpose for all standards when plotted against the ratio of the stimulus values to be compared, namely x/s.

Another, earlier generalization of Weber's law was proposed by Fechner (1860). It is formulated as

$$JND(s) = ks + c,$$

where $k > 0$. A variant is:

$$\sigma(s) = ks + c,$$

where σ is the standard deviation of the psychometric function. Some of these various forms of Weber's law have been studied with duration discrimination, in man by Treisman (1963) and by Allan and Kristofferson (1974), in animal by Church and co-workers (1976) and Gibbon (1977).

A MODEL FOR SIGNAL DETECTION. A model for signal detection can be applied to animal discrimination studies since, for subjects faced with a standard and a comparison stimulus, one of the stimuli can always be considered as the "signal" and the other as the "noise". These were the basic notions of the theory of signal detectability developed by Peterson and co-workers (1954). This theory was originally designed for the detection of weak signals against a noise background. The language and concepts may, however, be applied by analogy to any context in which the sensory input is ambiguous. The theory assumes that:

the relevant information available to the subject as a result of stimulation can be summarized by a number;
repeated presentations of a stimulus produce not the same number but a distribution of numbers;
the subject behaves as if he knew the distribution associated with each stimulus.

The relevance of these assumptions to animal subject has been discussed by Boneau and Cole (1967). With this model, the aspects of sensitivity, as the result of perceptual processes, and bias, as the result of the subject's decision rule, are estimated as different parameters (d' and β, respectively). The "classical" theory made strong assumptions about the distribution of the effects of the noise and signal; these were supposed normal and of equal variance. But recent research has developed non-parametric measures for sensitivity and bias.

The main indices of sensitivity are Grier's A' (1971), and Frey and Colliver's SI (1973). The same authors also propose bias indices: B" and RI, respectively. The parameters reflecting sensitivity and bias are computed from two conditional probabilities; $p(Y|S)$, the probability of giving the response associated with the presence of the signal ("yes"), also called probability of *hit*, and $p(Y|N)$ the probability of giving the "yes" response in the presence of noise, or the *false alarm* probability. These two probabilities are estimated from the data by proportions ("rates") of hits and false alarms. When these values are plotted against one another for any possible degree of bias, the curve obtained is called the *isosensitivity* or *Receiver operating characteristic* (ROC) curve. When hit and false-alarm rates are plotted against one another for any possible level of sensitivity, they form an isobias curve (Swets, 1973).

An example of the application of signal detection theory to duration discrimination is given by Stubbs' (1976) study in which ROC curves were obtained from three pigeons. Bias was varied by manipulating the relative reinforcement rate for correct

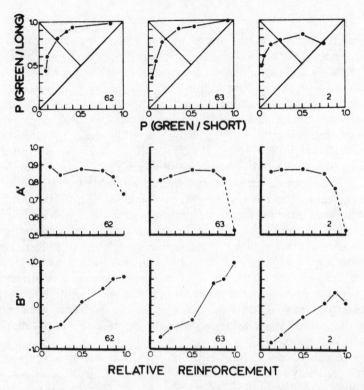

FIG. 2.24. Measures of sensitivity and bias in three pigeons (62, 63, 2) according to signal detection theory (Stubbs, 1976). The duration discriminated range from 11 to 15 s for short stimuli and from 16 to 22 s for long stimuli. All long stimuli are considered as "the signal". Responses in green stand for "yes" responses. Top row shows ROC plots; the hit and false-alarm rates are plotted against each other for different values of relative reinforcement rate for green-key correct responses. Middle and bottom rows show the values taken by the non-parametric index of sensitivity A', and the non-parametric index of bias B'' with increasing relative reinforcement rate (From Stubbs, 1976. Copyright 1976 by the Society for the Experimental Analysis of Behavior, Inc.)

responses. The hit rates were given by the proportions of correct responses on the coloured key associated with long stimuli—which in this case represented the signal. The false alarm rates were given by the proportions of error responses on the same key. Figure 2.24 presents, along with the ROC curves, the non-parametric indices A' (sensitivity) and B'' (bias) as a function of relative reinforcement rate.

ALTERNATIVE MODELS. Models other than signal detection could be used for estimation of sensitivity and bias parameters. These models have already received application in different discrimination contexts (see Laming, 1973, for detailed description). Models specifically devoted to time estimation have been developed with human subjects (Eisler, 1975; Treisman, 1963). These models could also be applied to animal time discrimination.

B. Discrimination of Complex Temporal Properties of External Events

Time may be integrated in more complex characteristics of stimuli, namely rhythm and velocity.

a. Rhythm

Rhythm is involved in stimulus control by click-rate. The discrimination is based on the difference between the intervals separating the clicks that form an auditory pattern. The methods most commonly used in click-rate discrimination belong to the "generalization-type" experiments defined earlier. Farthing and Hearst (1972) and Williams (1973) used the multiple VI-extinction schedule with pigeons, but other authors associated a schedule of negative reinforcement, i.e. with shock postponement as the reinforcement, with one of the click-rates, while the other rate was correlated with no shock (Sidman, 1960; Hearst, 1965). A comparative study for rats and humans was set up by Mostofsky and co-workers (1964), with a "go-no go" response requirement (see 2.3.2.Ac). The subjects were first trained with a positive stimulus value (e.g. a rate of 10 Hz) and a negative value that was inferior (e.g. 2 Hz). After a test for generalization with three intermediate durations never followed by reinforcement, the subjects received additional discrimination training with the same positive rate but with a different superior rate for the negative stimulus (e.g. 18 Hz). Then another generalization test took place. Assessment of stimulus control by click-rate was also achieved by experiments where conditioned suppression of a positively reinforced behaviour resulted from repeated presentations of a warning stimulus (a given click-rate) followed by unavoidable shock. Generalization gradients were based on the proportion of response suppression recorded during each test click-rate (Hendry and others, 1969; Winograd, 1965).

Other experiments consider rapid rhythm of visual flashes up to the point of perceptual fusion into steady light (Critical fusion frequency, CFF). Such studies are in fact concerned with visual perception, as previous results with humans have shown that CFF varies as a function of stimulus intensity. However, the discrimination of flickering versus steady light is based on the presence versus absence of a perceptible interval between two stimulations; an absolute threshold for time can be found at the flashing rate which leads in 50 per cent of the cases to a behaviour similar to the one controlled by the steady stimulus. The variation of this threshold with the intensity of the stimulus has been measured in pigeons (Hendricks, 1966) and in budgerigars (Ginsburg and Nilson, 1971).

b. Movement Velocity

The perception of the velocity of a motion integrates temporal as well as spatial elements since the time required by the target to cover a given distance may be used as a cue. In Hodos and co-worker's (1976) experiments, pigeons discriminated the presence versus the absence of motion in visual targets; in one experiment radii on a circle, in the other horizontal stripes on a film tape (Fig. 2.25). The stimuli were presented on a central display. The subjects matched location of the two side keys (right or left) with the stimulus displayed (motionless or moving). Training consisted of repeated presentations of various speeds in decreasing order. The absolute threshold for movement was defined as the speed for which a 75 per cent correct choice is found. The study was completed by non-parametric indices of sensitivity and bias computed according to a signal-detection model.

"Apparent movement" speed also possesses discriminative properties. By means of polarized lights rotating behind sensibilized pictures, Siegel (1970) simulated various

FIG. 2.25. Displays for the discrimination of velocity of movement in pigeons. Top: rotating radii. Bottom: vertically moving stripes. The discs on left and right of the displays are response keys. The food aperture is just below the stimulus panel (From Hodos and others, 1976. Copyright 1976 by the Society for the Experimental Analysis of Behavior, Inc.)

motions (horizontal, vertical, clockwise, etc.) with controllable speed. Some of his pigeons were reinforced according to a VI-30 min schedule during presentation of "moving" pictures at a rate of 4 cps, and were never reinforced during presentation of motionless pictures. For the other pigeons, the reverse conditions prevailed. Generalization tests were performed, some of them including presentation of high speeds to assess whether these speeds would lead to the same rates of responding as the motionless pictures.

PART II
BASIC RESULTS

III. *Temporal Parameters and Behaviour*

3.1. TEMPORAL REGULATION OF THE
SUBJECT'S OWN BEHAVIOUR

3.1.1. Properties of Spontaneous Regulation

A. Absolute Delays and Relative Timing Behaviour

The typical pattern of behaviour generated by the classical FI schedule has been repeatedly obtained along a wide range of values. Dews (1965b) has submitted pigeons to intervals from 500 s to 100,000 s, Schneider (1969) to intervals from 16 to 512 s; Lejeune (1971a), working with cats, has explored intervals from 2 to 15 min (Fig. 3.1). These are only a few examples. The post-reinforcement pause that develops spontaneously is in all cases a function of the total interval duration. Whatever the index used to measure the performance, behaviour is controlled by relative, not absolute

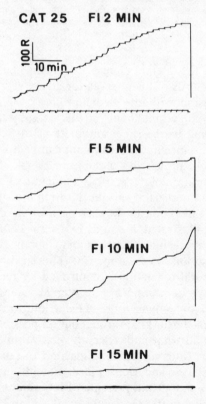

FIG. 3.1. Cumulative records of a cat in a Fixed-Interval schedule with interval values increasing from 2 to 15 min. (After Lejeune, 1971a.)

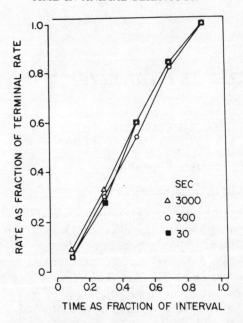

FIG. 3.2. Response rates in five successive segments of intervals of various length (30, 300, 3000 s). The mean number of responses in each segment is expressed as a fraction of the mean number of responses in the final segment. (From Dews, 1970. Copyright 1970 by Appleton Century Crofts, Inc. and Prentice Hall, Inc. By permission.)

time. The pause extends as the interval increases. In Schneider's study, the breakpoint was located at about two-thirds of the interval, for all six values explored (from 16 to 512 s, in geometric progression, with intermediate values of 32, 64, 128, 256). The slope of the line relating the average breakpoint of the six subjects to the values of the interval was 0.67. According to the same author, and to Schneider and Neuringer (1972), 99 per cent of the variance in the breakpoint can be accounted for by the value of the interval, very little being left to the subject variable. Similarly, Shull (1971a) showed that the median duration of the pause is a function of interval duration (for FI of 30, 60 and 300 s). When intervals of various duration are subdivided into an equal number of parts, response rates coincide for each part, whatever the absolute length of the interval. This is clearly illustrated in a study by Dews (1970, Fig. 3.2) and has been shown in less classical situations by La Barbera and Church (1974). General activity in a Fixed-Time situation also follows the same rule (Killeen, 1975).

That spontaneous temporal regulation is controlled by relative not absolute time holds true for a wide range of situations and interval values. However, it does not follow that the function relating any measure of behaviour to the temporal parameters of the schedule is of a simple linear kind throughout all possible values of the interval. Indeed, there is evidence that while still reflecting relative time, performances become less adjusted to time as the interval increases beyond certain values. This is clearly evidenced by the Fry, Kelleher and Cook index. For instance, two cats studied by Lejeune (1971a) on FI 2, 5 and 10 min presented FKC values (ten-session average for each of the three FI values) of 0.637, 0.630 and 0.545 and 0.710, 0.642 and 0.562 respectively (computed on eight subdivisions, the index could fluctuate between

−0.825 and +0.825). Still more convincing are the observations reported by Stubbs and others (1978) on pigeons, for FI 10 s, 20 s and 100 s. The FKC index (with maximal value of ±0.75, since four subdivisions were used) decreased from 0.6 or 0.5 to 0.4 or 0.35 when passing from FI 20 to FI 100 s. Though other variables such as the duration of reinforcement might partly account for these results, the decline of FKC index at high values of the interval suggests that temporal regulation, though obviously present throughout a wide range of interval durations, is best within a given range for a given species (or individual). The function relating behaviour to interval length is not a simple logarithmic linear function.

Striking evidence for the control of behaviour by relative rather than by absolute time is also found in Harzem's studies of *Progressive-Interval* (PI) schedules (Harzem, 1969). In these schedules, the duration of the interval increases from interval to interval. In one version of these contingencies, the progression is arithmetic, that is, a constant value is added to the value of the preceding interval. Starting with 60 s, and adding a constant amount of 60 s, successive intervals will be 120, 180, 240 and so on. As one progresses in such a series, the relative increase from one interval to the next gets smaller and smaller. In another version, the progression is geometric. Starting with 60 s, a constant proportion of 20 per cent is added from one interval to the next. The relative difference between successive intervals remains constant throughout the series. Subjects adjust to these progressive increases of the interval, but they do so much more regularly to the second version (geometric PI) than to the first. Typical results of two rats are shown in Fig. 3.3. The mean time elapsed until the second response was emitted is plotted against the interval duration on double logarithmic co-ordinates. The values obtained are located on a diagonal with a slope of 1.016. If one ignores the discrepancies between the subject with geometric progression and the subject with arithmetic progression for the shortest interval value, both subjects' data appear well approximated to the best fit line for the range from 3 to 50 min.

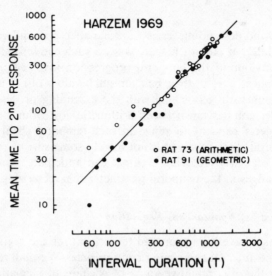

FIG. 3.3. Adjustment to progressive interval schedules, arithmetic and geometric, as a function of interval duration. See text. (Data from Harzem, 1969, as plotted by Gibbon, 1977. Copyright, 1977 by the American Psychological Association. Reprinted by permission.)

A similar adaptation to changing intervals has been demonstrated by Innis and Staddon (1971) in pigeons, with a cyclic interval schedule. Each cycle consisted of seven intervals of increasing duration followed by seven intervals of decreasing duration. The duration of the intervals increased from $2t$ to $8t$ or decreased from $8t$ to $2t$. t values of 2, 4, 10, 20 and 40 s have been explored. Two conditions were compared: in

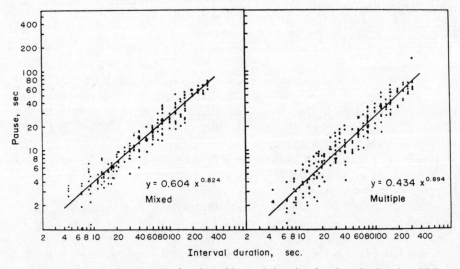

FIG. 3.4. Post-reinforcement pause as a function of interval duration for the mixed and multiple cyclic interval schedule. Both scales are logarithmic. Equations for the regression line are indicated on the figure. (Modified, from Innis and Staddon, 1971. Copyright, 1971 by the Society for the Experimental Analysis of Behavior, Inc.)

one, the ascending and descending series were signalled by a different coloured stimulus (multiple schedule); in the other, there was no such cue (mixed schedule). Subjects adjusted their behaviour to the temporal progression involved in the contingencies (Fig. 3.4). Performances were at their best for small values of t (2 and 4 s) and when a discriminative stimulus was associated with the ascending or descending series. The same study includes comparison between arithmetic, logarithmic and geometric progressions. The subjects' capacity to adjust to such temporal changes is quite remarkable. The behavioural clock involved is not merely some sort of recycling device that can be tuned to a regularly recurring event. It can anticipate, in a very sophisticated manner, lawful changes in the temporal parameters characterizing the contingencies.

B. The Determinants of Spontaneous Regulation

Several hypotheses have been proposed to account for the regulation of behaviour observed with Fixed-Interval and similar contingencies. We shall review three of them. The first relies on the discriminative value of a chain of mediating behaviours. The second emphasizes the delay to reinforcement and the third the time estimation from a point of origin, for instance, the reinforced response.

The Chaining Hypothesis

According to this hypothesis (Anger, 1963; Dews, 1962), the temporal regulation evidenced by the pause-activity alternation would be due to the subject discriminating its own behaviour. Intervals would be filled with a chain of *mediating behaviours*, be it the operant responses themselves, or any other behaviour occurring during the pause or between operant responses. This hypothesis, that is akin to the explanation of temporal regulation by collateral behaviour (see Chapter 7) dispenses the organism with a timing mechanism proper. All what is needed is the capacity of a subject to react to its own overt behaviour. Attempts to explain behavioural data by purely behavioural variables, observable at the very same level, are not rare in psychology. They are heuristically valuable, even when experimental tests produce negative outcomes: we then know that we must look for the mechanism at another, possibly less accessible level.

In the present case, conclusive experiments initiated by Dews (1962) and extended by Wall (1965), McKearney, (1970b), Schneider and Neuringer (1972), Hienz and Eckerman (1974) definitely discard the chaining hypothesis. The procedure used is basically as follows: a discriminative negative stimulus never associated with the reinforcement is intermittently presented during the interval, alternating with phases of positive stimulus. For example, in a FI 500 s schedule, a bright house-light (S$^-$) is alternated with normal houselight (S$^+$) every 50 s, the last 50 s of the interval being associated with S$^+$. The S$^-$ presentations result in a decrease (eventually a suppression) of responding. However, the rate of responding in S$^+$ periods is unaltered, and depends upon the position of the period in the interval. The temporal pattern is preserved. This could not be the case if the operant responses played the role of mediating behaviour for the maintenance of temporal regulation (Fig. 3.5). This holds true when as few as two S$^+$ 50 s periods are provided in a 500 s interval (leaving 80 per cent of the total duration under S$^-$): the rate of responding is clearly a function of the temporal position of the period (Dews, 1966a). Data on pigeons have been confirmed by Dews in the squirrel monkey (1965a); they have also been explored under various experimental conditions (Dews, 1966b) and for a wide range of interval durations (from 500 to 100,000 s with S$^-$ periods extending up to more than $2\frac{1}{2}$ h) (Dews, 1965b).

Other studies cited above have obtained similar results, sometimes using response latency rather than response rate as a dependent variable. In all cases, the "strength of the response" changes as time elapses, no matter what is done to interrupt the chain of behaviours. It would seem that here, as in the case of motor sequences discussed by Lashley (1951), the organization of behaviour cannot be accounted for by a model that makes each portion of a sequence the necessary stimulus for the next portion. Though speed is not a critical argument in the case of spontaneous temporal regulation of the sort discussed here, we must look for a less peripheric and less indirect mechanism.

The Delay to Reinforcement Hypothesis

As formulated by Dews (1962, p. 373), this hypothesis states that "the progressive increase in rate of responding through a fixed interval would be based on a declining retroactive rate-enhancing effect of the reinforcing stimuli as the delay between response and reinforcement is increased". This gradient of delay to reinforcement, however, supposes temporal discrimination rather than explains it (for an opposite view,

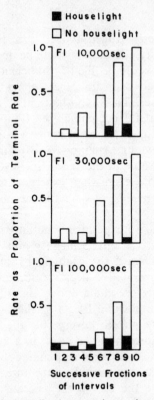

FIG. 3.5. Mean proportional rates of responding in successive tenth parts of FI. The numbers of responses in the individual segments are expressed as a fraction of the number in the terminal (tenth) segment of all intervals of a session and these ratios averaged to give the values shown in the graph. Houselight is the negative stimulus. (From Dews, 1965b. Copyright 1965 by the Society for the Experimental Analysis of Behavior, Inc.)

see Morse, 1966). As Jenkins (1970) demonstrates, after Dews (1962), delay to reinforcement must be discriminated by reference to a fixed origin point in order to exert control over behaviour. In other words, differential control by the delay to reinforcement would not be possible without estimation of the time from the beginning of the interval. It might then seem more economical to resort to such a temporal discrimination directly.

Discrimination of Duration Proper

Skinner (1938) relied on a process of generalization of responses to temporal stimuli throughout the interval to explain FI behaviour. A similar hypothesis is found in Spence (1956) concerning the regulation of behaviour in short and long alleys. The concept of temporal stimulus, however, is unclear. Time is a ubiquitous dimension of environmental events, and we do not know of any specific receptor for duration. We talk of temporal stimulus only in a metaphorical way. Straightforward as it is at the theoretical level, the estimation of time from the origin of the interval, independent of the delay to reinforcement, is not easy to demonstrate. A recent experiment by Stubbs and others (1978) shows that pigeons can discriminate the time elapsed from the origin of an interval—a fact that is by no means surprising in view of the performance in tasks involving discrimination of response latencies or of the duration of external events (reviewed below). Fixed intervals of 10, 20 or 100 s are intermittently interrupted by a choice procedure. Either early or late in the interval, two lateral keys are lit. Pecking on the left or the right key will be reinforced depending upon the time in the interval at which they are illuminated. Subjects adjust successfully to these contingencies. Control experiments have ruled out the possible discriminative control of FI responses. An experiment by Wall (1965) provides some convincing evidence regarding the issue. It aims at answering the following question: will we observe an increase in the probability of responding throughout the interval if responses *other* than the reinforced response have never been previously followed by reinforcement, be it delayed? A retractable lever is presented in the rat's cage every 60 s, and is withdrawn as soon as the reinforced response has been produced. After training in this procedure, the lever is introduced once during the interval and maintained for 5 s. Subjects in three different groups are presented with the lever at the 15th, the 30th and the 45th s of the interval, respectively. Subjects of another group are presented with the lever at all three points in time, randomly. Response latency (measured from the moment at which the lever is introduced) is shortest for the reinforced response. For other responses, it varies inversely with the time elapsed from the beginning of the interval (Fig. 3.6). The important point here is that this holds true for the very first presentations of the lever in the interval, that is to say for responses that have not been produced before in the same temporal relation to reinforcement. This cannot be accounted for by a gradient of delay to reinforcement. However, when the procedure is repeated, a steeper decrease in response latency during the interval is observed, possibly due to the retroactive effect of reinforcement. It therefore seems reasonable to conclude, after Jenkins (1970), that the stabilized patterns of responding in a FI schedule result from the interaction between a temporal discrimination of time elapsed from the origin of the interval and a delay to reinforcement effect. What are precisely the respective parts of these two factors is another unanswered question. A few experimental attempts to introduce a delay between the reinforced response and the delivery of reinforcement have produced drastic decrease in the rate of responding (Dews, 1969; Skinner, 1938). Thus, control by the reinforcing stimulus seems rather weak, as Zeiler (1977) observed, and it seems quite unlikely that it plays more than a secondary role, if any. Other experimental evidence bearing on the present issue is provided by

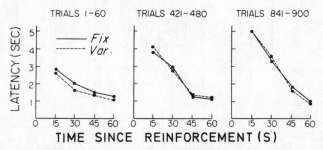

FIG. 3.6. Latency of responding as a function of time since reinforcement when a retractable lever is presented at one of three points in the 60 s interval (in addition to the presentation at 60 s) (Fixed condition) or at the same three points at random (variable condition). (From Wall, 1965. Copyright 1965 by the American Psychological Association. Reprinted by permission.)

studies investigating the function of external stimuli associated with the reinforcer in FI schedules (see Chapter 8).

C. Suprasegmental Properties of FI Behaviour

In the preceding sections, we were concerned with the characteristics of behaviour generated by FI contingencies at the level of the interval. Important as they are, they are not the only aspects of behaviour that deserve attention. It is interesting, for example, to look for possible effects of such periodic schedules throughout a series of intervals, over a whole experimental session, over a series of sessions, or after extended exposure to the contingencies. We shall call these and similar effects *suprasegmental* effects. They have been investigated far less than the segmental effects discussed above. However, a few suggestive results are worth mentioning.

Interval-to-Interval Fluctuations

The number of responses is not the same in all successive intervals. Dews (1970), studying what he has called *second-order effects* (a name derived from Skinner's *second order deviations*), has observed in a sequence of 200 180 s intervals, variations over a wide range (sometimes 100-fold, usually 50-fold—see Fig. 3.7). Intervals with many or few responses tend to occur in groups. There is no simple alternance of low and high

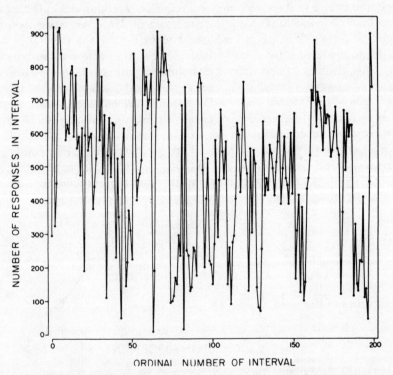

Fig. 3.7. The number of responses in consecutive intervals under FI 180 s. Abscissa: intervals in order of occurrence in session. Ordinate: number of responses in individual intervals (From Dews, 1970. Copyright 1970 by Appleton Century Crofts, Inc. and Prentice Hall, Inc. By permission.)

rates. This observation is at odds with the hypothesis that the number of responses in a given interval is a function of the number of responses in the immediately preceding interval (Herrnstein and Morse, 1958). The dynamic effect already observed by Skinner (1938) and further commented upon by Ferster and Skinner (1957) is clearly more complex. What is needed, if the cyclic properties of the interval to interval fluctuations are to be unveiled, is a mathematical analysis of the kind that is currently applied to periodic events in chronobiology.

Session-to-Session Fluctuations

Skinner (1938) described such fluctuations under the name of *first-order deviations*. Actually, we know very little about them. The variations of the rate of responding over successive sessions are not random, as Zeiler and Davis (1978) have recently shown. Analysing data from eleven pigeons on FI schedules ranging from 3 to 120 min, these authors have observed sequential dependencies somewhat analogous to those observed in the analysis of successive intervals: groups of sessions with high rates alternate with groups of sessions with low rates. They have found the same trends in their analysis of data obtained earlier by Gollub using rats (1964). The question arises as to whether these higher order cycles are produced by the FI contingencies themselves, or are caused by completely independent factors (such as some biological rhythm) and superimposed on the FI performance (see 5.6).

Effects of Prolonged Exposure

Several authors have reported that, after long exposure to the FI schedule, subjects tend to respond, in the activity phase of the interval, at a specific rate possibly selected by repeated reinforcement. Dews (1970) obtained similar terminal rates in both components of a Mult-FI 120 s FI 600 s schedule. The specific rate seems to generalize to the whole phase of activity, resulting in a typical two-state pattern after long exposure: the pause is immediately followed by a high and steady rate, a characteristic of the break-and-run pattern (Cumming and Schoenfeld, 1958; Schneider, 1969; Sherman, 1959; Skinner, 1938). This observation, however, has not been universally confirmed. Dews (1978) has analysed individual intervals in ten successive sessions on a FI 1000 s schedule using rhesus monkeys. The subjects had been exposed to the contingencies for about one year, totalizing more than 3000 intervals. Using several measures, Dews clearly found that the typical accelerated rate (scalloping) persisted after prolonged exposure (Fig. 3.8). Some data suggest that the break-and-run pattern would be more likely to develop with short durations of the interval (Schneider, 1969).

3.1.2. Properties of Required Regulation

The DRL schedule is by far the most extensively studied among the contingencies requiring temporal regulation of the subject's own behaviour as the condition for reinforcement. Therefore, most of the data and interpretive hypotheses presented in the present section will be drawn from experiments on DRL. Reviews concerning DRL performance have been published by Harzem (1969) and by Kramer and Rilling (1970).

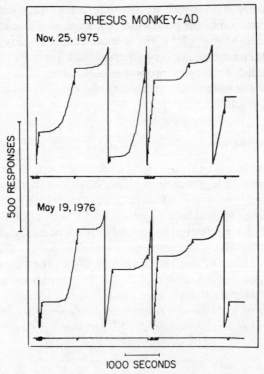

FIG. 3.8. Cumulative records of a monkey in a multiple schedule FI 1000 s—FR 50. Upper record illustrate pattern in two-hundredth sessions, lower record in the three-hundred-and-sixteenth session. Some 1160 FI had occurred between two records, but note similarity of pattern, with no greater tendency to break-and-run pattern in the lower record. The lower fixed pen in each record is downset during FR components (From Dews, 1978. Copyright 1978 by the Society for the Experimental Analysis of Behavior, Inc.)

A. Rate of responding, frequency of Reinforcement, and Temporal adjustment

DRL contingencies typically have two main effects on operant behaviour. One is a slowing down of the rate of responding. The other is the temporal adjustment to the critical delay value. One may ask: is the second a necessary correlate of the first, or is the first eventually a by-product of the second? A number of refined experiments have aimed at answering these questions.

That DRL contingencies reduce the rate of responding is clearly evidenced in several kinds of situations. Superimposing DRL contingencies on another schedule, for example, results in a decrease of rate. This was originally demonstrated by Skinner (1938) who required animals under a Fixed-Interval schedule to space their responses by 15 s and later by Ferster and Skinner (1957) in the Variable-Interval schedule and in the so-called tandem Variable-Interval-DRL schedule. Farmer and Schoenfeld (1964b) required spacing of responses by 1, 2, 4, 8, 16 or 24 s in a FI 30 s schedule. As rate of reinforcement in interval schedules is not modified by changes in rate of responding, the decrease in rate of responding when DRL contingencies are superimposed is not due to the reinforcement variable. A second source of evidence can be found in multiple or yoked schedules, where DRL performances are compared with behaviour under Sidman avoidance (Ross and others, 1962) or Variable-Interval schedules (Boakes and others, 1976; Ellen and others, 1978; Randich and others, 1978; Richardson, 1973, 1976). In experiments with yoked schedules, the rate of reinforcement in the compared schedule is equalized to the rate in the DRL schedule, so that it cannot account for the lower rate of responding in DRL (Fig. 3.9).

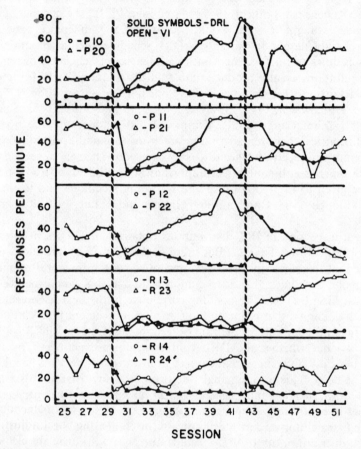

FIG. 3.9. Yoked and VI schedules. One subject in each pair is exposed to DRL 15 s, while the other is exposed to VI contingencies, and vice versa, depending on the experimental phase. Number and spacing of reinforcements in the VI are equalized with DRL. Contingencies were reversed in sessions 30 to 42, and reversed again (reinstating initial condition) in sessions 43 to 52. (From Richardson, 1973. Copyright 1973 by the Society for the Experimental Analysis of Behavior, Inc.)

The rate of responding in DRL is also inversely related to the value of the critical delay, as originally demonstrated by Wilson and Keller (1953) and by Zimmerman and Schuster (1962) and extensively documented in more recent studies. Gonzales and Newlin (1976) using a compound schedule (the details of which may be left out here), have compared delays of 10, 30 and 60 s. More convincingly, Richardson and Loughead (1974b) have explored delays of 1 to 45 min, and Richardson (1976) has explored delays of 0.5 to 300 s. As a rule, the longer the delay, the lower the rate of responding.

The temporal adjustment proper is classically analysed on the basis of the distribution of Inter-Response Times (IRT). These distributions are often bimodal (Conrad and others, 1958; Holz and others, 1963; Kramer and Rilling, 1970; Sidman, 1955), as already indicated in the methods section. The presence of a mode at very short values depends, to some extent at least, upon the species and upon the contingencies. Very short IRTs, typically obtained with pigeons, are less frequent in rats (Carter and Bruno, 1968; Kelleher and others, 1959; Marcucella, 1974). The peculiar status of the pecking responses might to some extent account for this difference between species (see Chapter 4 and Chapter 5). Variants of DRL schedule, such as situations with two responses, schedules of reinforcement of long latencies or schedules of reinforcement of response duration, usually produce unimodal distributions of IRTs or equivalent measure (Blackman, 1970; Catania, 1970; Cohen, 1970; Greenwood, 1978b; Lince, 1976; Logan, 1961; Macar, 1969; Molliver, 1963; Platt and others, 1973; Schwartz and Williams, 1971; Zimmerman, 1961). If bimodal distributions are the by-product of certain contingencies, there remains the problem of explaining the function of short IRTs. This point will receive further discussion later on. The role of secondary technical factors in generating bimodal distributions must also be considered. Special attention must be given to the type of response recording technique and to the finesse of time classes (Blough, 1963; Harzem, 1969; Kramer and Rilling, 1970; Ray and McGill, 1964).

Mere visual inspection of IRT distributions obtained in DRL schedules reveals a temporal adjustment of behaviour. But more evidence must be produced in order to demonstrate that this is characteristically generated by the temporal requirements of DRL and similar schedules, that is, to demonstrate that this cannot be accounted for by the general effect on rate of responding correlative of the decrease in reinforcement probability. It is known that reducing the rate of reinforcement in a Variable-Interval schedule results in a decrease in the rate of responding (Catania and Reynolds, 1968). One might wonder whether the DRL contingencies as experienced by the animal subject are basically different in that respect from those of the VI schedules. Various attempts have been made to disentangle the two effects involved, that is *reinforcement and response-rate reduction* and *timing behaviour*, and to show the specificity of DRL contingencies in controlling temporal adjustment.

A first and straightforward strategy consists in comparing the rate of responding and the IRT distribution in a DRL schedule and in a Variable-Interval schedule in which the reinforcements are delivered at exactly the same time and in equal number as in the DRL. This *yoked procedure* has been used by Richardson (1973) with rats and pigeons under DRL 15 s. In one experiment, different subjects were used in the DRL and in the simultaneous VI schedule, the yoked animal receiving reinforcement without any temporal restriction on the response each time the DRL subject produced a reinforced response. In a second experiment, the subject was its own differed yoked control: the record of reinforcement delivery under DRL was used to programme reinforcements under VI in a subsequent session. In a subsequent study, the same author (Richardson, 1976) applied the same principle in a single session, using a multiple schedule where VI and DRL components alternated. Delays used in the DRL ranged from 0.5 to 10 s in one group of subjects and from 10 to 300 in another. A similar procedure has been employed by Ellen and others (1978) with rats and delays of 10, 15 and 20 s. In all cases, as already mentioned (see Fig. 3.9), the rate of respond-

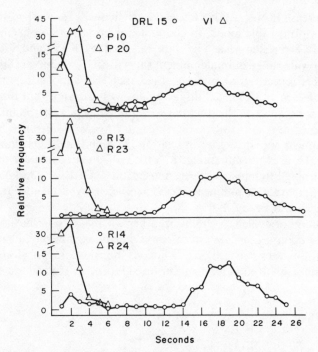

FIG. 3.10. IRT distributions in DRL 15 s (circles) and in yoked VI (triangles) from pairs of pigeons (P) or rats (R). The IRTs are plotted in 1 s increments (abscissa). (From Richardson, 1973. Copyright 1973 by the Society from the Experimental Analysis of Behavior, Inc.)

ing was lower in DRL than in VI. According to Richardson (1976), the *relative rate*, that is the rate in VI divided by the sum of VI rate plus DRL rate, is around 0.85 for almost all delays (3 to 300 s) that have been studied. Ellen and others (1978) found relative rate values of around 0.65. This clearly shows that DRL behaviour of rats and pigeons is not exclusively controlled by the rate of reinforcements. Moreover, IRT distributions are quite different in DRL and VI. As illustrated in Fig. 3.10, IRT distributions under VI are unimodal, with the mode at very short IRT, while they are bimodal under DRL, a second mode falling close to the critical delay value (at least for relatively short delays, as shall be discussed below).

Another experimental strategy applied to the same problem by Shimp and his co-workers consists of controlling the rate of reinforcement by a Variable-Interval contingency, and adding an IRT restriction to the delivery of reinforcement. For example, the reinforcement available according to the VI contingencies will be delivered only after an IRT of between 1.2 and 1.8 s has been emitted. Several versions of this *paced VI* schedule have been explored with various degrees of complexity that need not be detailed here (Moffitt and Shimp, 1971; Shimp, 1967, 1968, 1971 for example). Reinforcement of selected IRT classes has also been used in *percentile reinforcement schedules* or *interval percentile schedules*, where the contingencies are defined on the basis of the subject's recent behaviour (Alleman and Platt, 1973; Kuch and Platt, 1976; Platt, 1973). In all these studies, results show that reinforcing selected classes of IRT affects the IRT distributions (Fig. 3.11), and that this effect is quite distinct from the effect of rate of reinforcement. The reinforcement of selected IRTs produces an increase in the relative frequency of these IRTs, marked in the distribution by a clearly observable mode. In contrast, varying the rate of reinforcement has little effect on IRT distributions. The data reported by Kuch and Platt (1976) who have explored a wide range of reinforcement rates (from 30 to 360 per h) are especially conclusive in this respect. It should be noted that the demonstration of differential reinforcement of selected IRTs is in disagreement with the view that response rate can be accounted for by some molar relation to reinforcement rate (Baum, 1973; Herrnstein, 1970).

Another argument supporting a dissociation between timing behaviour and rate of responding can be drawn from studies in which an unescapable aversive shock, signalled by a warning stimulus, is superimposed on a DRL or similar schedule. The resulting change in rate is not necessarily paralleled by a change in the temporal regulation (Migler and Brady, 1964).

Despite the body of evidence summarized here, some authors have questioned the view that inter-response time is a dimension of behaviour amenable to reinforcement. They argue that this view is based on a fallacious interpretation, neglecting the mathematical constraints involved in the methods used for recording and analysing the data.

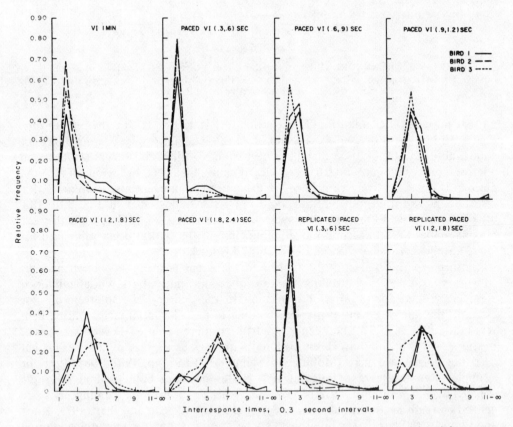

Fig. 3.11. Paced-VI schedule. IRT distributions for three pigeons. In each paced schedule, reinforcements were programmed only for responses terminating inter-response times between the two indicated values. (From Shimp, 1967. Copyright 1967 by the Society for the Experimental Analysis of Behavior, Inc.)

These objections have been developed by Reynolds and McLeod (1970). They do not deserve detailed discussion in regard to the accumulated facts in favour of an animal's capacity to estimate the duration of its own behaviour or, as what appears essentially the same, to estimate the time elapsed between two portions of behaviour. Not only are animals able to produce such spacing of their own behaviour, but they can use the produced intervals as discriminative stimuli to control further action. Reynolds (1966) designed an experiment with pigeons in which two properly spaced pecking responses on a red key changed the key-light to blue for a definite period; while the blue stimulus was on, pecks were reinforced by food according to a variable interval schedule if the two responses on the red key were spaced by 18 s or more (IRT). Though temporal differentiation of IRTs in the presence of the red stimulus did not clearly develop, the rate of responding in the VI component was clearly related to the duration of the IRTs that produced the blue stimulus. Another procedure requiring a single key-peck after red-key illumination following an inter-trial interval yielded the same results. A replication of this second procedure (Pliskoff and Tierney, 1979) confirmed the above relationship but revealed, however, a better temporal differentiation of red-key responses and a somehow weaker discriminative control over VI responding by the red-key response latencies (possibly due to the use of a very progressive learning up to the final red-key requirement), as shown in Fig. 3.12. Nelson

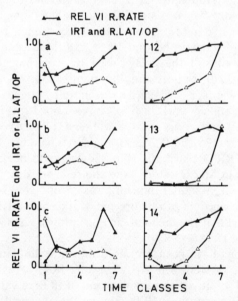

Fig. 3.12. Number of IRTs or response latencies per opportunity as a function of the duration of IRTs or response latencies emitted on a red key (and changing the colour of the key to blue for 30 s), and relative response rates on the blue key as a function of the preceding IRT or response latency on the red key. Data are plotted according to seven time classes (the first six classes cover the time up to the required IRT or response latency). Left: data from Reynolds (1966) with birds a, b, and c on a 18 s IRT requirement on the red key. Right: data from Pliskoff and Tierney (1979) with bird 12 on a 12 s and birds 13 and 14 on a 18 s latency requirement on the red key. The seventh time class gathers all response latencies greater than the required one: response latencies per opportunity equal therefore unity in this plot. (Redrawn after Reynolds, 1966, and Pliskoff and Tierney, 1979. Copyright 1966 by the Society for the Experimental Analysis of Behavior, Inc. Copyright 1979 by the Psychonomic Society.)

(1974) has also studied a choice behaviour controlled by the duration of the previous IRT in the same species.

As interesting as any critical approach can be, Reynolds and McLeod's viewpoint is but one example of a reluctance often found among behavioural scientists to endow laboratory animals with a real sense of time. Along the same lines and referring to Wilson and Keller's (1953) observation of what animals do during inter-response times, several authors have suggested that IRT distributions should be viewed as distributions of durations of *collateral behaviour* occurring between operant responses. Rather than estimating the passage of time, the subject would simply discriminate sequences of its own activities that happen to fill the appropriate time span. The apparent temporal regulation would in fact be mediated by such collateral behaviour, and one could dispense with internal timing devices. The importance of this hypothesis, both in terms of the number of experiments it has inspired and in terms of its theoretical significance that extends far beyond the DRL contingencies, deserves separate treatment. It will be discussed in detail in Chapter 7.

B. Absolute or Relative Time

Classical data on DRL performance indicate serious limitations as to the maximum delay to which typical laboratory animals can adjust. Pigeons perform poorly for delays longer than 20 s (Holz and Azrin, 1963; Holz and others, 1963; Kramer and Rilling, 1969; Reynolds, 1964a, 1964b; Reynolds and Limpo, 1968; Skinner and Morse, 1958; Staddon, 1965). Mice show little, if any, timing behaviour (as measured by IRT distributions) for delays beyond 10 or 15 s (Maurissen, 1970). Rats, cats and monkeys do better, but still the delays used by experimenters rarely exceed 60 s (Brown and Trowill, 1970; Malott and Cumming, 1964; Marcucella, 1974; Pliskoff and others, 1965; Richardson, 1976; Richardson and Loughead, 1974b; Skinner and Morse, 1958). Ferraro and co-workers (1965) have trained rats on a DRL 60 s; Hodos and co-workers (1962) and Weiss and co-workers (1966) have trained monkeys on DRL 21 or 20 s; Macar (1969, 1971b) has trained cats on a two-lever DRL 40 and 60 s. All describe IRT (or RRT) distributions as showing evidence of temporal regulation but in some cases (as in Macar's study), the choice of the delay indicates the limits of the subject's adjustment. A similar difficulty was encountered by Catania (1970) while training pigeons on a schedule of reinforcement of response latencies up to 48 s, and by Kuch and Platt (1976) on their already mentioned interval percentile schedule when reinforcing long IRTs.

Various explanations have been proposed for these limitations (Kuch and Platt, 1976). One suggests a lowered "sensitivity" of long IRTs to reinforcement. The problem of the differential sensitivity to reinforcement of IRTs of various duration was first discussed with regard to very short IRTs. Blough (1966) considered that in the pigeon, short IRTs (corresponding to *bursts* of responses and forming the secondary mode of IRT distributions) are not amenable to control by reinforcement. Mallott and Cumming (1964, 1966) talk about a *bias* in favour of short IRTs in rats, which means that some variable other than the contingencies arranged by the experimenters, possibly some species-specific variable, controls the emission of short IRTs. Millenson (1966) describes responses separated by short IRTs in pigeons as *stereotyped behaviour sequences* of which the sensitivity to reinforcement is different, compared with longer

IRTs. On the other hand, Shimp (1967) has shown that IRT classes of 0.3 to 2.4 s can be differentially reinforced. The same author (Shimp, 1968) has also demonstrated that long IRTs occur with a frequency equal to the frequency of shorter IRTs provided that the pay-off is increased in terms of relative reinforcement frequency. Shimp's studies, however, remain within the range of usual medium IRTs (long IRTs being of the order of 3.5 to 4.5 s) where temporal adjustment raises no problem at all. In fact, little is known about the low sensitivity to reinforcement of long IRTs, and if it were demonstrated (see for example Richardson, 1976), it would seem that it is something to be explained rather than an explanation. We should keep in mind that the timing mechanisms themselves could be responsible for the limitations observed.

Another explanation tends to account for the limitation of the DRL delay by a general loss of efficiency of reinforcement when the rate of reinforcement is reduced beyond a certain point. This is not a convincing argument, however, in face of the sustained control exerted by intermittent reinforcement in other kinds of schedules.

A third explanation is based on the analysis of the central tendency and dispersion of the IRT distribution (or distribution of some other appropriate temporal measure of responding). It has been shown that as the mode or mean increases, the dispersion also increases, and that the ratio between the index of central tendency and the index of dispersion is fairly constant, at least over a certain range of delays. Thus, Catania (1970) has reported a ratio of 0.30 for reinforced values of response latency of 0.6 to 36.4 s, Platt and co-workers (1973) ratios of 0.22 to 0.28 for response duration with reinforced values ranging from 0.8 to 6.4 s and Kuch (1974) reports ratios of 0.21 for response durations of 2.4 and 8 s. This increase in dispersion might reflect a reduction in discriminability of IRTs as the critical delay is increased that would in turn explain the limitation on temporal adjustment. This argument, however, is not very conclusive. First, resorting to a reduction in discriminability for long IRTs is just another way to state the fact that temporal adjustment, under these contingencies, deteriorates when the requirement exceeds a given value. Secondly, invariant ratios between dispersion and central tendency across a series of increasing delays may be viewed as an index of equivalent levels of adjustment. As already mentioned, those ratios may be considered as the equivalent of Weber's fraction at the level of temporal regulation of the subject's own behaviour. One may wonder nevertheless whether these ratios would remain constant for delays far longer than those currently explored. The lack of data precludes any present conclusion.

Part of the intricacies of the problem discussed here derives from the fact that we still lack an unequivocal index of temporal regulation in DRL and similar contingencies. This question has already been alluded to in the method section. It should be recalled here that the efficiency ratio is not necessarily correlated with the indices of central tendency drawn from IRT distributions, that these distributions can be characterized by central tendencies and/or by indices of dispersion or by the ratio between these two parameters and that transformation of these distributions to IRT/OP distributions provides for a more clear-cut appraisal of the temporal regulation. Conclusions concerning the ability of laboratory animals to adjust to the temporal requirement of the schedules are therefore largely dependent upon the measures chosen.

This should be kept in mind when looking at the data analyses proposed by various experimenters. In most studies designed to analyse the performance across a series of delays, indices of central tendency—mean, mode or median—are used. As a rule, when this kind of measure of behaviour is related to the requirement of the schedule, it can be verified that the organism adjusts to relative, not absolute time. Evidence for this law in rats and pigeons on classical DRL schedules can be drawn from Mallott and Cumming (1964) (delays: 1, 2, 5, 10, 15, 20, 50 and 100 s), Staddon (1965) (delays: 5.68, 10.8, 15.85, 21.10 and 31.50 s) and Richardson and Loughead (1974b) (delays: 1, 2, 3, 5, 7, 10, 15, 20, 30 and 45 min). Similar evidence for pigeons and schedules of reinforcement of long latencies is provided in Catania (1970) (reinforced latencies: 0.6, 2.75, 5.15, 7.5, 14.9, 24.4, 36.4 and 48.0 s). Additional evidence for rats and schedules of reinforcement of responses duration can be found in Platt and co-workers (1973) (response durations: 0.4, 0.8, 1.6, 3.2 and 6.4 s) and Kuch (1974) (response durations: 2, 4 and 8 s with limited hold equivalent to 25, 50 or 100 per cent of the duration requirement). In all these cases, the temporal measure of behaviour used is a power function of the IRT, response latency, or response duration requirement. The functions drawn from experimental data fit nicely with similar relations observed in other kinds of temporal regulation and estimation in animals in the frame of more general psychophysical laws. These similarities will be discussed at length below.

Elegant curves like the one reproduced from Richardson and Loughead's study (1974b) in Fig. 3.13 seem to confirm that a measure of central tendency (e.g. the mean of IRTs distribution) is perfectly appropriate for estimating the temporal adjustment of an organism. In addition, the fact that the relationship holds for delays much longer than those classically explored in the same species (that is, rats and pigeons) would indicate that the limitations on required temporal regulation tasks underlined by previous authors have been overcome and should therefore be assigned to experimenters rather than to the laboratory animals. A closer look at the data, however, reveals a very different picture. The power functions described by Richardson and Loughead are based on the mean of the IRTs distribution across a series of delay requirement ranging from 1 to 45 min. The distribution of relative frequencies of IRTs given for delays of up to 15 min (unfortunately not for longer delays) show very flat and stretched shapes except for the shortest delays (up to 3 min in rats, scarcely at 1 min for pigeons): there is no mode around the reinforced value (Fig. 3.14). The mean IRT is therefore not a correct estimate of the most frequent behavioural event, and cannot be taken as a valid index of the temporal regulation. The efficiency ratio improves as the delay increases in pigeons but it deteriorates in rats. In any case, it is not a reliable index of temporal adjustment either. A good efficiency in DRL can result from an oscillatory process where the operant behaviour is alternately extinguished (and thus produces reinforcement) and reconditioned (under the influence of the reinforcement obtained in the extinction phase). Clearly, we need more than the mean IRT to demonstrate control of behaviour by the delay in DRL and similar contingencies. Until more evidence is gathered, we shall retain the widely recognized observation that required temporal regulation appears only for short delays or temporal requirement, that is, depending upon the species, delays not exceeding 10 to 120 s, and that within the range of values where temporal adjustment is clearly demonstrated, the adjustment is made to relative, not absolute time. We have seen that this is confirmed in classical DRL, as well as in schedules of reinforcement of long latencies and response duration. It has also been found in more complex situations, some of which will be described here as illustrations of the refined capacities of organisms in making such temporal adjustments.

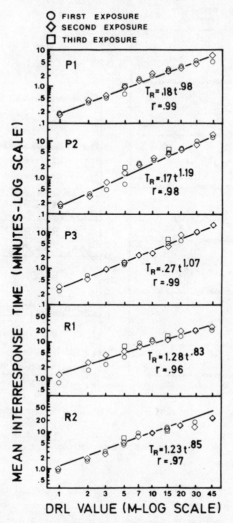

FIG. 3.13. Mean IRT during successive exposures as a function of DRL value in pigeons (P) and rats (R). Both axes are log scales. Each point is a mean of the last four or five days under a condition. The regression equations, computed using the method of least squares, are presented with each graph (r = product moment correlation coefficient. Tr = mean IRT in minutes, t = DRL value in minutes). (From Richardson and Loughead, 1974b. Copyright 1974 by the Society for the Experimental Analysis of Behavior, Inc.)

Logan (1967) trained rats on Mixed DRL schedules, in which two DRL components (5–30, 10–30, 15–30 or 20–30 s) alternated randomly without any signal being associated with the components. He found that the larger the difference between the two delays employed, the better the performance. Staddon (1969a) trained pigeons under contingencies in which the spacing requirement changed cyclically throughout a session (*cyclic Mixed DRL schedule*). The delays were in succession 8.5, 11, 16.3, 20.9 and 30.6 s (with a limited hold of 10 per cent of the delay value). Passage from one delay to the next took place every five min without any signal, a complete cycle consisting of one ascending and one descending series. Each session comprised two cycles and

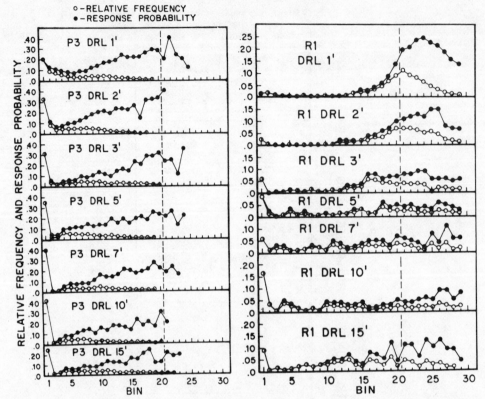

FIG. 3.14. Relative frequency (open circles) and IRT/OP (filled circles) distributions for one individual pigeon (P 3) and one rat (R 1) under different DRL values, from 1 to 15 min. Each bin width on the abscissa is one-twentieth of the DRL value. All IRTs to the right of the vertical, dashed line were reinforced. (From Richardson and Loughead 1974b. Copyright 1974 by the Society for the Experimental Analysis of Behavior, Inc.)

consequently, lasted 110 min. This experiment was designed to test the effect of the feedback of responses, and it will be discussed in Chapter 8. For our present purpose, only the baseline data before the feedback stimulus was introduced are of interest. One of the three animals tested exhibited a remarkable adjustment to the changing contingencies, the rate of responding and the modal IRT paralleling the changes in the delay. A second animal showed a similar, though less perfect adaptation, and the behaviour of the third did not show any lawful pattern correlated with the changing delay (Fig. 3.15). This kind of schedule, and the behaviour it generates, is to be compared with the progressive interval schedule studied by Harzem (1969) and the cyclic interval schedule studied by Innis and Staddon (1971) (see page 52) and leads to similar reflection as to the capacity of the timing device involved.

Different delays can be mastered not only in succession, but simultaneously. This was shown by Wilkie and Pear (1972) in rats, using a special schedule where inter-response times of 1.5 ± 0.5 s were the selected operant and reinforced according to a DRL 4 s schedule. In this situation, a spacing of two responses by a 1 to 2 s delay was followed by reinforcement only if it occurred at least 4 s after the previous operant IRT. The distribution of relative frequencies of inter-response times (that is, times between successive operant-IRTs) were bimodal with one of the mode at 4 or 5 s depending upon the individual subject, and the other at the shortest value, as is usual in DRL schedules.

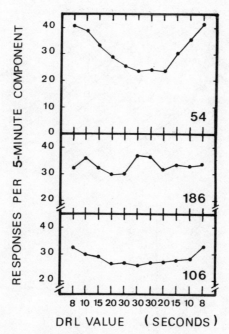

FIG. 3.15. Response rate in each component of a *Cyclic Mixed DRL schedule*, showing adjustment to cyclically changing delays. Abscissa: delay values of successive components, ordered as they appear in the cyclic schedule. Data averaged from seven sessions. (Redrawn, after Innis and Staddon, 1971. Copyright 1971 by the Society for the Experimental Analysis of Behavior, Inc.)

Finally, mention should be made of schedules in which sequences of behaviour satisfying specific temporal requirements are selectively reinforced. In such situations the whole sequence may be considered an operant response (Shimp, 1976, Zeiler, 1977). For example, sequences of *pause-activity* or *activity-pause* of given duration have been successfully conditioned (Staddon, 1972a; Wasserman, 1977). Or in a still more refined way, the contingencies may define precise spacing between the successive responses composing the sequence. This was done by Shimp (1973), in an experiment that also illustrated the control of behaviour by two different delays in succession, and by Hawkes and Shimp (1975) who reinforced increasing and decreasing rates of responding in pigeons. In this last study, during 5 s trials, the subject was required to pace its responses in such manner that the rate approximated 0 in the first second, 1 response in the second second, 2 responses in the third, 3 in the fourth and 4 in the fifth. As Hawkes and Shimp explained, in such a situation, the subject must emit key-pecks "at a rate that changes at a constant rate throughout the trial". The control exerted by these contingencies was clearly evidenced by comparison with the change in rate observed under a FI 5 s, as can be seen in Fig. 3.16.

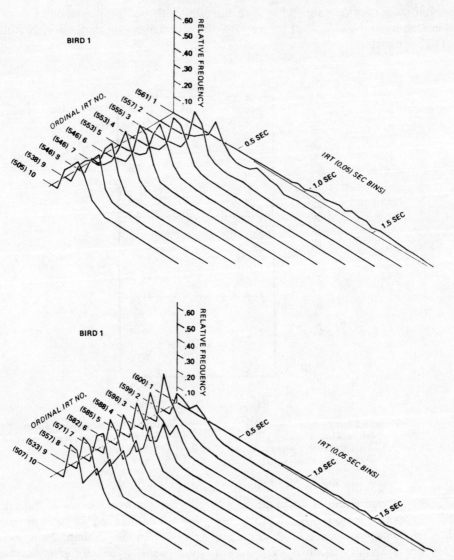

FIG. 3.16. Relative frequency distributions of IRTs for one pigeon for the first, second, third, ... tenth response in the two conditions of Hawkes and Shimp's experiment. Above, data from the schedule where reinforcement is contingent upon specified, increasing rates of responding. Below, control data from FI sessions. The numbers to the side of the ordinal—response—number axis are the frequencies of responses in the corresponding distributions. Points are averaged over last 600 trials of each condition. (From Hawkes and Shimp, 1975. Copyright 1975 by the Society for the Experimental Analysis of Behavior, Inc.)

C. Suprasegmental Traits in Required Regulation

As with spontaneous temporal regulation, less attention has been given to the fluctuations of behaviour throughout a session or from session to session than to the average pattern with respect to the critical delay. However, it would seem that sequential analysis can provide valuable information concerning the mechanisms of temporal adjustment. Several authors have been aware of this possibility, and have explored sequential dependencies using mathematical formulations of various degrees of sophistication (see method section). Farmer and Schoenfeld (1964a) originally demonstrated that in rats, the probability for two reinforced IRTs to occur in succession is higher than the probability of sequences of two IRTs one of which only is reinforced. This would suggest that temporal regulation is improved after a reinforced response. This was not fully confirmed by Ferraro and co-workers (1965), who found a high probability for an IRT of a given value to be followed by another IRT of the same value. Other analyses show that reinforced IRTs tend to occur in long sequences, interrupted by sequences of non-reinforced short IRTs, pauses or bursts (Ferraro and others, 1965; Kelleher and others, 1959; Malott and Cumming, 1964; Weiss and others, 1966). Bursts are infrequent after reinforced responses (Kramer and Rilling 1970; Sidman, 1956), but frequent after IRTs just too short to be reinforced (Kramer and Rilling, 1970). Conversely, reinforced IRTs occur frequently after a burst of responses (Ferraro and others, 1965). This suggests that bursts have some function in temporal regulation, the nature of which remains to be elucidated. Finally, it should be noted that these sequential dependencies between bursts and other IRTs have not been demonstrated in pigeons (Kramer and Rilling, 1970). Until more data become available concerning sequential dependencies, only tentative conclusions can be drawn from the few analyses performed. There is no evidence that temporal regulation is the result of a step-by-step correcting process in which a too short IRT would induce a slightly longer IRT, and so on until a reinforced IRT is produced, maintained for a while, shortened, and again corrected. Long sequences of IRTs (or response latencies, or response durations) present a different pattern (Fig. 2.12): large phases of long, eventually reinforced values alternate with phases of contrasting short values, suggesting wide oscillations across a session rather than a step-by-step adjustment. Long series should be analysed in order to detect possible periodic characteristics of these oscillations and to identify their causes and functions.

Prolonged exposure to DRL or similar contingencies is reported to improve performance (Ferraro and Grilly, 1970; Laties and others, 1969; Staddon, 1965; Wilson and Keller, 1953) though in some cases it may lead to a deterioration of behaviour, and eventually to emotional by-products in and outside the experimental situation (Macar, 1971a).

3.2. CONTROL OF BEHAVIOUR BY TEMPORAL PROPERTIES OF EXTERNAL EVENTS

3.2.1. Generalization

The general sensitivity of animals to temporal dimensions of external events is evidenced by the gradients of generalization obtained along duration (Reynolds and Catania, 1962) and click-rate continua (Hearst, 1965; Sidman, 1961). The shape of these gradients seems independent of the absolute values of stimuli: gradients established for different ranges of values in the same subject superpose when plotted as a function of relative click-rate (Mostofsky and others, 1964) or duration (Catania, 1970, illustrated in Fig. 3.17).

Another important feature is that temporal generalization gradients do not appear spontaneously: the mere exposure of an animal to a stimulus of fixed duration or rate

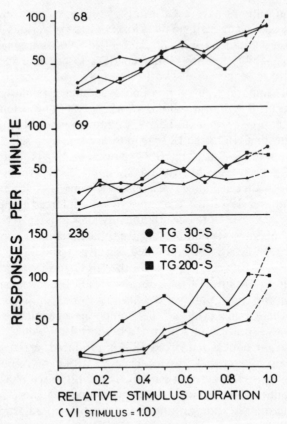

FIG. 3.17. Superposition of generalization gradients for duration in three pigeons (68, 69, 236). The ranges of values studied were 3–30 s (circles), 5–50 s (triangles), and 20–200 s (squares) with the longest duration as the positive (VI) stimulus. Rate of responding is plotted against relative stimulus duration. (Redrawn, after Catania, 1970. Copyright 1970 by Appleton Century Crofts, Inc., and Prentice Hall, Inc. By permission.)

does not make it sensitive to that dimension of the stimulus (Elsmore, 1971a; Farthing and Hearst, 1972; Hearst, 1965; Williams, 1973). It is necessary to give a preliminary training both with a positive and a negative temporal stimulus to obtain differential rates of responding in the presence of test stimuli. A supplementary condition is that both training stimuli belong to the same continuum; although gradients were obtained by Farthing and Hearst (1972) when the negative stimulus was a 0.0 Hz click-rate (i.e. background noise only), these gradients were often bimodal and showed irregularities, especially when the positive stimulus was located at midrange of the part of the continuum under study. Williams (1973) observed flat gradients when a period of silence was contrasted with a 2.45 Hz click-rate. Such results are consistent with the concept of the "attending hierarchy" of stimulus dimensions developed by Baron (1965). This hierarchy is based on the complexity of the training necessary to achieve stimulus control over behaviour. The highest level is occupied by stimulus dimensions that spontaneously give rise to generalization gradients, e.g. wavelength in pigeons (Guttman and Kalish, 1956). At a lower level are dimensions that require preliminary differential training before generalization can develop, for example, auditory frequency in pigeons (Jenkins and Harrison, 1962). A third, lower level could be added, occupied by dimensions that require differential training with the additional constraint that the stimuli used in training belong to the same source continuum; temporal stimuli could be placed at this level.

3.2.2. Thresholds

Some duration discrimination experiments were designed for the evaluation of a just noticeable difference (or differential threshold) between two temporal intervals. The minimal degree of discriminability offers a numerical basis for comparison studies (Richelle and Lejeune, 1979) or for assessment of the effect of the source variable, such as response requirement (Perikel and others, 1974) or nature of the stimulus (empty versus filled: Mantanus, 1979). The most important information derived from threshold experiments concerns the relationship between the value of the threshold and the duration of the standard stimulus. This problem refers to the more general framework of psychophysics. The verification of Weber's law requires that the ratio between the value of the threshold and the value of the standard stimulus (i.e. the Weber fraction) remains a constant whatever the absolute value of the standard. This requirement was met in an experiment by Stubbs (1968). Three pigeons successively discriminated the durations of three ranges of ten stimuli, the extremes being 1–10, 2–20 and 4–40 s, and the increasing step being of 1, 2 and 4 s respectively. The first five values in each range constituted the class of short stimuli, the other durations being the long stimuli. The individual psychometric curves superposed when plotted against relative stimulus value. The cut-offs or 0.5 response probabilities varied between 45 and 55 per cent of the longer duration in all cases, and a constant Weber fraction was found. The author mentioned an indicative 0.25 value for this fraction in three other birds with a similar 1 to 10 s discrimination where correct choice was reinforced intermittently. It should be pointed out that other experiments found Weber fractions very close to this value (Catania, 1970; Perikel and others, 1974) using different procedures.

The question of whether functional relationships other than Weber's law could describe the variation of the differential threshold with the standard stimulus was raised recently. In a study using rats Church and others (1976) tested the adequacy of two models describing the relationship between difference threshold and duration of the standard stimulus. These models are the *Counter Model* and the *Weber Model*.

The basic assumptions of the Counter Model are that pulses are emitted randomly in time and counted, and that the subject forms an "internal representation of time" equivalent to the number of pulses that have been emitted during presentation of the stimulus. If this model holds, the variance of the psychometric function is a constant proportion of the duration of the signal. According to the Weber Model, the standard deviation, and not the variance of the psychometric function should be a constant proportion of the duration of the stimulus. The psychometric function is supposed to be a normal ogive, and it is assumed that factors of variability related to stimulus onset and termination (Allan and Kristofferson, 1974) also intervene in both models. If these factors are labelled v, and s is the duration of the standard stimulus, the Counter Model may be formulated as

$$[JND(s)]^2 = ks + v \quad \text{where } k \text{ is a constant}$$

and the Weber Model as

$$[JND(s)]^2 = ks^2 + v \quad \text{where } k \text{ is a constant.}$$

To identify which exponent of s (1 or 2) described the relation between the squared JND and the duration of the standard, Church and others placed three rats in a two-lever situation with auditory stimuli. The standard values successively presented were 0.5, 1, 2, 4 and 8 s. The value of the comparison stimulus leading to a 75 per cent correct responses was determined by a titration method and thus only one point of the psychometric curve obtained (the difference limen, JND was calculated with reference to the standard stimulus). The squared difference limen was plotted as a function of the standard stimulus (Counter Model) and as a function of the squared standard stimulus (Weber Model). As shown in Fig. 3.18, the squared difference limen was linear with respect to the squared standard values, thus confirming the Weber Model. When the squared difference limen was plotted against the value of the standard (Counter Model) the amount of variance not accounted for by the linear model was greater than for the Weber Model plot.

Though experimental results seem to favour Weber's law, the constancy of the Weber fraction is not an absolute rule in animals. With very short standard stimuli, Weber fractions may take different values; in the experiment by Church and others (1976), just mentioned and illustrated in Fig. 3.18, the Weber fraction decreased with increasing values of the standard stimulus, but the function reached an asymptote at the larger values (4 and 8 s) since stable values were recorded (0.18 to 0.50, depending on the subject). It is reasonable to assume, as the authors did, that a variability term (v in the models mentioned above) could account for the differences in Weber fractions, but it remains to be established just how far methodological factors influence the relative importance of that term.

Measures of temporal discriminability other than differential threshold are supplied by the sensitivity indices derived from the model for signal detection presented in the preceding chapter. These indices appear roughly constant whatever the degree of response bias induced by manipulations of stimulus probability (Elsmore, 1972), re-inforcement probability (Kinchla, 1970; Stubbs, 1976) or schedule of reinforcement of correct responses (Stubbs, 1968). It would be interesting to check this constancy of sensitivity indices when temporal variables are involved; for instance would a constant relative difference between long and short stimuli lead to equal d' or A' indices when the absolute duration of the stimuli is varied? This would lead to a more complete description of temporal accuracy since above-threshold levels would be taken into account.

3.2.3. Scaling

A functional relationship may be established between temporal stimuli and measures of timing behaviour. The relationship can be described by the general term "temporal scaling". It will be noted that scaling takes the form of a direct relationship between temporal characteristics of stimuli and temporal characteristics of responses in all situations where temporal intervals are produced by an organism; e.g. scaling of latencies (Catania, 1970), lever-press durations (Platt and others, 1973), inter-response times (Richardson and Loughead, 1974b). With duration discrimi-

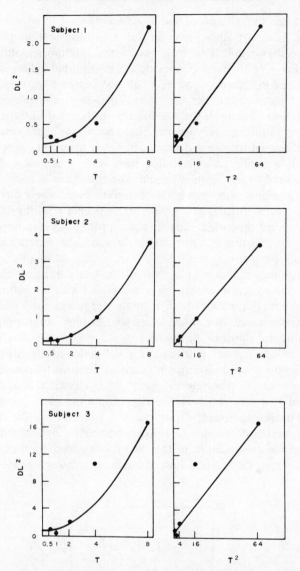

FIG. 3.18. Testing the counter Model versus Weber Model (see text): the squared difference limen (DL^2) of three rats (subjects 1, 2, 3) is plotted against the duration of the standard (T), according to Counter Model (left column), and against the squared duration of the standard (T^2), according to Weber Model (right column). The best linear fitting are found for the Weber Model. (From Church and others, 1976. Copyright by the American Psychological Association. Reprinted by permission.)

nation, scaling is indirect, as it is with any psychophysical situation in which the stimulus and the response do not share a common property; as pointed out earlier, duration discrimination studies deal with behavioural measures that do not possess temporal properties.

An experiment done by Church and Deluty (1977) offers a good example of how duration discrimination data provide information about the kind of transform temporal stimuli may be assumed to undergo when perceived by the organism. Church and Deluty transposed the method of bisection, widely used in human psychophysics, to animal subjects. Their hypothesis may be stated as follows; if an animal, trained to discriminate between two durations in a two-manipulandum situation, is presented with a new stimulus such that responding on either manipulandum is equally likely, this stimulus may be considered subjectively half-way between the two training values. It is labelled "the bisection point" of the interval between training values and may be expressed as a function of them. If the bisection point is equal to the arithmetic mean of the two training durations, this implies that the psychological transformation of time is linear with respect to stimulus duration. If the bisection point is equal to the geometric mean, this implies a logarithmic transform of stimulus duration. If the bisection point is equal to the harmonic mean, the transform is reciprocal. Eight rats were trained to discriminate between pairs of stimuli: successively 2–8, 1–4, 4–16 and 3–12 s. Following training with a given pair, intermediate test durations were introduced. Five values were presented with an equal probability and were geometrically distributed between the training values. The three middle durations were approximately equal to the harmonic, geometric and arithmetic mean of the training durations. No responses were reinforced after any of the test durations. The test duration that led to 50 per cent responses on the lever associated with long stimuli was defined as the bisection point. For each pair of training durations, the bisection point was close to the geometric mean. This result, illustrated in Fig. 3.19, supports the hypothesis of a logarithmic transform of duration.

This result is also consistent with earlier data collected by Stubbs (1968) in a two-manipulandum duration discrimination context. Pigeons trained to discriminate 1 and 10 s stimuli "bisected" the interval at about 3 s (which is approximately the geometric mean of 1 and 10) when eight intermediate durations were presented in a subsequent phase of the experiment. Other pigeons, trained from the start to discriminate duration ranges (1 to 5 versus 6 to 10 s) showed an equal probability of responding between 5 and 6 s, the location of the "cut-off" defined by the contingencies of reinforcement. However, these birds showed a tendency toward equal accuracy levels

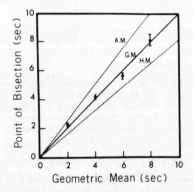

FIG. 3.19. Bisection of temporal intervals by rats. The bisection points of four ranges of stimulus durations (1–4, 2–8, 3–10 and 4–16 s) are plotted against the geometrical mean of the shortest and longest stimuli of each range. The vertical lines represent the standard error of the mean (eight subjects). The lines labelled AM, GM and HM refer to the arithmetic, geometric, and harmonic mean of those durations. It can be seen that the bisection points fit very closely to geometric mean at each value considered. (From Church and Deluty, 1977. Copyright 1977 by the American Psychological Association. Reprinted by permission.)

at logarithms of stimulus durations located at equal distances from the cut-off (like 3 and 10 s). This result was confirmed when the same birds were run in a task where durations were selected for symmetrical "logarithmic distance" from the cut-off. The short stimuli were 1 to 5 s in 1 s steps (as in the previous experiment) and the long durations were 6, 8, 10, 15 and 30 s. The proportions of responses on the key associated with long stimuli were close to perfect reciprocity for values whose logarithms were equidistant from the cut-off (see Fig. 3.20). Stubbs concluded that "accuracy was a function of relative rather than absolute difference of a duration and the cut-off" (p. 229).

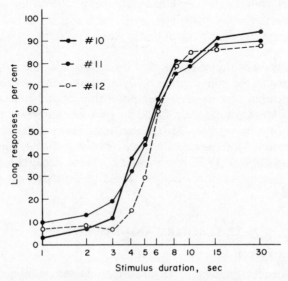

FIG. 3.20. Symmetry of the left and right parts of psychometric functions when the short and long stimulus values are matched for logarithmic distance from the cut-off between 5 and 6 s. This figure is to be compared with Fig. 2.21 which shows the performance of the same three birds when stimulus values were matched for arithmetic distance from the cut-off. (From Stubbs, 1968. Copyright 1968 by the Society for the Experimental Analysis of Behavior, Inc.)

Stubbs' results also show how contingencies of reinforcement can affect scaling. In the examples cited the bisection (or 0.50 response probability) points are either 3 or 5.5 s depending on the training conditions, i.e. with two or ten durations presented, respectively. In the latter case, the arbitrary definition of a cut-off for reinforcement at equal arithmetic distance between the two extreme values resulted in another scale of durations extending beyond the 10 s stimulus. This scale was estimated when longer values of stimuli up to 30 s were used; the psychometric functions, plotted against logarithms of stimulus durations, appeared more complete at upper asymptote, and symmetry was found between right and left parts of the S-shaped curves. Logarithmic scaling of time was also observed on two-valued multiple FI schedules with pigeons (Stubbs, 1976). Different behavioural measures co-varied at the time corresponding to the geometric mean between long and short FI values.

The consequence of scaling on aspects of behaviour other than the discriminative response are shown in an experiment by Snapper and others (1969). In this two-lever design, and using rats, responses during stimulus presentation had no consequence, so the subjects consistently pressed first the lever associated with the short stimulus, and then switched to the lever associated with the long stimulus if no reinforcement had appeared in the meantime. These extra-responses also reflected scaling, as a severe drop in responding on the lever associated with short stimuli ("short" lever) occurred at the geometric mean between the stimuli (1 and 5 s). It can be shown that this drop in responding that preceded switching to the other lever corresponds, on the average, to the 95th percentile of the distribution of responses on the "short" lever.

The few examples reviewed here already offer good evidence in favour of a logarithmic relationship between temporal stimuli and behaviour. Such a relationship can only evoke Fechner's law, as cited by Guilford (1954) who gives the following formulation:

$$R = C \log S$$

where R is the measure of the response, S is the measure of the stimulus, and C is a constant. This remarkable convergence of data stresses the suitability of duration discrimination procedures for achieving temporal control and collecting psychophysical data. Highly similar results were found in spite of important procedural differences, such as sensory modality of the stimulus, response requirement, and method of presentation of stimuli. Moreover results may be collected over long periods of time since, as Church and others (1976, p. 309) noted, animal subjects show "considerable enthusiasm for a task which for a person would be an extremely dull and tedious psychophysical effort".

3.3. SUMMARY AND DISCUSSION

3.3.1. One Clock or Several

Looking at the results obtained in various experimental situations, one is struck by the discrepancy between performances under contingencies such as Fixed-Interval schedules on the one hand and DRL schedules on the other. Paradoxically, behaviour seems to adjust to time more effectively when the temporal regulation is what we have termed "spontaneous", than when it is required as a condition for reinforcement. The temporal parameter of a Fixed-Interval schedule is usually defined in terms of several minutes or occasionally several hours, while it is defined in terms of a few seconds or exceptionally minutes, in a schedule of differential reinforcement of low rates. The same animal that spontaneously pauses for several minutes after each reinforcement in a FI schedule scarcely adjusts one out of three or four of its responses to the critical delay of a DRL 30 or 60 s. This is even more surprising as in the first case, temporal regulation has no consequence on reinforcement (at the scale of the experimental situation, one could say that it has no survival value for the subject), while it has in the second.

Part of the difficulties encountered by members of some species, for instance pigeons or mice, might be related to species-specific characteristics that would limit their performance in DRL and similar situations. It is true that pigeons and mice compare favourably with rats, cats, or monkeys under FI contingencies, but do not master DRL delays longer than 10 to 20 s while these other species do. For example, it has been argued on the basis of recent research questioning the arbitrariness of the pecking response in pigeons that the very nature of the selected operant is an obstacle to control by certain contingencies. Cross-species differences will be described and discussed at length in Chapter 4. For our present purpose it suffices to say that whatever argument might be used to account for the relative inferiority of certain species with respect to timing behaviour, it would leave unanswered the paradox described here: within the same species, temporal adjustment seems much easier in some situations than in others, and is typically easier where it is not required. Obviously, we must turn to an analysis of the situations to which laboratory animals have been exposed and try to find out which properties of some situations induce "spontaneous" temporal regulation, and which properties of other situations make required regulation so difficult. Our survey of methods and our summary of the main properties of temporally regulated behaviour do not provide us with all the necessary elements for a detailed analysis of the various situations, as they are defined by the experimenter and as they are experienced by the subject. At present, we shall only outline some hypotheses (Lejeune, 1978) that will be further documented in the following chapters.

FI and DRL schedules differ as to the kind of operant upon which the reinforcer is made contingent. Though at first sight the same key-peck or lever-press is defined as the response in both situations, actually what is reinforced in a DRL schedule is a time interval between two motor responses or inter-response time (IRT). This is a quite different class of operant, possibly more difficult to bring under the control of reinforcement. In DRL, as in schedules of reinforcement of response duration, the subject must discriminate the temporal properties of its own behaviour while in stimulus duration discrimination tasks, it must discriminate the temporal properties of external events and in schedules of reinforcement of response latencies, it must do something of both or something in between. Such attention to the duration of events, external or in the subject's own behaviour, is not involved in Fixed-Interval schedules.

We have noted that temporal regulation in FI schedules cannot be accounted for by chains of behaviours, the successive moments of which would be discriminated by the subject, thus relieving it of a real time estimation. By contrast, *collateral behaviours* are often observed under DRL contingencies and in many cases have been shown to facilitate temporal adjustment (see Chapter 7). Supposing, as does one hypothesis, that collateral behaviour provide all the necessary information for spaced responding, one might wonder why a subject would not resort to it under FI contingencies and why spontaneous regulation is still much better with respect to the magnitude of the delay.

One possible explanation, as will be seen in Chapter 7, is that collateral behaviours do not actually provide a mediating mechanism or a substitute for time estimation proper, but that they yield a behavioural output compensating for the strains involved in the schedule requirement. In this hypothesis, it is assumed that temporal regulation whatever its kind, implies some sort of active inhibitory mechanism (see Chapter 9). For this mechanism to function for a certain period of time, it would require compensatory phases of activity. Various situations would put various loads on this inhibitory mechanism and provide more or less occasions for compensation. Looking at FI versus DRL schedules from this point of view, it can be argued that FI schedules exhibit features quite distinct from DRL. They control behaviour in a manner analogous to the Pavlovian *time conditioning*, as several authors have noted (Fraisse, 1967; Malone, 1971). In fact, when plotted in cumulative curves, salivary responses to periodic unconditioned stimuli present the same scalloped shape, so familiar to operant conditioners using FI schedules. Moreover, the rhythm induced by FI contingencies and reflected in the typical patterning of the cumulative record is clearly seen to persist for a while when a subject is abruptly transferred to a higher value of the delay. It could be said that the subject has to make no effort to cope with time, it just let time model its behaviour. The level of temporal organization involved would be more primitive but, also, FI contingencies would allow for an alternance of pause and activity phases, providing for the compensatory mechanism hypothetized above. This is in fact what is observed: animals under FI produce a number of responses at the end of the interval. Only exceptionally do they give a single response at the very moment the reinforcement is made available and unique responses emitted a certain time after the delay is over are typical of FI with long delays (and in this latter case for reasons that do not contradict the general interpretation outlined here).

Temporal patterning of behaviour under FI contingencies could be explained by the highly regular and periodic structure of the schedule that contrasts with the lack of periodicity in the actual distribution of reinforcements under DRL. The structure of different types of contingencies will be analysed in more detail in Chapter 8, and the role of stimuli associated with reinforcement in FI schedules in inducing and maintaining temporal regulation will be described. The power of periodic events in shaping behaviour is further evidenced by experiments with Fixed-Time schedules. In this kind of situation, food (or some other reinforcer) is delivered at regular intervals irrespective of the subject's behaviour. It has been shown that responding occurs under these conditions, and that it has the same general pattern as under FI while this is not the case when non-contingent reinforcers are delivered according to a Variable-Time schedule (that is at variable, unpredictable intervals) (Zeiler, 1968). This strengthens the analogy with Pavlovian time conditioning.

If the main contrast is between behaviour under FI (or more generally spontaneous regulation) and behaviour under DRL, DRLL, or Response-Duration schedules (or required regulation), more subtle differences between various situations of the same general category should not be overlooked. The opposition outlined above between two different classes of operant (IRTs between two successive responses versus single motor responses) is too crude to account for the diversity of facts. Some data suggest that estimating the duration of one's own behaviour might not be equivalent to estimating the duration of external events. Pigeons are better at discriminating between external events differing by duration than at spacing their own motor responses: they discriminate fairly accurately visual stimuli up to values of 40 s or so (and possibly more) while they hardly master delays of 15 to 20 s in DRL.

On the other hand, increasing the constraints of the motor response used as the operant increases the limitation of temporal regulation. Thus cats, which show excellent temporal regulation in Fixed-Interval schedules of 10 or 15 min, show great difficulty in spacing two brief responses by more than 40 to 60 s and demonstrate yet greater difficulty when maintaining a motor response for a given duration (Greenwood, 1977). This is not merely a matter of being able to keep the paw in the same position for a long time, since prolonged responding can be maintained if a stimulus signals the moment at which the animal must release the lever (see Fig. 3.21). The difficulty does not come from motor control proper; it seems inherent to time estimation under specific task requirements.

Future experimental work and theoretical analysis should therefore aim at identify-

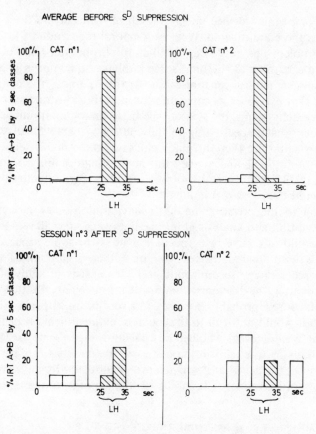

FIG. 3.21. Distributions of response durations (IRT A → B) in two individual cats when the availability of reinforcement on lever B is signalled by an auditory stimulus (above) and when it is unsignalled (below). (After Greenwood, 1977, unpublished data.)

ing the variables that are responsible for the observed differences in performance. Following the line of thinking proposed for explaining FI versus DRL behaviour, it would seem advisable to look at the structural particularities of the contingencies and their structuring power over behaviour and to evaluate their status with respect to the inhibitory load they impose on the subject's behaviour and the compensating mechanisms they allow for. In the example cited above, the response duration requirement would be considered as drastically limiting the range of possible collateral behaviour and consequently, possible compensation for inhibition.

Whatever insight further research will bring to the questions discussed here, the differences in temporal adjustment with varying situations raises another important problem. How should an individual's or a species' capacity to master time be characterized? Obviously, no single situation can be assumed to reflect the true *Timing Capacity* of an organism. This capacity can be defined only in a relative manner, as a function of the particular situation in which the performance is measured and it will be defined very differently depending upon the situation used. There remains another related problem: what do all these different performances have in common? Is there a

common unifying timing device underlying all of them, or are there as many timing
mechanisms as there are different types of temporal regulation?

A similar problem has been long familiar to chronobiologists. In the same organism, they find a number of rhythms, some in phase, some not in phase, and rhythms
being sometimes in phase, sometimes not. Are these under the control of a unique
internal clock that expresses its capacities through the constraints of various biological systems and subsystems, modulating the basic timing information? Or are there a
number of clocks, each specializing in the control of part of the systems exhibiting
periodicity? Or again are there as many clocks as there are systems, in other words,
are time bases ubiquitous and at the limit not distinguishable from what is taken as
their expression? This problem is not yet solved in chronobiology, nor is it in the
study of acquired temporal regulations.

The solution to the problem, in our behavioural framework, will require more
extensive description and analysis along the lines suggested above. Psychophysiological research would of course bring decisive answers to some aspects of the problem,
though results accumulated up to now are rather disappointing (see Chapter 6).
Another research strategy recommends itself. It consists of studying individual performance in a variety of different situations. Curiously enough, this simple approach
has not been favoured, probably because of its routine and time-consuming characteristics. Though it would not lead to the location or identification of the timing device
(or devices), it would reveal whether an individual's performance ranks similarly in
various situations or whether, on the contrary, the quality of its performance in one
situation does not predict what it will achieve in another. Highly homogeneous individual profiles would give more plausibility to the hypothesis of a common device.

3.3.2. The Relativeness of Behavioural Time: Weber's Law

It appears from the preceding section that the different experimental situations and
their particular methods of measurement lead to results that give the impression of a
large variety of data. This in turn gives rise to hypotheses of a multiple basis for
temporal control. This point of view, however, may be the result of a superficial survey
of the results, and differences could conceal some essential common traits. A closer
look at the data brings out characteristics that readily appear common to all types of
temporal regulation. Particularly good evidence is offered by experiments where organisms are submitted to several values of the same schedule; the behaviour of the
subjects reflects a same adaptative pattern at all delays. Examples are found in "spontaneous" and "required" regulations as well as in discriminations of duration. These
will be briefly reviewed.

The FI schedule provides evidence about the constancy of the relative length of the
post-reinforcement pause (Shull, 1971a), the constancy of the scallop (Dews, 1970), and
the constancy of the relative location of the breakpoint (Schneider, 1969).

Relative adaptation to time is also reflected in other aspects of behaviour, such as
the latency with which animals approach a lever inserted for a few seconds at different
points in the interval (Wall, 1965). In FT schedules, the distribution of "interim" and
"terminal" activities is proportional to the length of the inter-reinforcement interval
(Killeen, 1975). Similar data are found using DRL schedules and related contingencies
(latency, duration, and Sidman avoidance schedules). Histograms can often be super-

posed when class width is proportional to the delay (e.g. Catania, 1970, with latencies). Moreover, several authors agree on a power function to relate the central tendency of the produced intervals to the schedule parameter, thus establishing the form of a psychophysical scale for time in animals (Catania, 1970; Platt and others, 1973; Richardson and Loughead, 1974b; Sidman, 1953). The ratio between the dispersion and central tendency measures of interval distributions appears constant in schedules reinforcing latencies (Catania, 1970) and lever-press durations (Platt and others, 1973). An immediate parallel can be drawn between these schedules and the duration discrimination experiments, as the Weber fraction is equivalent to this ratio. Consequently, two very different procedures find a common measure in spite of other technical discrepancies. Besides constancy of the Weber fraction (Church and others, 1976; Stubbs, 1968), duration discrimination results confirm relative timing in other respects: when subjects are submitted to different sets of stimulus durations, proportional generalization gradients (Catania, 1970), and a constant rule for bisection (Church and Deluty, 1977) can be found.

Considering such striking similarities, one would be tempted to postulate that all kinds of temporal adaptations bear more similar than distinctive features, and to build a general model accounting for the common mechanism or "internal clock" that regulates behaviour in a basic identical fashion whatever the temporal schedule. Indeed, this has been done. The most sophisticated model has been proposed by Gibbon (1977). It accounts not only for the functioning of the hypothetic "clock", as derived from relative timing data, but also for the "translation" of time estimates into behaviour.

The first assumption in Gibbon's model is that any interval or duration gives rise to an estimate. All time estimates are simply proportional to the estimation of a time unit. As a consequence, the dispersion and central tendency of the distribution of estimates are in the same ratio whatever the length of the schedule parameter. This constitutes a broadened vision of Weber's law applied to all temporal schedules of reinforcement. The second assumption concerns the modulation of behaviour by a motivational parameter or "expectancy" that increases as a reciprocal function of the estimated time to reinforcement. It is compared to an overall level of expectancy that depends on external conditions such as the degree of deprivation of the animal, the amount and rate of reinforcement, etc. When the ratio between increasing and overall expectancies reaches a given threshold, responding is released; in FI schedules the subject breaks pausing, in DRL schedules it terminates an IRT or a latency. In duration discrimination schedules, a comparison is made between expectancy of reinforcement for a "short" report and expectancy of reinforcement for a "long" report but choice is also based on the value of a critical ratio. In Gibbon's model, the motivational variable explains more or less "biased" translations of temporal estimation into behaviour, i.e., the propensity to respond sooner or later.

The hypothesis of a unique mechanism for temporal estimation, after Gibbon (1977), accounts for a large number of results, some of which were mentioned at the beginning of this section. Some problems, however, should be mentioned. Although this model combines coherence and simplicity in characterizing adaptations by the sole coefficient of variation measure, it does not provide an account of certain important aspects of temporal regulation, like response bursts: as mentioned by the author (p. 284), anticipatory behaviour related to the beginning of the interval is not considered. The model does not sufficiently justify the differences in adaptation between FI, DRL, and duration discrimination schedules. The problem here lies in the lack of an unequivocal characterization of the expectancy concept, which intervenes as a correcting factor in the application of the general law to the particular situations.

Besides the problem of modelization, there is a persistent lack of unequivocal support for the hypothesis of a unique internal clock and "profiles" of temporal regulation in successive procedures with crossed controls are not yet available, as has been pointed out above. Only recently was an experimental design set up to investigate the properties of the hypothetical "clock". In a series of experiments, Roberts and Church (1978) attempted to "stop the clock" in different groups of rats run either on FI or on duration discrimination schedules. The sequence of intervals or stimuli was interspersed with blackouts which in the experimental groups ("STOP" groups) actually interrupted the schedule, while having no consequence on the schedule in control ("RUN") groups. The results are not, however, unequivocal, as the data only involve group reactions over blocks of sessions. They seem to reflect a more general adaptation to the modifications of the contingencies of reinforcement (e.g. lengthening of the intervals, splitting of the signals) rather than subtle changes in timing behaviour, which would be expected to appear on individual trials.

IV. Comparative Studies

4.1. STUDIES OF VARIOUS SPECIES

Most examples used throughout this book concern one of the few animal species traditionally favoured in conditioning laboratories: on the Pavlovian side, dogs; on the operant side, rats and pigeons, and less frequently, monkeys or cats. Despite this limited range of species, important differences in temporal regulation have been observed. The most striking of these differences opposes pigeons to monkeys, cats or even rats in DRL schedules. It suggests that various species do not adjust equally well to temporal properties or requirements of controlling contingencies. As for any other aspect of species-specific characteristics, the question arises: what is the biological logic of the differences? How and why did they emerge and evolve? If one wishes to answer this kind of question, one needs much more information from a reasonable variety of species than is now available. An exhaustive inventory of performances of all species is of course unthinkable. Nor is it necessary to obtain a coherent picture of the way species compare with each other with respect to acquired temporal regulations. But half a dozen favoured species is notoriously insufficient, even if two or three dozen occasional attempts with other species are added.

A review was made of about 175 papers published in specialized journals between 1953 and 1977. The inquiry was restricted to behavioural studies of temporal regulation, using the most classical contingencies and their main variants (that is Fixed Interval, DRL, and discrimination of duration of external events). Psychophysiological and psychopharmacological studies were left out of the account. This gives the category "other species" a more favourable position, as subjects of psychophysiological and psychopharmacological studies are generally rats, pigeons, cats or monkeys. Despite this restriction, the place occupied by "other species" is still very limited, as can be seen from Fig. 4.1, 4.2 and 4.3, and one would not venture any detailed comparative analysis on such grounds. Pigeons and rats are by far the most widely studied species, with about 79 per cent of the total number of experiments. It is worth noting that rat studies are prevalent in FI and DRL, and pigeons studies dominate discrimination of event duration. This means that even comparisons between these familiar species are biased by the unequal quantity of information in various situations. The range of delays explored under the different categories of contingencies is roughly indicated in Fig. 4.1, 4.2 and 4.3. It does not reflect the caprice of experimenters, but the contrasts already discussed between required and spontaneous regulations on the one hand, and between regulation of one's own behaviour and discrimination of duration of external events on the other. Within one category, the range of delays explored reflects the cross-species differences.

Let us now take a closer look at studies assembled under the category "other species". It must be noted that part of these are avowedly exploratory attempts, sometimes with one single subject. Many of them give only raw data in the form of cumulative records or the like.

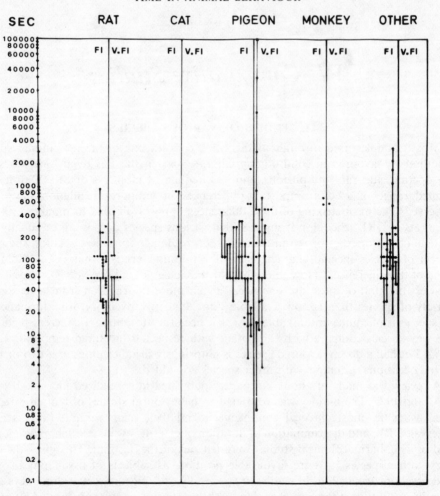

FIG. 4.1. *Comparative studies on FI schedules.* Each point indicates for one species or group of species the value of the interval (log scale in seconds). When several values have been explored in the same study, the points are connected by a vertical line. For each species or group of species, the vertical trait from top to bottom separates studies on classical FI schedule (left), and studies on variants of FI (right). (From Richelle and Lejeune, 1979.)

Studies of mammals include: gerbils in DRL (Beatty, 1978); mice in FI and DRL (Maurissen, 1970); prairie dogs in FI (Todd and Cogan, 1978); a horse in FI (Myers and Meskers, 1960); guinea pigs in FI, DRL and FT (Hauglustaine, 1972; Urbain and others, 1979); Syrian hamsters in FI and DRL (Anderson and Shettleworth, 1977; Beatty, 1978; Godefroid and others, 1969); bats in FI (Beecher, 1971); beagle dogs in FI (Waller, 1961); rabbits in FI (Rubin and Brown, 1969); prosimians: hapalemur in FI and DRL (Blondin, 1974) and Potto in FI (Lejeune, 1976); chimpanzees and baboons in FI (Byrd, 1975), besides numerous studies using different monkey species (squirrel or rhesus monkeys for example); raccoon in FI (King and others, 1974b). Studies of birds include: chickens in FI (Lane, 1961; Marley and Morse, 1966); budgerigars in FI (Ginsburg, 1960); black vultures in FI (Witoslawski and Anderson,

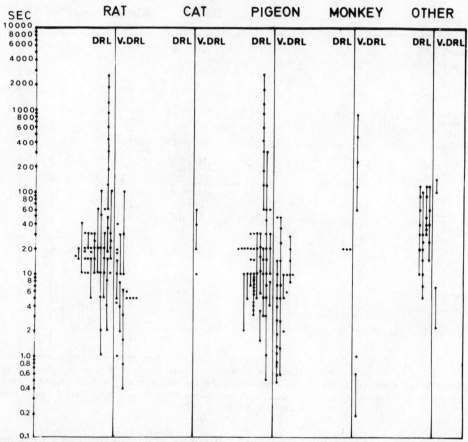

FIG. 4.2. *Comparative studies on DRL schedules.* Each point indicates for one species or group of species the value of the delay (log scale in seconds). When several values have been explored in the same study, the points are connected by a vertical line. For each species or group of species, the vertical trait from top to bottom separates studies on classical DRL schedule (left), and studies on variants of DRL (right). (From Richelle and Lejeune, 1979.)

1963); crows in FI and DRL (Powell, 1971, 1972a, 1974); ravens in FI (Haney and others, 1971); quails in FI (Cloar and Melvin, 1968; Reese and Reese, 1962).

A few studies of fish include: tilapia (*Tilapia macrocephala*) in FI (Eskin and Bitterman, 1960; Gonzalez and others, 1962); gourami fish in FI (Wolf and Baer, 1963); goldfish in FI (Rozin, 1965) and in DRL (Scobie and Gold, 1975).

There is one study of honey bees in FI by Grossman (1973).

We shall review the main results of most of these studies, some of which remain unpublished. For species listed above and not considered in more detail below, available results are either too fragmentary or extremely similar to those of the usual laboratory species.

4.1.1. Mammals

In an unpublished study, Maurissen (1970) conditioned mice in FI and in DRL, using a specially designed lever and a dry-powder food reinforcer. A total of twenty-

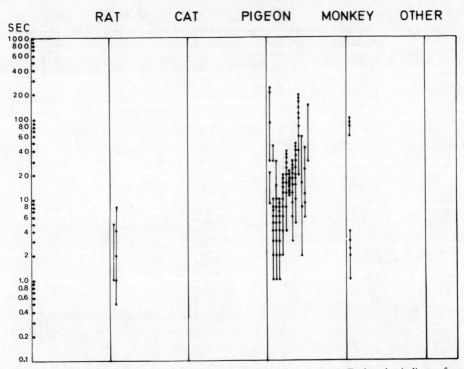

F<small>IG</small>. 4.3. *Comparative studies on discrimination of duration of external events.* Each point indicates for one species or group of species the value of the duration of the external event (log scale in seconds). When several values have been explored in the same study, the points are connected by a vertical line. (From Richelle and Lejeune, 1979.)

six subjects, belonging to strains C3H and NMRI, were submitted either to a FI of up to 4 min or to DRL 30 s or 100 s schedule. The number of sessions for a given individual in a given schedule amounted to a minimum of 60 and a maximum of 175 (a session lasted 60 or 90 min depending upon the particular experiment). The study furnished two major findings.

First, the performance of mice on FI schedules compares favourably with that of rats, pigeons, or cats. Fig. 4.4 shows the curvature index as a function of the delay for three subjects. Fig. 4.5 shows the relative frequency of responses in the successive parts of the interval for three different delays in one individual subject. Secondly, perform- ance in DRL is very poor beyond delays of 10 s or so. Fig. 4.6 shows the IRT distributions for two sample sessions at different stages of exposure to DRL 30 contingencies for four individual subjects. For two of the subjects (mice 3 and 7), there is no sign of temporal regulation. For the other two the shape of the distribution suggests temporal control, but the mode is still far from the reinforced value and the proportion of reinforced responses (Efficiency Ratio) remains very low. These subjects were transferred abruptly after shaping to the final value of the delay, that is 30 s. Progressive training from 5 to 30 s in small steps did not result in better final performances.

Hauglustaine (1972) successfully conditioned more than twenty guinea pigs, a species rarely used in operant conditioning research. On the basis of careful observa-

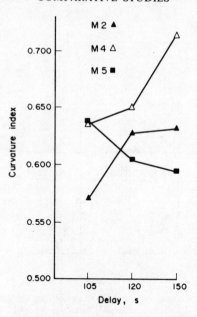

FIG. 4.4. Curvature index as a function of the fixed interval in three individual mice. (After Maurissen, 1970, unpublished data.)

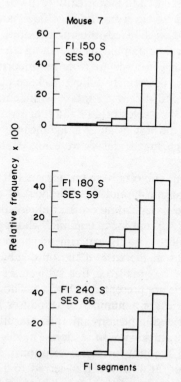

FIG. 4.5. Distribution of responses in consecutive segments of the interval, for three values of the interval, in one individual mouse. (After Maurissen, 1970, unpublished data.)

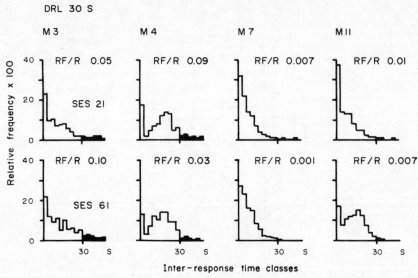

FIG. 4.6. IRT distributions from two sessions on DRL 30 s at different stages in training in four individual mice (M 3, M 4, M 7, and M 11). Efficiency ratios RF/R are given in the figure. (After Maurissen, 1970, unpublished data.)

tions of their feeding habits, weight and motor activity, he designed a lever terminated by a 3.5 cm wide disc located 5.5 cm from the cage floor, and used carrot juice as a reinforcer. All subjects trained on a Fixed-Interval schedule, with delays of up to 80, 140 or 200 s, produced typical scalloping as shown in the samples of cumulative records presented in Fig. 4.7. The records are from non-deprived animals: preliminary observation had shown that it is extremely difficult to keep guinea pigs in good health when they are food deprived, and that water intake is difficult to control if the animal is allowed to consume fresh vegetables (a guarantee of balanced diet). Control tests with water-deprived animals did not produce a significantly different picture. Haug-lustaine suggested that, though longer delays were not explored, his subjects would probably have adjusted.

In a second experiment, four subjects were trained on DRL. Delay values were 30, 36, 40, and for one of the subjects, 49 and 90 s. The first of these values was reached either abruptly or progressively, depending on the individual. Exposure to each delay lasted fifteen to forty sessions. Except for one animal whose performance deteriorated with the 36 s delay, the subjects in this experiment showed good adjustment with delays of 36, 40 and even 49 s in the case of the only subject trained with this last delay. Samples of cumulative records from this subject are shown in Fig. 4.8. IRT distributions from two subjects are presented in Fig. 4.9. The mode is typically close to the reinforced value, and there are a number of responses in the first IRT class—a phenomenon classically described in pigeons and rats. The efficiency ratio ranges from 0.20 to 0.70, with variations from session to session. The degree of water deprivation, as control tests show, cannot account for these variations. Subject 18, which showed very good temporal regulation at 49 s, was transferred to DRL 90 s. The last histogram in Fig. 4.10 shows the subject's failure to adjust to this delay even after thirty training sessions.

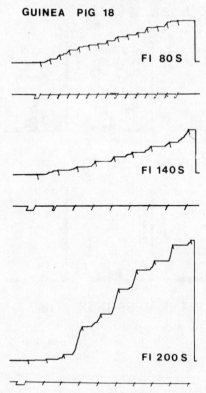

FIG. 4.7. Cumulative records from one individual guinea pig on FI schedule, with various interval values. (After Hauglustaine, 1972, unpublished data.)

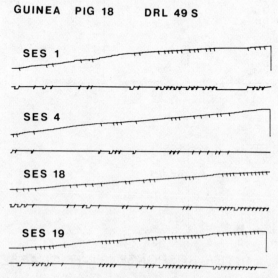

FIG. 4.8. Cumulative records from one individual guinea pig on DRL 49 s, at different stages of training (from top to bottom). The bottom pen on each record is downset when the critical delay is over and the reinforcement is available. (After Hauglustaine, 1972, unpublished data.)

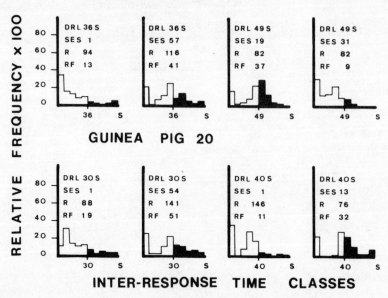

Fig. 4.9. IRT distributions for two guinea pigs on DRL. Delay values, rank number of session using a given delay, response and reinforcement numbers are shown above each graph. (After Hauglustaine, 1972, unpublished data.)

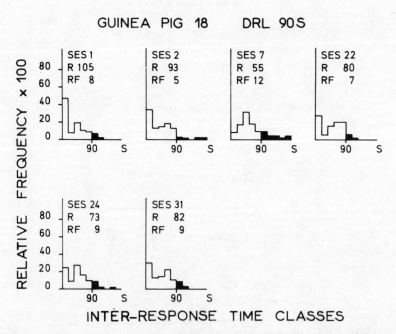

Fig. 4.10. IRT distributions of one guinea pig on sample sessions on DRL 90 s showing failure to adjust to that delay. Compare with performances shown in Fig. 4.9 for shorter delays. (After Hauglustaine, 1972, unpublished data.)

If these few subjects are representative of the species, it would seem that guinea pigs reach a limit of possible adjustment to DRL somewhere between delays of 50 and 90 s.

Rubin and Brown (1969) also employed water as a reinforcer in conditioning rabbits with FR, VI or FI contingencies. Under FI schedules, the duration of the post-reinforcement pause was a function of the interval (inspection of cumulative records shows that it practically reduced to nothing in FI 30 s, and that it extended over more than three-quarters of the interval on FI 180 s). No scalloping, but an abrupt break-and-run pattern was observed. This, as well as the rather high rates of responding, was probably due to the nature of the response: subjects grasped the lever with their teeth and depressed it with head movements.

Following a first successful attempt to condition lever-pressing responses in non-deprived Syrian hamsters (Richelle and others, 1967), and a year-long study of hoarding behaviour as measured by operant responding (Godefroid, 1968), Godefroid and others (1969) submitted two animals of this species to Fixed-Interval contingencies. The subjects were housed in large cages subdivided into a nesting-place, a toilet corner, and a storing-place for food. The home cage was connected through a tunnel with the experimental chamber to which the subject had access for 45 min five days a week between 5 and 7 p.m. The choice of this time schedule was the result of earlier observations: Godefroid had found that the hamsters, being crepuscular animals, would engage in high rates of operant responding for food that they would not eat but stock towards the end of the day. Most of the grains received as reinforcement were carried to the store-place rather than immediately consumed. The subjects had free access to the hoarded food, the amount of which always largely exceeded their need. After three months of exposure to continuous reinforcement, the FI was introduced and progressively increased to 2 min by 10 s steps. Only five sessions for one animal and ten for the others were run with the final 2 min delay. The observed performance might not reflect the maximal capacities of the subjects. However, both exhibited temporal regulation. One produced 40 to 50 per cent of its responses during the last quarter of the interval and its performance improved throughout training; the other gave comparable results up to the 50 s delay, but the index dropped to 34 per cent when the subject was transferred to the 2 min delay. Variation in the number of responses was not systematically related to the duration of the delay.

Another study of the behaviour of Syrian hamsters under FI and FT schedules has been recently published by Anderson and Shettleworth (1977). The duration of the interval employed was 30 s. The authors concentrated on observing and analysing various classes of behaviour that were produced during the interval between food presentation, bar pressing being one of them in the subject exposed to Fixed-Interval schedules. Some behaviours occur more and more frequently towards the end of the interval, paralleling lever pressing in FI, and are characteristically terminal activities. Others are typically produced in the beginning or the middle of the interval (interim activities). There is clear evidence that all these activities are temporally patterned. The dynamics of their alternation has been discussed by several authors (see Falk, 1977; Killeen and others, 1978; McFarland, 1974; Rachlin and Burkhard, 1978; Silby and McFarland, 1974; Staddon, 1977; Staddon and Simmelhag, 1971), and we shall mention some of their hypotheses in Chapter 7 and Chapter 9. Unfortunately, Anderson and Shettleworth did not give quantitative estimations of the temporal regulation of their subjects' behaviour. Inspection of their graphs, however, does reveal scalloping for lever-pressing responses and some terminal activities (Fig. 7.6).

Attributing Boice and Witter's (1970) failure to condition prairie dogs to their having used water as reinforcement, Todd and Cogan (1978) successfully trained eight subjects on FR, VI or FI schedules, with solid food as the reinforcer. They reproduce cumulative records of FI 30 to 150 s for three animals, and these exhibit classical scalloping, post-reinforcement pauses and response rates inferior to rates on FR. The performance compares with behaviour generally described under FI contingencies.

Though still exploratory, data obtained by Blondin (1974) on a prosimian species are interesting in two respects: they bear on a non-domesticated organism, and on a species of mammals that belongs to the most primitive category of primates. Using the exceptionally adequate facilities of the Prosimian Research Laboratory of the Musée National d'Histoire Naturelle at Brunoy (Paris), Blondin conditioned several indivi- duals of the species *Hapalemur griseus griseus*, a member of the family *lemuridae* that lives in Madagascar. The study was aimed at designing a convenient operant con- ditioning situation, using positive reinforcement, that could be used as an efficient tool in analysing aspects of the behaviour of this species (for example psychophysical functions). Preliminary to and concurrent with operant training, careful direct obser- vation of the subjects guaranteed that the experimental procedure would not go counter to basic aspects of the natural repertoire.

Subjects were housed in pairs (one male, one female) in large living-spaces, provided with the necessary items for survival in captivity (branches, a close nesting-place). They were not manipulated during the experiment. Instead, the experimental enclos- ure was connected to the living space for the time of each session. This ethological approach, exceptionally respectful of the natural behaviour, makes the results gath- ered by Blondin valuable in spite of the limited number of subjects (five, of which one was trained only on Fixed Ratio), and of the fact that she did not intend to produce a systematic picture of temporal regulations as such.

One female subject was trained under FI 5 min for twenty-six sessions after an FR (2 to 100) training which extended over 200 sessions. One male subject was trained under FI 1 to 5 min after exposure to DRL contingencies for more than 100 sessions. One naïve male was trained under Fixed Interval with delays increased from 10 s to 20 min and later to 60 min (total number of sessions: 240). Performances under DRL were explored in the male subject later trained under FI, and in another female subject that did not undergo any further training. The maximal delay was 60 s for the male, and 40 s for the female, and was reached in 5 s steps in each case, with stabilization of the performance at 20 and 40 s.

The subject initially trained under FI showed a performance typical of that schedule until the delay reached 30 min. The index of curvature varied between 0.32 and 0.47 (for a subdivision of the interval into eight temporal classes, the maximal value is 0.875). These values are inferior to those obtained with rats, cats, monkeys and mice. Interestingly, they are not a function of the duration of the interval for delays from 40 s to 20 min, as shown in Fig. 4.11. Deterioration of performance at 30 min and beyond might indicate a limit of adjustment or the effect of other kinds of variables (possibly length of session).

The two subjects trained on FI after exposure either to DRL or to FR contingencies showed the same kind of general pattern with comparable values of the index of

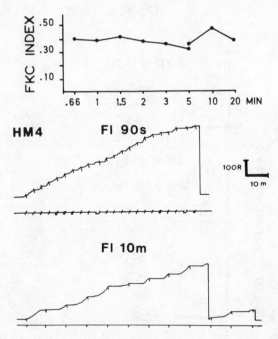

FIG. 4.11. Fixed-interval performance in a male hapalemur. Above: curvature index (FKC) as a function of the interval. Below: samples of cumulative records under FI 90 s and FI 10 min. (After Blondin, 1974, unpublished data.)

curvature, but with rates of responding plausibly influenced by previous training. The subject transferred from DRL to FI produced almost five times fewer responses than the subject transferred from FR to FI and four times fewer than the subject trained exclusively on FI. Cumulative records in Fig. 4.12 allow for comparison of the individual performances. Temporal regulation under DRL developed slowly but reached very good quality for DRL values of up to 60 s, as illustrated by the IRT distributions in Fig. 4.13. Sequential analysis of IRTs indicates that short IRTs tended to occur either after another short IRT or after an IRT that fell short of the reinforced value. Reinforced IRTs tended to occur after IRTs of comparable size. We shall retain for a later discussion the fact that *Hapalemurs*, though they do not show temporal regulation of the same quality as mice or pigeons in FI, do nevertheless master much longer delays in DRL.

Another exploratory study on operant conditioning of prosimians has been published by Lejeune (1976). Two subjects, one female and one male, belonging to the species *Perodicticus potto edwarsi* (from Gabon) were trained on CRF, Fixed-Ratio, and FI schedules. The conditioning sessions took place in the home cage, in order to avoid undesirable manipulations of the animals. A special manipulandum adapted to the particularities of the hand of this species was designed, and milk was the reinforcer. Though pottos are rather slow-moving organisms, under Fixed Ratio 20 they produced high rates of responding (averaging 11.4 responses per min). The rate was much lower under FI 2 min and FI 1 min, but the temporal regulation was poor. Figure 4.14 shows the relative frequencies of responses in the four successive parts of

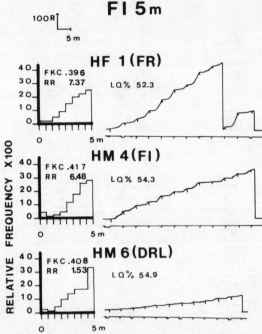

FIG. 4.12. Performance under FI in three individual hapalemur with different histories of conditioning. Left: Average relative frequencies distributions of responses in consecutive segments of the interval (with values of the curvature index, FKC, and response rate per minute, RR, indicated above each graph.) Right: cumulative records. The abbreviations following the identification of each subject stand for the schedule experienced before. The percentage of responses in the last quarter of the FI (L.Q. %) is also indicated. (After Blondin, 1974, unpublished data.)

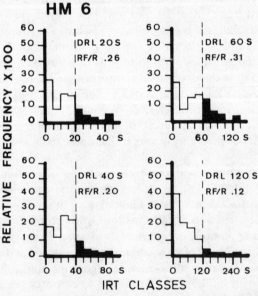

FIG. 4.13. Individual performance under DRL in hapalemur: IRT distributions for increasing values of the critical delay. (After Blondin, 1974, unpublished data.)

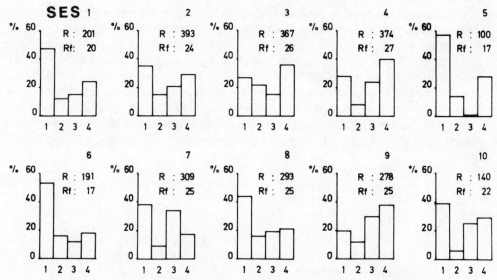

FIG. 4.14. Performances of a female Potto under FI 2 min. Relative frequencies distributions in the four successive segments of the interval for ten consecutive sessions. (From Lejeune, 1976.)

the 2 min interval for the female subject: after ten sessions of exposure to the final contingencies (following ten sessions of progressively increased delays), the classical FI pattern did not develop and many responses were still produced in the first quarter of the interval. This may be due to the previous exposure to the Fixed-Ratio schedule, but the male subject did not perform better on a 1 min delay. Considering the drastic change of environment for such a species transferred from the African forest to a laboratory cubicle, the number of sessions was far too small for conclusions to be drawn about the limitations of the potto's capacities for temporal regulation.

Data on temporal regulations in apes are very scarce. In a pharmacological study, Byrd (1975) submitted two chimpanzees and three baboons to a multiple FI-FR schedule. The interval in the FI component was 10 min. During sessions without drug, subjects of both species exhibited good temporal regulation, with a quarter-life index ranging from 57 to 88 per cent.

4.1.2. Birds

Ginsburg (1960) obtained scalloped curves in five budgerigars on FI 30, 60 and 120 s. The response was a brief vocal emission (chirping).

Pecking responses can be conditioned in quails, and brought under various forms of schedule control, as first shown by Reese and Reese (1962). These authors furnish a record of one subject of the species *Coturnix coturnix japonica* trained on FI 1 min: the rate of responding is very low (averaging 1.4 response per min) and the response is often emitted after the interval has elapsed. These exploratory findings indicate, as the experimenters suggest, that with more training, these birds might show precise temporal discrimination. This has not been confirmed, however, in a subsequent and more detailed study by Cloar and Melvin (1968). Subjects belonging to two different species, *Coturnix coturnix japonica* and *Colinus virginianus*, failed to develop temporal regulation. The rate of responding remained high, and though there was some scalloping or break-and-run pattern depending upon the species (see Fig. 4.15), the post-reinforcement pause was short. Long exposure to the schedule did not result in improved performance in this respect. Possibly some artifact (unsuspected signals from the equipment) might be responsible for Reese and Reese's results, but the differences between the two experiments with regard to the deprivation variable offer the most plausible explanation. Reese and Reese maintained their subjects at their normal body weight and deprived them for less than 24 h, while Cloar and Melvin resorted to the reduction to 80 per cent of free-feeding body weight widely used with laboratory animals.

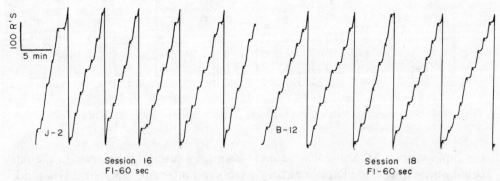

FIG. 4.15. Selected cumulative records of responses on a FI 60 s schedule for a Japanese quail (*Coturnix coturnix Japonica*) and a bobwhite quail (*Colinus virginianus*). (From Cloar and Melvin, 1968. Copyright 1968 by the Society for the Experimental Analysis of Behavior, Inc.)

Members of the *Corvus* family have been exposed to Fixed-Interval schedules. Haney and co-workers (1971) have observed in the white-necked raven (*Corvus cryptoleucus*) a performance comparable with that of familiar laboratory species for delays of 4 min. Powell (1972a) exposed crows (*Corvus brachyrhynchos*) to FI 60, 120 and 240 s. He obtained mean response rates much lower than with FR, VR or VI schedules and scalloping or break-and-run patterns with a quarter-life index approximating 70 per cent and pause duration extending over 50 to 70 per cent of the interval. Referring to numerous data published on pigeons, Powell concludes that crows can discriminate time more effectively than pigeons. Powell also exposed crows to DRL, in a systematic comparative study that will be reviewed in the present chapter under 4.3.

4.1.3. Fish

Fish of the species *Tilapia macrocephala* (African mouthbreeder) initially trained on FI 1, 2 or 4 min were maintained on FI 2 min in an experiment aimed mainly at studying the effect of pre-feeding on behaviour under FR and FI schedules (Eskin and Bitterman, 1960). Pre-feeding had no significant effect on the rate of responding in FI, while it reduced the rate in FR. No detailed analysis of FI performance is given, but the authors make this important statement:

"Records of the FI subjects, except for a lower over-all rate of responses, are not readily distinguishable from those of the FR subjects; brief pauses after reinforcement are common, but no indication of scalloping appears. It cannot be concluded, however, that this absence of scalloping in the fish reflects some functional difference between fish and higher animals. Whether scalloping can be depended upon to appear spontaneously in higher animals—which once was taken for granted—seems now to be in doubt".

Other experimenters, however, have obtained scalloping and temporal regulation in fish of other species. Wolf and Baer (1963) reported results from one gourami fish under FI and Rozin (1965) from five goldfish (*Carassius auratus*) exposed to FI 1 and 2 min for thirty to sixty sessions of 30 min each (see Fig. 6.31 and 6.32).

Scobie and Gold (1975) studied four goldfish on DRL 10, 20, 40 and 80 s. They observed an inverse relation between response rate and length of delay, as is classically described in familiar laboratory species. IRT distributions showed evidence of a temporal regulation for delays of 10 and 20 s, but not for 40 and 80 s.

4.1.4. Insects

There are, unfortunately, very few studies on operant conditioning of insects and an experiment by Grossman (1973) on honey bees (*Apis mellifera*) is, to our knowledge, the only one that makes use of a temporal schedule. Grossman designed an ingenious device in which bees had to enter a plexiglas tube in order to suck a sugar solution. The situation was fairly "natural", bees visiting the experimental device from a neighbouring hive. FI schedules with delays of up to 20 and 90 s produced rates of responding clearly lower than FR schedules, but no temporal discrimination developed; there was no scalloping and no post-reinforcement pause, as can be seen from the cumulative record in Fig. 4.16.

4.2. PROBLEMS OF COMPARISON

Scarce as they are, these data indicate that spontaneous temporal regulations develop with delays of from half a minute to 10 or 15 minutes with very similar perform-

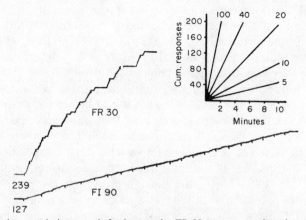

FIG. 4.16. Representative cumulative records for bees under FR 30 (upper record) and under FI 90 s (lower record) schedules. Both records are from the seventh day of training on the schedule (From Grossman, 1973. Copyright 1973 by the Society for the Experimental Analysis of Behavior, Inc.)

ances not only in familiar laboratory species such as rats, pigeons, cats and monkeys, but also in most other species that have been submitted to Fixed-Interval contingencies. Exceptions, some of them possibly due to various aspects of the experimental situation rather than to species limitations, are bees, *Tilapia macrocephala* fish, one prosimian, and quails.

By contrast, interspecific differences clearly show up in DRL. Roughly, species for which data are available (unfortunately, there are very few) would rank as follows according to the quality of temporal regulation and the magnitude of the delay to which the subjects adjust (in decreasing order): monkeys, cats, rats, then clearly much lower on the scale, pigeons, mice, fish.

Ranking species along a behavioural dimension such as temporal regulation raises a number of extremely difficult problems classically encountered in all comparative studies of behaviour and repeatedly identified by experimenters who have attempted to determine a hierarchy for learning capacities or aptitude for problem-solving (see Bitterman, 1960). Are apparently identical situations really comparable when applied to different species? How can we evaluate the equivalence of measures of performances taken with species possessing various morphological characteristics and completely different natural repertoires? Is a cat pressing a lever for food comparable to a rat pressing a lever for food? And is a pigeon pecking a key comparable to a rat pressing a lever or a monkey pulling a chain? A reinforcement pertaining to the same category, say food, and distributed according to the same schedule, may very well not have an equivalent reinforcing value in different species with very different feeding habits (carnivorous predators for instance can space their meals as the hazards of prey-hunting demand and eat big quantities at a time, while small rodents have a more regular feeding schedule and eventually hoard food in their nest). The same reasoning could be evoked concerning the responses, the discriminative stimuli and the contingencies: though many experimenters and theoreticians of learning have long tended to ignore this point, the status of a given reinforcer, of a given response, or of a given stimulus is not equivalent for all species (see for example Dunham, 1977; Mackintosh, 1977; Schwartz and Gamzu, 1977).

Looking at it carefully, the problem is very tricky indeed. It is tempting to claim that very similar performances from different species can rightly be compared, and to use the similarity of the observed behaviours as a criterion for comparability. A famous case is the juxtaposition by Skinner (1956) of three cumulative records obtained under multiple Fixed-Interval-Fixed-Ratio schedules from a pigeon, a rat, and a monkey. True enough, such a comparison demonstrates that an experimenter can arrange contingencies in such a way as to obtain similar control over the behaviour of different species, and this is obviously what Skinner meant to do. But this does not tell us whether the compared species are equivalent with respect to the behaviour under study. In order to demonstrate that, one should first demonstrate equivalence of the independent variables used. One main variable in studying temporal regulation is for instance the magnitude of the delay. Pigeons, mice, rats and cats compare favourably in DRL if the delay is 10 s, but differences become evident with longer delays. Contingencies work similarly as long as species-specific limitations do not oppose their control. Of course when we attempt to compare species, the point at which limitations are observed is more important than the point at which all species (or group of species) perform equally well.

Thus, we are faced with a twofold problem. One aspect is the need to demonstrate that dissimilarities observed in the temporal regulation of different species do not merely reflect the experimenter's ignorance of species-specific characteristics; that he did not use the right response, the right reinforcer, etc., for revealing the animal's capacity. The other aspect is making sure that similarities are really significant, that we have not just picked up by chance and possibly assisted by theoretical inclination, the most favourable point for comparison. With few exceptions, studies reported so far in the present chapter bear on one single species, and comparison with results of other species, mainly familiar laboratory animals such as rats or pigeons, refers to experiments done by other authors. This means that some details—and often very important details—of the experimental situation always differ, and that consequently, the twofold problem formulated above is never dealt with rigorously. To do that, we need studies which systematically compare two or more species in equalized experimental settings. Such studies are extremely rare.

4.3. COMPARATIVE STUDIES

A good example is provided by Lowe and Harzem (1977) who explicitly addressed themselves to the detailed comparison of rats' and pigeons' performances under FI and FT schedules. The delays in this study were 30, 60 or 120 s. Subjects were first submitted to eighty FI sessions, followed by fifty FT sessions and another series of forty FI sessions, all at the same delay. As expected, the classical temporal regulation developed, but a number of distinctive features also appeared. Responding was not (or was little) maintained in FT in rats whatever the delay, while it persisted at the same rate as in the previous FI schedule in pigeons, at least for delays of 30 and 60 s. Cumulative records shown in Fig. 4.17 illustrate this effect. Also, when FI contingencies were reinstated after exposure to FT, rats did not respond at the same overall rate as they had in previous FI sessions but showed a much lower rate.

Detailed analysis of pause duration and of rate during the activity phase reveals other differences between species. Following the method proposed by Killeen (1975, see 2.3.1.), the authors fitted normal curves to response rates plotted as a function of time since previous reinforcement (Fig. 4.18) and plotted the midpoint of the normal curve obtained against the interval value (Fig. 4.19). The data for rats (first and second exposure) fell along straight lines with slopes of 1.16 and 1.32 respectively, while those for pigeons had slopes of 0.95 and 0.96. Supposing this mathematical treatment provides, as Killeen contends, a correct estimation of temporal judgement in animals, it must be admitted that because the psychophysical function has different values in both species the capacity for time judgement is different. Such a conclusion is tempting, since it would add to the well-established difference between rats and pigeons in DRL, an unsuspected difference in FI performance. As Lowe and Harzem rightly warn, however, this conclusion should not be adopted too readily. Rather than reflecting species' basic differences in their capacity for temporal adjustment, the differences described here might very well be accounted for by the status of the operant response. Key-pecking in pigeons is both food-producing and food-induced, that is both operant behaviour and respondent, elicited behaviour, as the numerous experiments on auto-shaping have shown. Lever-pressing in rats is more exclusively an operant. This would explain why responding can be maintained for delays as long as 60 s in Fixed-Time

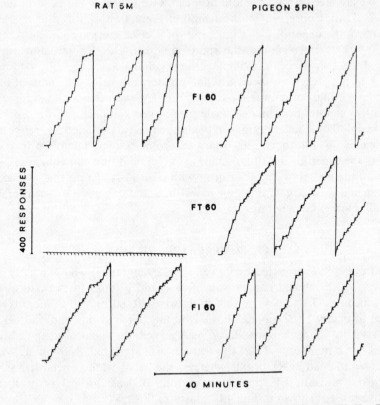

FIG. 4.17. Cumulative records of one rat and one pigeon from the last session on FI 60 s, FT 60 s and re-exposure to FI 60 s schedules. (From Lowe and Harzem, 1977. Copyright 1977 by the Society for the Experimental Analysis of Behavior, Inc.)

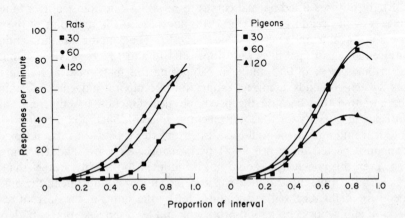

FIG. 4.18. Rate of responding by rats and pigeons in successive tenths of the interval, in the last four sessions of first exposure to the FI schedules. Data points are averages (geometric means) across animals (eleven rats and twelve pigeons) in each group, for the first nine-tenths of each interval. (From Lowe and Harzem, 1977. Copyright 1977 by the Society for the Experimental Analysis of Behavior, Inc.)

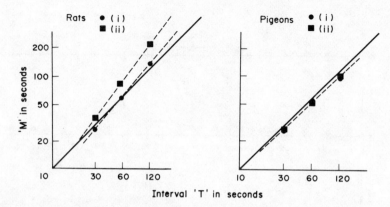

FIG. 4.19. The relation between interfood interval (T) and midpoint of the normal curve (M). The data are averaged across animals in each group, taken from the first (filled circles) and second (filled squares) exposure to the FI schedule. (From Lowe and Harzem, 1977. Copyright 1977 by the Society for the Experimental Analysis of Behavior, Inc.)

schedules with pigeons but not with rats, and why the slope of the function relating the midpoint of the normal curve to the length of the interval is not the same in each species (for a discussion of the response variable see Chapter 5, Section 3).

Equally cautious conclusions as to the significance of observed differences were drawn by Richardson and Loughead (1974b) from their data on DRL with long delays (1 to 45 min) in rats and pigeons. Rats had lower response rates and higher reinforcement rates than pigeons at all DRL values; pigeons had more very short "bursting" responses than rats and the more so as the DRL delay was increased, which was not the case in rats; the correlation between mean log inter-reinforcement time and mean log DRL value was higher in rats than in pigeons, etc. Though all these differences furnish support for a species difference in temporal regulation, one should not rule out the possibility that some secondary variable such as the type of response used, rather than timing-capacity proper, is responsible for it.

A study by Powell (1974) is exemplary in that it compares in rigorously equalized situations a familiar laboratory animal, the pigeon, and a wild species, the crow (*Corvus brachyrhynchos*). Four crows and three pigeons were trained under DRL over a large number of sessions with delays increased from 10 to 120 s according to an efficiency criterion. Pigeons were not trained beyond 10 or 30 s because of their poor efficiency. Crows performed very efficiently with delays of 30, 60 or 120 s. In the absence of a limited hold, they usually produced very long IRTs as can be seen from Fig. 4.20 but still showed evidence of temporal regulation. Introducing a limited hold (10 to 2 s in DRL 20, or 5 to 2 s in DRL 10) resulted in improved temporal discrimination (see Fig. 4.21).

In a control experiment, the end of the critical delay was signalled by a change in colour of the response key. This external clock (see Chapter 8) made it unnecessary for the subject to estimate time. Under this procedure, pigeons performed just as well as crows, and obtained more than 80 per cent of the reinforcements for delays of up to 120 s. Powell also reported the presence of many collateral behaviours (mainly in the form of pecking other parts of the chamber) in his crows under both unsignalled and

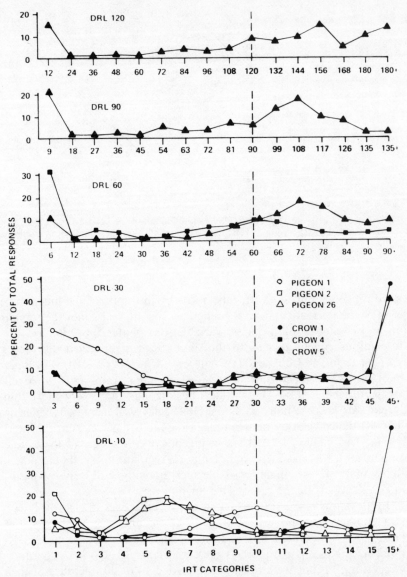

FIG. 4.20. Relative frequency distribution of inter-response times (IRT) at selected schedule values. (The numbers on the abscissa represent the upper limit for each category. The data were averaged over three consecutive sessions during the last five sessions at the schedule requirement. The dashed vertical line indicates the minimum IRT required to obtain reinforcement). (From Powell, 1974. Copyright 1974 by the American Psychological Association. Reprinted by permission.)

signalled DRL, while such behaviours were observed in pigeons only under signalled DRL.

Powell interprets his findings in terms of the higher position of his crows on the avian evolutionary scale. He emphasizes the fact that the behaviour of corvids in the wild is less specialized than that of pigeons and that the brain development is more advanced in corvids than in other avian species.

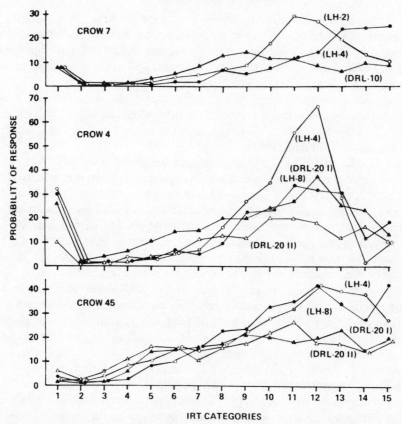

FIG. 4.21. Conditional probability curves for three crows under baseline conditions (DRL 10 and DRL 20 s triangles) and at two limited hold requirements (LH). The greater-than-15 category is not presented, since it always yields a probability of 0.0 or 1.0. (From Powell, 1974. Copyright 1974 by the American Psychological Association. Reprinted by permission.)

A recent study by Urbain and others (1979) compares rats and guinea pigs on Fixed-Time schedules. The delay between non-contingent food reinforcement was increased from 7 to 120 s or decreased from 120 to 7 s. Polydipsia was observed in rats but not in guinea pigs, while interim lever presses were produced only by those guinea pigs exposed to the descending delay values. Rats and guinea pigs exposed to ascending delays emitted a negligible number of such responses. Unfortunately, the authors give no data on the temporal distribution of interim activities, so that this comparative study adds little to our knowledge of temporal regulation.

4.4. TEMPORAL REGULATION AND EVOLUTION: SOME HYPOTHESES

Assuming that the inter-species differences observed are not merely artefactual, several hypotheses may be proposed to account for them (see Richelle 1968, 1972). Data presently available are insufficient to give definite support to any of these, however. We shall therefore present them as being equally plausible, and leave it to future research to discover convincing evidence favouring one of them.

The Evolutionary Hypothesis.   Temporal regulation of behaviour might be more and more refined and efficient as we progress along the phyletic scale from simple to more complex organisms. According to this view, one should find better performance in apes than in monkeys, in monkeys than in prosimians, in primates than in rodents, in mammals than in fish, in vertebrates than in invertebrates, and so on. Whatever its physiological basis, temporal regulation of behaviour would improve with biological evolution. According to Razran, Soviet data on delayed conditioning "suggest that the maximum CS–US delay which an organism can master is a positive function of evolutionary ascent" (Razran, 1971).

Operant conditioning studies also indicate that the limitations on the delay that can be mastered differ between species and on the whole, the differences observed would not contradict an evolutionary hypothesis (see for instance Powell's study that compared crows with pigeons). They do not suffice, however, since limits have not yet been systematically explored in many species. A systematic study should bear on several different situations and measures of behaviour since, as we have seen, no single situation or measure can be considered as reflecting the true capacity of the timing mechanism or mechanisms—the nature of which remains completely elusive.

The magnitude of delays that can be mastered or the degree of precision of time estimation might not be the relevant dimension along which the evolutionary hierarchy will appear. More complex aspects of temporal regulation should possibly be looked at, aspects which have drawn very little attention up to now: for instance, the capacity to master simultaneously the temporal dimension of several events, or of several actions, or both; or the pasticity of temporal regulation, making for rapid adjustment and readjustment to changing parameters. Along these lines, Bolotina (1952, in Dmitriev and Kochigina, 1959) noted that monkeys showed better adjustment to new delays than dogs or animals ranking lower on the phyletic scale.

The Reductionist Hypothesis.   As long as the evolutionary hypothesis is not more strongly documented, one might just as well advocate the opposite view, that temporal regulation is basically similar in all animals. The idea that some universal fundamental timing device is at work behind all varieties of performances is of course suggested by the ubiquitousness of rhythms in living systems, as evidenced in chronobiological research. Periodicity is observed in plants as well as in animals, in unicellular as well as in complex organisms. It seems to be a very basic character of living matter and for some authors like Richter (1965), the evolutionary trend is not toward enrichment of rhythmic properties, but rather toward liberating the organism from the constraints of so primitive a mechanism.

Reducing all acquired temporal regulations to some basic mechanism common to all animals (possibly the same mechanism underlying biological rhythms) leaves us with the problem of accounting for the differences in performance in different species. Some of these differences might, as we have seen, be more apparent than real, and be due to the non-equivalence of some of the variables (refer for instance, to the data produced by Lowe and Harzem, 1977, discussed above). Some might be explained by the experimenter's failure to make the designed contingencies effective. In a more sophisticated version of the reductionist hypothesis, real differences might be assigned not to the temporal mechanisms *per se* but to the kind of use the organism makes of them given its sensory-motor equipment. In other words, there would be no basic difference in timing capacities between fish, rats and monkeys, but only differences in levels of motor inhibitory processes, in levels of perceptual attentional processes and the like, and consequently these processes would integrate temporal information in the context of varying degrees of complexity.

The Ethological Hypothesis. The third hypothesis, already suggested by Richelle (1972), proposes that differences between species with respect to temporal regulation are not related to some general evolutionary trend, but that they should be explained by reference to the specific properties of the natural repertoire. Temporal regulation would have been selected by environmental pressure as any other aspect of the morphology or of the behaviour of a species. To take an admittedly oversimplified example, cats, as carnivorous predators, would have developed watching-behaviours involving capacities for temporal regulation reflected in laboratory situations such as DRL, while mice, having evolved in the opposite way, have developed flight responses certainly more effective for potential prey—and exhibit poor performance in DRL situations.

Among the authors mentioned in the present chapter, Grossman (1973) has implicitly favoured this hypothesis in his discussion of the behaviour of bees under schedules of intermittent reinforcement. He draws a parallel between experimental and natural contingencies, pointing to the fact that bees visiting flowers do not always find the expected reinforcement: some flowers do not contain nectar because they are already empty or not yet filled. He does not suggest any explanation, however, for the failure of experimental bees to produce scallops on a FI schedule (but discussion of other aspects of bees' behaviour, p. 138, could throw some light on this point).

Some of the comparative results reported above might at first sight fit better with this hypothesis than with the other two (and especially the first *evolutionary hypothesis*). Thus, quails do not perform like pigeons under FI, while pigeons perform very much like rats or monkeys. Quails and pigeons are closer on the phylogenic scale than pigeons and monkeys. Pottos—at least if Lejeune's (1976) results are confirmed—do not adjust to FI as do hapalemurs, their fairly close neighbours on the evolutionary scale, and they are hardly better than bees or quails, which are biologically far away.

It would seem more logical to look for an explanation of these dissimilarities and similarities in the species-specific characteristics, rather than in evolutionary position.

This hypothesis has a number of features to recommend it. It emphasizes the wide diversification of species and of species-specific behaviour and by so doing it discards oversimplifications of biological evolution. It is in line with current thinking in experimental behaviour analysis. Many aspects of experimentally controlled animal behaviour are now being interpreted in the framework of natural repertoires, rather than exclusively as expressing general laws valid for all species or for large groups of species. Significantly enough, Collier and others (1977) devote one half-page of *The Handbook of Operant Behavior* (Honig and Staddon, 1977) to discussing the analogy between DRL performance and the lion's prey-watching. Finally, it might seem a more adequate hypothesis to account for the data available up to now. It also raises difficult problems, not the least of which is how to make a convincing demonstration of its validity. Relating precise measurements of one aspect of behaviour in the laboratory to species-specific patterns of behaviour in the natural environment is an extremely difficult task and at any rate, it requires more than the loose, intuitive comparison of the kind outlined above. The joint expertise of experimental psychologists and of ethologists is essential for collecting factual arguments which will eventually lead to the validation of the theory.

V. *Factors Influencing Temporal Regulations*

5.1. CONTINGENCIES OF REINFORCEMENT

Temporal regulations are largely a function of the contingencies of reinforcement. As the reader probably has already observed, this is one of the central themes of this book, and the basis for most of the discussions presented in the previous chapters as well as those to come in the following chapters. Thus it is only for the sake of logic in presentation that we mention it under this heading, and list it first among the factors that influence temporal regulations.

The main contrast between spontaneous and required regulations, as exemplified by the typical pattern of behaviour generated by FI and DRL schedules respectively, need not be commented upon again. Only as a reminder shall we point to the difference, in terms of the magnitude of the delay that can be mastered, between these two categories of behaviour. At another level, we have also encountered differences in performance depending upon the fact that the contingencies require a temporal estimation of an external event or a temporal adjustment of the subject's own behaviour. The best documented example here is the capacity of pigeons to estimate the duration of visual stimuli of 30 or 40 seconds while they can hardly space pecking responses by more than 10 to 15 seconds. More subtle differences are also observed within more restricted classes of situations. Changing the stimulus from continuous light of a given duration to an "empty" interval of time of the same duration between two brief flashes, produces different performance in a task of discriminating the duration of external events in pigeons (Mantanus, 1979). Spacing two brief motor responses by a given delay seems easier for cats than maintaining the paw on a lever for the same period of time (Greenwood, 1977).

Further along the same line, it can be shown that minor changes in the contingencies defining a given schedule affect the temporal regulation. As such minor changes can be arranged *ad infinitum*, and have in fact been arranged by experimenters among a wide range of variations, an exhaustive enumeration would appear fastidious and serve only to provide an inventory.

If minor changes in the contingencies may result in differences in performance, it must be noted that, conversely, important changes may not prevent striking similarities. A case in point is the Fixed-Interval schedule in which an electric shock, rather than food, is contingent upon the first response emitted after the delay has elapsed. Strangely enough, operant responding can be maintained under such conditions. Most experiments have been performed on squirrel monkeys or rhesus monkeys, except for one study with cats (Byrd, 1969). Subjects have been submitted to FI schedules of shock presentation after previous exposure to variable interval contingencies with food reinforcement (Kelleher and Morse, 1968a), after exposure to a schedule of shock escape (McKearney, 1974a), or after previous shock avoidance learning (Barrett and Glowa, 1977; Byrd, 1969, 1972; Kadden, 1973; Kelleher and Morse, 1969; McKearney, 1968, 1969, 1970a; Malagodi and others, 1973b, 1978; Morse and Kelleher, 1970;

Stretch and others, 1968, 1970). In some experiments, the behaviour controlled in this way has been maintained for hundreds of sessions (Malagodi and others, 1973b).

Most studies report a patterning of behaviour very similar to that obtained under FI using food as a reinforcer: there is a pause after the shock, then a phase of activity with responses emitted at an accelerated rate until the next shock is delivered. Fig. 5.1 shows this effect (for a detailed analysis, see Morse and Kelleher, 1977 and also McKearney and Barrett, 1978). Several authors give quantitative characterization of their subjects' performances. Byrd (1972) has obtained quarter-life values between 0.55 and 0.60 in squirrel monkeys under FI 8 min of shock presentation. Malagodi and others (1978), using two subjects of the same species, reported quarter-life value of 0.74 and 0.77 respectively under FI 6 min. McKearney (1969) observed little change in the quarter-life index after shifting from FI 10 s to FI 60 s.

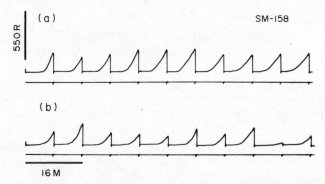

FIG. 5.1. Cumulative records of responding in one squirrel monkey (SM-158) during final sessions of FI of shock presentation when (a) the first response occurring after 8 min produced an electric shock and (b) a 1 s delay without a stimulus change intervened between the first response occurring after 8 min and delivery of a shock, and a 30 s time-out followed shock delivery. (After Byrd, 1972. Copyright 1972 by the Society for the Experimental Analysis of Behavior, Inc.)

Variables controlling behaviour under FI with food reinforcement seem to control behaviour under shock presentation as well. Rate of responding is related to the length of the interval and to the magnitude of the reinforcement (that is the intensity of shock) as with food reinforcement (McKearney, 1969; Malagodi and others; 1973b, DeWeeze, 1977). Delaying the reinforcement results in a decrease in response rate, without altering the general pause-activity pattern, as is the case with food reinforcement (Byrd, 1972. See Fig. 5.1b). That responding is truly under the control of shock presentation is demonstrated by the fact that behaviour extinguishes if the shock is no longer delivered, and is resumed if the shock is reinstated (McKearney, 1969). Furthermore, behaviour has been successfully maintained in second-order schedules where FI components of stimulus presentation are maintained under FR of shock presentation (Byrd, 1972) and on VI or FR schedules of electric shock presentation (Kadden, 1973; McKearney, 1970a, 1972, 1974b). However, continuous reinforcement fails to sustain behaviour (Kelleher and Morse, 1968a; McKearney, 1972). The question remains unanswered, as to how long-lasting the control of behaviour would be with different schedules of electric shock presentation.

A no less strange case of what appears a "perversion" of the biological function of reinforcers is the schedule of food postponement in all respects similar to Sidman non-discriminated avoidance, (see 2.3.1) except for the fact that the electric shock is replaced by food presentation. Experimental subjects produce responses that have the consequence of postponing food presentation. Such behaviour has been obtained from rats after training under continuous reinforcement (Smith and Clark, 1972) and in the squirrel monkey after Sidman shock avoidance (Clark and Smith, 1977). Besides poss- ible species-specific factors, the experimental history seems more important here than for the schedule of FI of shock presentation in the emergence of a temporal patterning of responses, but temporal patterning typical for Sidman avoidance was observed by Clark and Smith (1977) in the squirrel monkey. Control exerted by food postpone- ment has been demonstrated with extinction and reconditioning, as can be seen on Fig. 5.2. Moreover, stable behaviour has been described for a period of about one year.

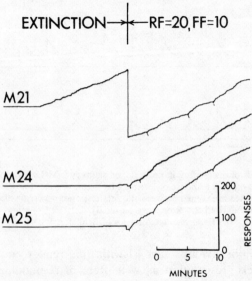

FIG. 5.2. Food-postponement schedule in squirrel-monkeys. Cumulative records of three individual subjects for a session during which the food-postponement contingencies were reinstated after extinction. The record reset at the end of the extinction period. Only 20 min before and 20 min after that point are shown. (From Clark and Smith, 1977. Copyright 1977 by the Society for the Experimental Analysis of Behavior, Inc.)

These intriguing experiments throw some new light on the concept of positive versus negative reinforcement. It would seem impossible to assign any particular physical stimulus to one or other classes of reinforcers in any absolute way. The same physical event, say food or electric shock, may be a positive or a negative reinforcer, depending upon the contingencies and upon the individual's previous history (Morse and Kelleher, 1977). The power of certain contingencies in inducing temporal regula- tion of behaviour is once more demonstrated when one considers how temporal patterning persists despite such unusual manipulation of the reinforcing event.

We shall now turn to more classical aspects of the reinforcement variable in tem- poral regulations.

5.2. THE REINFORCER

The notion of contingencies of reinforcement refers mainly to the set of conditions that relate reinforcement, responses, and discriminative stimuli. The same contingencies, however, may be applied using different reinforcers, in terms of quality or quantity. The nature of the reinforcer may vary, not only according to the classical opposition of positive versus negative (for instance food versus electric shock), but within each of these classes: not only food can be presented as positive reinforcer, but also water, intracerebral electrical stimulation, habit forming drugs, etc. The magnitude of the reinforcer may also vary: the amount of food, the intensity of shock, the concentration of a drug are easily controlled independent variables. It is of course interesting to look at temporal regulations under various quantitatively or qualitatively different conditions of reinforcement.

5.2.1. The Nature of the Reinforcer

Though food under various forms is by far the most widely used of possible reinforcers, intracerebral self-stimulation (Olds and Milner, 1954) has been used in FI schedules (Brown and Trowill, 1970; Cantor, 1971; Pliskoff and others, 1965; Sidman and others, 1955) and in DRL schedules (Brady and Conrad, 1960a; Cantor, 1971; Couch and Trowill, 1971; Owens and Brown, 1968; Pliskoff and others, 1965; Terman, 1974).

Some experimenters have used drugs such as cocaine or pentobarbital in FI schedules (Balster and Schuster, 1973; Dougherty and Pickens, 1973; Goldberg and Kelleher, 1976; Johanson, 1978; Pickens and Thompson, 1972 for example). Behaviour has also been maintained in second-order schedules where FI components of stimulus presentation are maintained under FR of cocaine presentation (Goldberg and others, 1975; Kelleher, 1975; Kelleher and Goldberg, 1977), and FR components under FI of cocaine or morphine presentation (Goldberg, 1973, 1975, 1976; Goldberg and others, 1976; Goldberg and Tang, 1977).

As a rule, it seems that performances under a given schedule are quite similar whatever the reinforcer. However, few studies have been designed to provide for a real comparison of different kinds of reinforcers. An already ancient study by Brady and Conrad (1960a) is one of these exceptions. In rhesus monkeys, they observed similar IRT distributions under DRL 20 s contingencies using either food or intracerebral self-stimulation as a reinforcer (Fig. 5.3). Another possibility for comparison in one single experimental context is offered by Balster and Schuster (1973). They used rhesus monkeys and a multiple schedule. The successive components of the schedule were Fixed-Interval 9 min with a limited hold of 3 min and Time-Out periods of 15 min (mult. FI 9 LH 3 TO 15 min). One out of four FI components was reinforced with food, and the other three components were reinforced with a drug injection (cocaïne). This experimental design was not aimed at a comparison between reinforcement, but at studying the long term effects of drug injections. However, it provides an opportunity for the kind of comparison we are interested in here (Fig. 5.4). As far as one can judge from the reported data, it seems that the general FI pattern was found for drug as well as for food, though the rate of responding was higher for drug as dose increased. Barrett and Glowa (1977) and Morse and Kelleher (1966) have obtained analogous patterns of responding in squirrel monkeys under Fixed-Interval contingen-

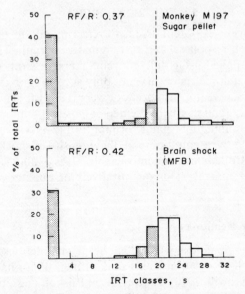

FIG. 5.3. IRT distributions in DRL 20 s for one individual rhesus monkey, with food (top) or electrical cerebral self-stimulation in the medial forebrain bundle-MFB. (Redrawn and modified, after Brady and Conrad, 1960a. Copyright 1960 by the Society for the Experimental Analysis of Behavior, Inc.)

cies reinforced either with food, by escape from a stimulus associated with brief electric shock presentation, or by shock presentation.

Comparing reinforcers differing in nature raises a difficult problem of equalizing them quantitatively, since we know—as we shall see in the next paragraph—that the magnitude of the reinforcement affects performance. This difficulty has been especially dealt with by researchers who have attempted to compare food reinforcers with electrical self-stimulation. Some felt that the duration of the consummatory response for food should be matched by the extension of the cerebral stimulation over a certain period of time (Brown and Trowill, 1970; Hawkins and Pliskoff, 1964; Pliskoff and Hawkins, 1967; Pliskoff and others, 1965). It is doubtful, however, that reinforcers can really be equalized in this way (Mogenson and Cioe, 1977). The question is further complicated by two other main factors, namely the intensity of stimulation (or more generally the physical parameters characterizing the stimulation current) and the locus of the stimulation. Thus, when electrical self-stimulation seems to produce different performances as compared with food reinforcement it might well be due to an imperfect equalizing of the two different reinforcers rather than to a basic difference related to their nature.

5.2.2. The Magnitude of the Reinforcer

Changing the reinforcer quantitatively—that is changing the amount, the concentration, the dose, or the intensity, depending upon the kind of reinforcer used—has effects upon the rate and upon the temporal pattern of responding under Fixed-Interval contingencies. We shall consider first those studies using food as a reinforcer.

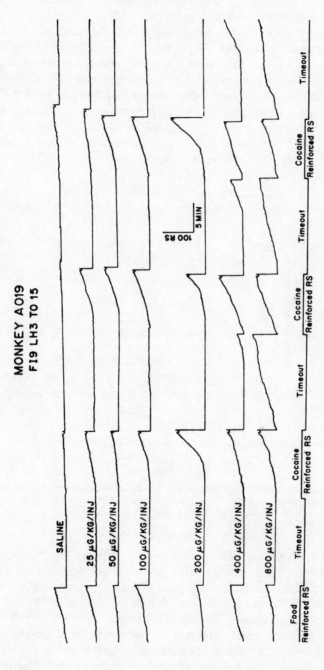

FIG. 5.4. Cocaine or food as a reinforcer on FI 9 min in a rhesus monkey. Portions of the cumulative record for the last day of each test dose. The pen reset at the completion of each fixed interval or limited hold and also after each time out. (From Balster and Schuster, 1973. Copyright 1973 by the Society for the Experimental Analysis of Behavior, Inc.)

First, it should be recalled that there is a direct relationship between response rate and absolute reinforcement frequency or, stated in another way, total amount of food earned over a session (Catania and Reynolds, 1968; Collier and Myers, 1961; Schneider, 1969; Skinner, 1938; Stebbins and others, 1959). The rate of response is related to the quantity of the reinforcer, but in a way that depends upon some critical aspects of the experimental procedure. When different quantities of food are given over successive blocks of sessions, the rate of responding shows a direct relationship to the quantity of the reinforcer as shown in Fig. 5.5 (Collier and Myers, 1961; Collier and Willis, 1961; Guttman, 1953; Hutt, 1954; Meltzer and Brahlek, 1968, 1970; Stebbins and others, 1959). One exception to this rule is reported by Keesey and Kling (1961) in a study on pigeons; it might be accounted for by the fact that rate measures were undertaken only after day-to-day stabilization of response rates with each quantity of reinforcement, which has not always been the case in other studies.

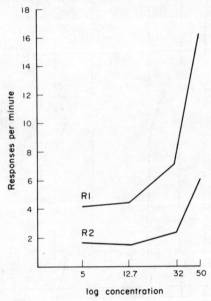

FIG. 5.5. Response rates as a function of percentage of sucrose concentration in two individual rats under FI 2 min. The concentration values are plotted on a logarithmic scale. (Redrawn after Stebbins and others, 1959. Copyright by the Society of the Experimental Analysis of Behavior, Inc.)

When different quantities of food are tested during the same experimental session, an inverse relationship is observed between the rates of responding and the quantity of the previous reinforcement (Madigan, 1978; Meltzer and Howerton, 1973, 1975, 1976; Staddon, 1970a) (Fig. 5.6). Results obtained by Jensen and Fallon (1973) and by Lowe and others (1974) provide exceptions. The effects obtained with this second procedure are sometimes termed *context effects* (Harzem and others, 1975; Madigan, 1978), and have been recently analyzed as a function of the interval value.

Indices of temporal regulation are also affected in different ways depending upon the procedure. When a given quantity of food is constantly used throughout a block of sessions, most authors report no change in the distribution of responses in successive

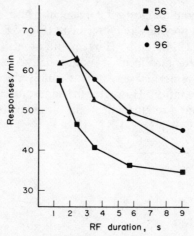

Fig. 5.6. Response rate over the interval following five reinforcement durations on FI 60 s in pigeons. Curves are for three individual subjects, and the values plotted are averaged across the last five days of exposure to the contingencies. (Redrawn after Staddon, 1970. Copyright 1970 by the Society for the Experimental Analysis of Behavior, Inc.)

parts of the interval (Meltzer and Brahlek, 1970) nor in the duration of the post-reinforcement pause (Harzem and others, 1975) (see Fig. 5.7). However, Stebbins and others (1959) describe a U-shaped relationship between the quality of temporal regulation and the sucrose concentration in rats. When different quantities of the reinforcer

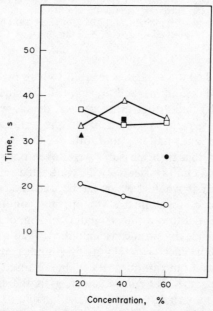

Fig. 5.7. Mean durations of post-reinforcement pauses in FI 60 s as a function of sucrose concentration in the liquid reinforcement (milk), when each concentration is used during a whole block of sessions. Data points are from the baseline stage in each concentration. Filled data points show redetermination on the first concentration experienced by each subject. Each symbol is for one individual rat. (From Harzem and others, 1975. Copyright 1975 by Academic Press, Inc.)

are tested in the same session, the post-reinforcement pause varies as a direct function of the magnitude of the previous reinforcement (Fig. 5.8), as shown by Staddon (1970a), Lowe and others (1974) and by Harzem and others (1975). The relation holds even when the primary reinforcement is occasionally omitted, though associated stimuli are presented in the normal way (this is simply the zero point on the continuum of reinforcement quantity). However, the index of curvature was not significantly altered as a function of reinforcement magnitude or of reinforcement omission in a study by Jensen and Fallon (1973).

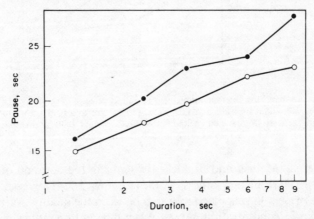

FIG. 5.8. Post-reinforcement pauses as a function of the duration of reinforcement in FI 60 s. Abscissa: duration of access to grain dispenser (subjects are three pigeons). The duration of reinforcement is varied throughout each session. Data points are averaged over the last five days (filled circles) or the first five days (open circles) across the three birds. (From Staddon, 1970a. Copyright 1970 by the Society for the Experimental Analysis of Behavior, Inc.)

It must be noted that the kind of studies referred to here involve peculiarly intricate experimental designs and conditions, and apparent discrepancies among results from different studies might be understood if procedural details were taken into account and the notion of reinforcement magnitude itself was seriously re-examined. Discrepancies between Jensen and Fallon's and others' studies, for example, may be accounted for by the fact that their measures were taken only after behaviour in their multiple FI 45 s FI 45 s TO 60 s schedule with contrasted reinforcement magnitudes had stabilized. This analysis would lead us far beyond the scope of this book, into subtleties that might be of great interest to operant conditioners, but would not contribute anything essential to our understanding of temporal regulations.

The problem is, of course, by no means simplified when a drug is used as a reinforcer, since it has a double effect as a reinforcing stimulus and as a variable affecting subsequent behaviour, as a function of dose, of time elapsed, and the like. It is not surprising that data are sometimes contradictory, as is the case with cocaine in the studies quoted above. One crucial variable is the latency of action, and the rate of metabolizing the injected dose. Thus Balster and Schuster (1973), using different infusion speeds, have obtained different effects. One could expect that procedures with discrete trials separated by long Time-Out periods, as Balster and Schuster (1973) employed, or very long FI s as Pickens and Thompson (1972) employed (30, 60 and

90 min), will create conditions and produce results quite different from those characterizing procedures with shorter delays or uncontrolled intertrial intervals, as in
Dougherty and Pickens' study (1973). Considering these complexities (see Johanson,
1978), and the limitation of most studies to rhesus or squirrel monkeys (Dougherty
and Pickens have used rats), a general coherent picture is difficult to draw.

Finally, a study by Pliskoff and others (1965) explored the effect of varying the
amount of electrical self-stimulation (in terms of number of stimulations available at
the end of each interval) on behaviour generated by FI schedules ranging from 10 s to
10 min. Unfortunately, this study does not provide any parametric analysis of the
relationship between reinforcer magnitude and temporal regulation.

Another simple though indirect way to manipulate the reinforcement variable is to
vary the degree of food deprivation. Satiation reduces the rate of responding, and
eventually results in suppression of operant responses. Conversely, increased deprivation produces an increase in rate of responding (Carlton, 1961; Skinner and Morse,
1958). The quality of the temporal regulation, however, is not affected or only little
altered, except in cases of extreme satiation or deprivation. Thus, Powell (1972b) has
shown in pigeons that chronic modification of body weight (from 72 to 93 per cent)
influences the rate of responding but does not affect the duration of the post-reinforcement pause nor the quarter-life. Pre-feeding the birds with amounts varying from 0 to
15 grammes, or depriving them for periods ranging from 24 to 72 h, has no significant
effect on the three behavioural measures being considered (Fig. 5.9). The absence of
effects of acute pre-feeding or deprivation on response rate in this case is at odds with

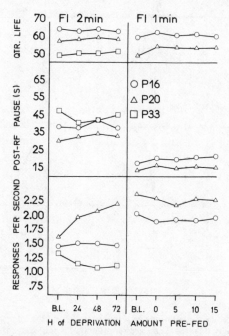

FIG. 5.9. Mean performances for three pigeons as a function of the number of hours of food deprivation
(left) or of the amount of food given prior to the experimental session (right). Data points corresponding to
baseline (B.L.) on FI 2 or 1 min represent the mean performance over ten sessions. (Redrawn after Powell,
1972b. Copyright 1972 by the Psychonomic Society, Inc.)

the general rule stated above. Powell points to possible explanations in terms of species differences (rats versus pigeons) or methodological details such as the duration of the baseline before deprivation or satiation effects are assessed. Finally, Johanson (1978) reports no effect of cocaine deprivation (inter-session times ranging from 1 to 45 h) on a fixed-interval 5 min schedule of cocaine reinforcement in rhesus monkeys, the absence of effect being due to the short-duration drug action and to the inability to induce dependence.

In DRL schedules, increasing the amount of food per reinforcement (from 1 to 4 pellets) results in a reduction of the mean IRT and a lowering of the efficiency ratio, as shown by Beer and Trumble (1965) using a 18 s delay with rats. A similar relation was observed in pigeons during the acquisition phase of a DRL 20 s by Lejeune and Mantanus (1977), when a 3 s access to the grain magazine was compared with the delivery of one single grain of vetch per reinforcement. This inverse relation between

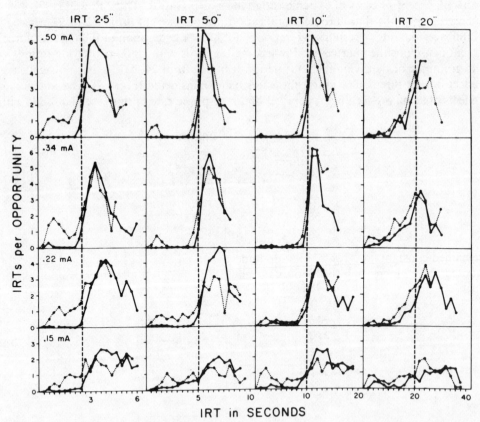

FIG. 5.10. IRTs per opportunity distributions as a function of stimulus intensity across four IRT schedules (from 2.5 to 20 s), for one individual rat. Data are represented for four current amplitude values which were used across all schedules. Each column presents the results for a different IRT schedule, with the vertical dashed lines denoting the criterion intervals. Solid curves represent IRTs following a reinforced response; dotted lines represent IRTs following a non-reinforced response. (From Terman, 1974. Copyright 1974 by Pergamon Press, Ltd.)

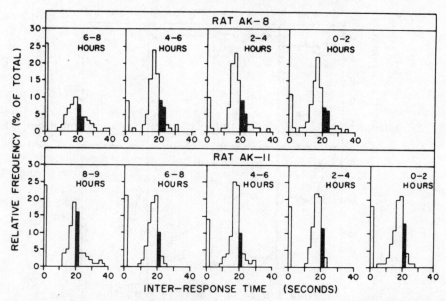

FIG. 5.11. Inter-response time distributions at decreasing stages of progressive satiation during a 10 h test session of limited-hold DRL. Black sections indicate reinforced inter-response times. (From Conrad and others, 1958, 1958. Copyright 1958 by the Society for the Experimental Analysis of Behavior, Inc.)

quantity of reinforcement and the quality of temporal regulation has not been confirmed for intracerebral stimulation. Terman (1974) has performed a systematic analysis of IRT and IRT/OP-distributions as a function of the amplitude of hypothalamic self-stimulation in rats. For the various values of delay used in this study (from 0.6 to 20 s), a direct monotonic relation was found between the amplitude of the stimulating current and the probability of producing a reinforced IRT (Fig. 5.10) A refined analysis of results based upon signal detection theory has confirmed this conclusion. The difference between food and electrical stimulation could be due to the difference in the duration of the reinforcing event. In Terman's study, the duration of the stimulation remained constant (at 0.5 s) for all amplitudes, while in the two experiments with food quoted above, increasing the amount of reinforcement meant also increasing the time spent eating.

Satiation paradoxically results in better temporal regulation of DRL performances, up to a point where the subject stops responding. This improvement is a by-product of a reduced rate of responding. This has been shown in rats (Aitken and others, 1975; Carlton, 1961; Conrad and others, 1958), monkeys (Conrad and others, 1958) and pigeons (Reynolds, 1964a; Skinner and Morse, 1958). Conrad and others (1958) provide IRT distributions from successive parts of a 10 h session under DRL 20 LH 4 for two rats. Before the animals ceased responding because of satiation, the proportion of long IRTs increased and that of short IRTs decreased, as evidenced in Fig. 5.11.

Increasing the degree of food deprivation produces an opposite effect, the magnitude of which is a function of the degree of deprivation. Thus Conrad and others (1958) have described an increase in rate of responding and a progressive reduction of long IRTs in one monkey with deprivation up to 9 h before the beginning of the sessions, and in one rat with deprivation up to 21.5 h. Longer deprivation periods

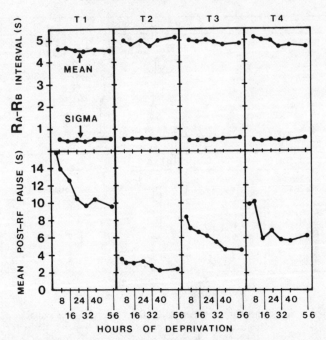

FIG. 5.12. Effects of deprivation upon waiting time and post-reinforcement pause in a temporal discrimination procedure. Each of the four columns of graphs represent the data for one rat. (Redrawn, after Mechner and Guevrekian, 1962. Copyright 1962 by the Society for the Experimental Analysis of Behavior, Inc.)

failed to further change the performance. However, if experimental conditions provide for a dissociation between crucial and secondary variables with respect to temporal regulation, deprivation can be shown not to affect the timing mechanism itself. This was elegantly demonstrated by Mechner and Guevreckian (1962) with rats, using a two-lever DRL schedule (see 2.3.1). The rats were reinforced whenever a response on lever B followed a response on lever A by a 5 s delay. Changing the deprivation level did not alter the average Response A to Response B interval, but did change the post-reinforcement pause, that is, the interval separating a reinforced response on B and a following response on A (Fig. 5.12). Finally, it must be noted that occasional omission of reinforcement results in an increased rate of responding, as is the case in FI schedules (Davenport and others, 1966).

5.3. THE RESPONSE

In the operant methodology, it has been considered for years that the behavioural units or responses selected for study are arbitrary, that is they are but samples of behaviour chosen for their ease of recording. However, in the late sixties important discoveries evidenced the particularities of the pecking response in the pigeon. Naïve birds presented with a light and food association spontaneously pecked at the light source (autoshaping: Brown and Jenkins, 1968), even when pecking prevented food appearance (negative automaintenance: Williams and Williams, 1969). These phenomena cast serious doubt on the arbitrariness of the key-peck in pigeons, recalling earlier findings by Breland and Breland (1961) concerning the interference of species-specific characteristics of behaviour with contingencies of reinforcement. Since then, the nature of the operant response has become a critical variable, the importance of which has been investigated in many situations, such as those of stimulus control (Hemmes, 1973; McSweeney, 1975; Scull and Westbrook, 1973), pharmacology (Fontaine and others, 1977), cerebral lesions (Huston and Ornstein, 1975).

In tasks involving temporal regulation, the type of performance also depends upon the selected operant. Most of the studies have been conducted with FI and DRL schedules. With negative schedules of reinforcement (such as Sidman avoidance), the interest has focused more on the acquisition of the response than on its temporal characteristics.

One aspect of the response possibly responsible for differences in performance is the position of the response in the natural behavioural repertoire of the species. Some operants present a high degree of similarity to the consummatory response evoked by the reinforcement (Jenkins and Moore, 1973), and thus they are not free of the constraints linked with feeding or drinking behaviour. For instance, key-pecking in the pigeon is closely related to grain consumption, and is contaminated by the characteristics of frequency (Staddon, 1972b), duration (Schwartz, 1977) and strength (Topping and others, 1971) which constitute the fixed attributes of the peck for food in the natural repertoire of grain-eating birds. These characteristics are of particular importance in the contingencies that require temporal spacing of the response, such as the DRL schedule of reinforcement.

The choice of a behavioural unit morphologically different from the consummatory response has generally led to a better adaptation. In the pigeon, treadle-pressing with the foot has been compared to key-pecking in DRL (Mantanus and others, 1977a). Treadle-pressing reaches more quickly an efficiency criterion for increasing the delay, as shown in Fig. 5.13, and the proportion of reinforced IRTs is superior to that obtained with key-pecks (Hemmes, 1975). The existence of a natural association between the reinforcer and the response also influences the rate of responding under chained DRL schedules; key-pecking is more sensitive to the rate-lowering effect of the secondary reinforcement than is treadle-press (Mantanus and others, 1977b).

In a species in which a non-consummatory response is commonly used, it is possible to worsen temporal performance by choosing a less arbitrary—more consummatory —operant. Rats using a tube-licking response on a DRL schedule with a liquid reinforcer exhibit a poorer performance, in terms of percentage of reinforced responses, than when they use a lever-press response (Kramer and Rodriguez, 1971). In avoidance schedules, the choice of an operant in the appetitive repertoire, that is key-pecking in pigeons, can alter the very acquisition of avoidance, because the emission of the operant is incompatible with the reflexive behaviours elicited by the shocks. Bolles (1971) argued that responses close to species-specific defence reactions ought to be selected for avoidance schedules. In the pigeon, treadle-pressing proved quite efficient in this respect (Foree and Lolordo, 1970; Smith and Keller, 1970). In the rat, some advantages of the head-poking response over the lever-press were found (Carrigan and others, 1972) but further investigation stressed the adaptive similarities between the two operants (Ayres and others, 1974).

Responses that are "arbitrary" with respect to consummatory behaviour may nevertheless bear constraints at some other level. The complexity of the motor execution

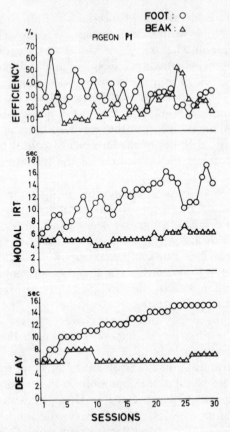

FIG. 5.13. Compared performances under DRL using key-pecking or treadle-pressing as the response in one individual pigeon. The DRL value was increased according to a performance criterion (20 per cent of reinforced responses and modal IRT ⩾ to criterion IRT). It can be seen that training progressed more quickly and with much better temporal regulation with treadle pressing. (From Mantanus and others, 1977a.)

may interfere with the temporal regulation of a response. In this case, the development of timing behaviour may be obscured for some period during which learning to execute the response takes place. For example, cats trained to maintain a paw on a lever for a given number of seconds needed many more sessions to reach a given delay than other cats trained to give two brief presses spaced in time; however, the final distribution of emitted durations and intervals respectively did not differ significantly (Greenwood, 1977). Moreover, subtle constraints exist within a given class of operants. Glick and Cox (1976) found side-preferences in rats. Given the availability of two identical levers for responding on a DRL schedule, the rats displayed preferences for one or the other, the strength of which was directly related to the percentage of reinforced responses on that lever.

The question raised by the comparisons of operants on temporal schedules concerns the relative validity of the operants for the expression of an organism's timing ability. Some early results suggested somewhat paradoxical adaptations in the same subjects, as if the organisms reacted differently to the contingencies according to the type of operant they were using. A counterbalancing trend in results appeared soon thereafter, as some experiments were designed to restore the key-peck response with some of its value, for instance, its generality under DRL schedules (Richardson and Clark, 1976), its sensitivity to duration-controlling schedules (Ziriax and Silberberg, 1978), or its ability to condition to avoidance schedules (Todorov and others, 1974).

These opposed and somewhat confusing results warn as to the necessity of analyzing the nature of the differences encountered when two operants are compared. Some aspects involved in their topographies may induce different response outputs that to some extent mask a similar patterning of the responses in time. For instance, response bursts are more probable with key-pecks than with treadle-presses, and their occurrence may affect the percentage reinforced responses. On the other hand, execution of an operant with greater muscular involvement can create problems of stability which compel the animal to pace its responses; this automatically lengthens the interresponse times without improving timing. The top row of Fig. 5.14 provides an illustration of an apparent discrepancy in the temporal regulation of two responses, a lever-press and a complete run in a circular cage (a technique designed by Fontaine and others, 1966). Lejeune (1974) compared these two responses in two groups of twelve rats on an FI 2 min schedule. The top row shows the average rates of responding during the successive quarters of the interval. In the bottom row, the differences between responses disappear when a relative measure of responding (the proportion of responses per quartile of interval) and the value of the index of curvature (see 2.3.1) are considered.

It must be pointed out that a limited number of operants have been studied up to now. Although a complete inventory of all possible operants would be inconceivable;

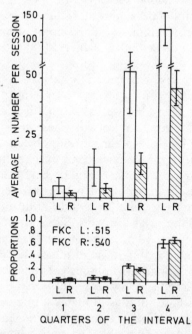

FIG. 5.14. Lever-press versus running response in FI 2 min with rats. Striped blocks: run; white blocks: lever. Each couple of data, from left to right, correspond to successive quarters of the total interval (see text). (From Lejeune, 1974, unpublished data.)

some careful selection has to be done. The position of a response with regard to the
natural repertoire seems of special importance as far as it imposes clear-cut limitations
on the temporal adjustment of that response. Similarly, the difficulty of executing
responses should be considered when a facilitative, pacing effect can be suspected.
Anticipating Chapter 9, we shall suggest that, in choosing a response for temporal
schedules, one should take into account an aspect almost totally ignored until now,
namely, the degree of inhibitory control to which the operant response is amenable.
This degree may vary as a function of the status of the response in the species-specific
repertoire (see Chapter 4), of the motor sequence involved, of the energy required, and
the like.

5.4. DEVELOPMENTAL AND HISTORICAL FACTORS

Temporal regulations are largely a function of the environmental conditions to
which the subject is presently or has been recently exposed, in other words of the
contingencies of reinforcement. It would seem important to ask whether they are also
a function of the moment in the individual's life and of its previous history.

5.4.1. Developmental Factors

Though the ontogenetic approach has for some time drawn the attention of chronobiologists, students of acquired temporal regulations seem to have ignored it completely. In fact, surprising as it may sound, their ignorance reflects a general neglect of
the developmental dimension among specialists of operant conditioning. This situation also matches the scarcity of developmental studies in the psychophysics of time
with humans—while ontogenetic studies on the concept of time have been numerous
(for example Fraisse, 1948; Fraisse and Orsini, 1958; Piaget, 1946, 1966; Zuili and
Fraisse, 1969).

Technical difficulties are no real justification for this situation. It is true that experiments on temporal regulations usually extend over long periods of time—several
months being more common than a few weeks—so that it might seem just hopeless to
expect a developmental description with animals whose growth period is shorter than
the minimal period required for the experiment. This objection might be sensible with
regard to mice or rats; it does not hold for other species, such as cats or monkeys, let
alone species less familiar to experimenters but who have a sufficient life-span.

Exploring the early stages of development before weaning is of course extremely
difficult, if not unfeasible with the usual procedures of operant conditioning, at least
with some of the most traditional laboratory species such as mice and rats (though
some attempts have been made; see Amsel and others, 1977; and Brake, 1978). Larger-
sized species, born with a more mature level of motor development, would be more
amenable to that kind of developmental approach. The problem is also easily solved
with species of birds in which the young leave the nest immediately after hatching.
Marley and Morse (1966), whose study is probably a unique exception to the statement made at the beginning of this section, have conditioned newly hatched chickens
under a multiple FR-FI schedule. They obtained nice scalloped behaviour after a few
sessions in the FI 2- or 3-min component of a MULT FI-FR schedule (Fig. 5.15). They
also furnished a cumulative record of a two-month-old chick for a MULT FR 25 FI 7

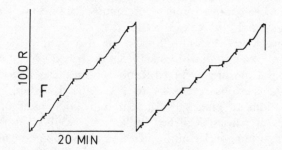

Fig. 5.15. Cumulative record obtained on the tenth day of life in a young chick trained from his third day under a multiple FR 30 FI 3 min schedule. Alternating FR and FI periods are clearly seen on the record, showing both temporal regulation in the FI component and discrimination between the white and the red light signalling the two components. (From Marley and Morse, 1966. Copyright 1966 by the Society for the Experimental Analysis of Behavior, Inc.)

min schedule. Marley and Morses' purpose was a purely practical one: they were looking for rapidly conditioned, easily housed, and inexpensive subjects, rather than searching for developmental aspects of temporal regulation.

Until researchers have undertaken experimental work along these lines, we can only emphasize the theoretical importance of ontogenetic analysis for a thorough under-standing of temporal regulations. The main question here is whether the capacity exhibited by a given species or individual is already present in the young organisms, or whether it evolves at a certain stage of development and is progressively built through growth. In the latter case, one will wish to elucidate the variables at work, and to disentangle the possible respective parts of endogenous maturation and en-vironmental influences. Not only would this cast some light on the status of temporal regulation in the organization of behaviour, but it could suggest important hypotheses concerning the mechanism involved.

5.4.2. Historical Factors

What has preceded, recently or formerly, the exposure to temporal contingencies in an animal's life may be of some consequence for the quality of its performance, or for the effect of other independent variables on the performance.

An illustrative case of the effect of recent history is found in the comparison of various procedures used to reach the final contingencies. For instance, the operant response can be obtained through systematic progressive shaping by the experimenter, or the animal can be immediately placed in the cage under the control of automatic equipment and have to find the effective response by itself. It has been shown that handshaping is not necessary for normal adjustment to DRL or FI contingencies (Linwick and Miller, 1978; Miller and Linwick, 1979).

If the final contingencies are defined by a given delay, let us say 20 s, the subject can be brought very progressively to the final delay, or it can be abruptly exposed to it as soon as the operant response has been acquired. Though the performance can develop similarly in both cases, cerebral lesions can produce different effects, as was shown by Schmaltz and Isaacson (1966a) for hippocampal lesions in rats. This specific interven-tion did not affect animals abruptly introduced to the final delay, but it did affect those who had first been trained under continuous reinforcement, that is reinforcement of each response irrespective of the IRT.

Previous exposure to other contingencies can also modify the usual pattern of temporal regulation. Skinner and Morse (1958a) described effects induced on FI behaviour by previous exposure to Fixed-Ratio contingencies and induced on DRL behaviour by previous exposures to FI or FR. After 10 h under FR 10, rats on DRL show alternations of phases characterized by high rates of responding and phases of low rates or pauses (Kelleher and others, 1959). Conversely, as could be expected, exposure to temporal contingencies can have consequences on later performance under other contingencies. Richelle (1968) has reported the transfer of an FI pattern under FR contingencies in cats with a very long previous exposure to FI. Post-reinforcement pause was curiously lengthened from the beginning to the end of a session and from one session to the next, eventually up to the point where only two or three reinforcements were obtained or operant responding was completely extinguished (Fig. 5.16). Another, transitory effect of previous history is also observed, occasionally, when an animal is transferred abruptly from one delay of FI contingencies to a longer delay. The previously acquired temporal patterning still occurs now and then.

Fixed-Time schedules offer an interesting situation for the study of the effects of previous history on temporal regulations. It will be recalled that in Fixed-Time schedules some reinforcing stimulus is presented at fixed intervals of time, irrespective of the animal's behaviour. Periodic presentation of non-contingent reinforcement is of course the definition of Pavlovian temporal conditioning. If this is done in an experimental space designed for operant conditioning, equipped with a response lever or the like, it eventually induces lever responses. These can be explained by referring to the process of adventitious reinforcement identified by Skinner (1948) and proposed as an elementary paradigm for superstitious behaviour: any bit of behaviour that happens to be emitted in temporal contiguity with the reinforcement is likely to become conditioned and hence repeated; becoming more frequent, it has more chance to occur again in temporal contiguity with the reinforcement, thus strengthening the association. What is needed, as is always the case in operant conditioning, is that the response be produced at least once for some reason, so that it can be selected by the reinforcement. Close analysis of superstitious behaviour has led recently to reconsider and complicate its interpretation in terms of adventitious reinforcement. Current hypotheses take into account the possible respondent—Pavlovian—nature of such behaviours, their similarity to displacement activities as defined by ethologists, and a number of other phenomena described as *adjunctive behaviour* (that is behaviour that is not expected nor required by given contingencies, but still systematically occurs, as a by-product of some schedules. See Reberg and others, 1977, 1978; Staddon, 1977; Staddon and Ayres, 1975; Staddon and Simmelhag, 1971 for more details).

Whatever the nature and function of these behaviours, we are concerned here with their temporal patterning. This is largely a function of the previous history of the subject. If it has been exposed to DRL contingencies, low steady rates will be observed under Fixed Time. Positively accelerated rates will be observed after exposure to Fixed-Ratio contingencies (Alleman and Zeiler, 1974). If the FT procedure is initiated after reinforcing the first three responses produced, responding at fairly high rates will be observed *after* food presentation, followed by a pause extending over the rest of the

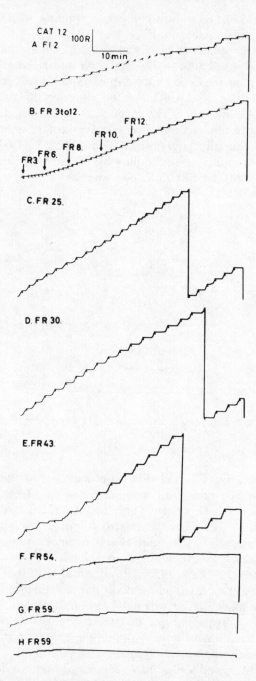

FIG. 5.16. Transfer of FI pattern in FR training in a cat with a long history of exposure to FI 2 min. A: control cumulative record under FI 2 min; B: first session under Fixed-Ratio schedule, with progressive increase of the number of responses required from three to twelve; C to H: FR with increasing ratios. Note the typical FI pattern persisting under FR, and the deterioration of behaviour at high values of FR. (After Richelle, 1968.)

interval (Neuringer, 1970). By contrast, previous training with a FI schedule will produce post-reinforcement pauses in FT.

Most studies of FT behaviour after previous FI training used pigeons, except Edwards and others (1968), Lattal (1972), and Lowe and Harzem (1977) who studied rats. These studies outline the importance of species specific factors and present discrepancies demanding further research. Lowe and Harzem (1977) report very low and even rare responding in rats on FT 30, 60 or 120 s as compared with pigeons, after eighty sessions of forty reinforcements on equal valued FI schedules. In contrast, Lattal (1972) describes sustained response rates on a MULT FT 60 s VT 60 s schedule for twenty-five and even fifty sessions following thirty-five to forty sessions of fifty reinforcements on the MULT FI 60 s VI 60 s schedule.

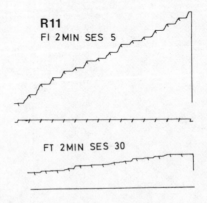

FIG. 5.17. Cumulative records from one individual rat in Fixed Interval 2 min (FI 2 min) and in Fixed Time 2 min (FT 2 min).

The similarity between FT and FI performance is, however, not complete (Fig. 5.17). First, response rates are lower in FT than in FI (Appel and Hiss, 1962; Edwards and others, 1968; Herrnstein, 1966; Lattal, 1972; Lowe and Harzem, 1977; Shull, 1971b; Zeiler, 1968). Secondly, the activity phase following the pause is not always as clearly patterned as it is under FI. It resembles more the bitonic pattern (pause, activity, pause) observed under conjunctive Fixed-Time-Fixed-Ratio 1 schedules (see Chapter 2), (Shull, 1970b, 1971b; Shull and Brownstein, 1975). These differences concerning the activity phase make the resemblance concerning the post-reinforcement pause all the more striking (Lowe and Harzem, 1977; Shull, 1971b; Zeiler, 1968) and once again indicate that these two aspects of Fixed-Interval (or Fixed-Time) performance, namely the pause on the one hand and the activity phase on the other, are controlled by different variables (Elsmore, 1971b; Farmer and Schoenfeld, 1964b; Marr and Zeiler, 1974; Morgan, 1970; Neuringer and Schneider, 1968; Shull, 1970b, 1971b; Shull and others, 1972; Schneider and Neuringer, 1972; Staddon and Franck, 1975).

Trapold and others (1965) have studied the effect of previous exposure to Fixed-Time reinforcement on FI performance. A group of rats previously exposed for twelve sessions to a Fixed-Time 2 min reinforcement schedule reached stable adjustment to

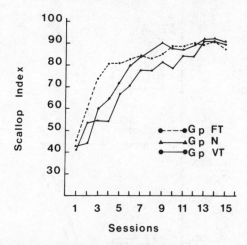

FIG. 5.18. Effect of pre-training in Fixed Time (filled circles, dotted line) or in Variable Time (filled circles, solid line) as compared with no pre-training (filled triangles, solid line) on the training in Fixed Interval 2 min. Abscissa: successive sessions under FI; ordinate: mean scallop index (each point = averages from six rats). The scallop index is a proportion of bar presses in a session which occur in the second minute following reinforcement. (Redrawn, after Trapold and others, 1965. Copyright 1965 by the Psychonomic Society, Inc.)

an FI 2 min schedule faster than a group exposed to variable time reinforcement or a group not submitted to any pretraining (Fig. 5.18).

Previous experience is also an important factor in Fixed-Time schedules where the non-contingent periodic event is not a positively reinforcing stimulus such as food but an aversive stimulus such as electric shock. Azrin (1956) has described reduced response rates and reversed scallops when free shocks are superimposed at periodic intervals upon a food reinforced VI baseline in pigeons (Appel, 1968, has described the same effect for fixed-interval punishment in the rat). Other effects of the Fixed-Time shock paradigm have been described after previous escape (McKearney, 1974a) or avoidance learning. First, when Fixed-Time shocks are superimposed upon an avoidance baseline, response rates are enhanced. Second, when the escape or avoidance schedule is discontinued, the free shock presentation alone controls responding and the typical Fixed-Interval pattern, that is a pause followed by an activity phase, usually develops (Hutchinson, 1977, p. 427; McKearney, 1974a; Malagodi and others, 1978). Sustained response rates, sometimes slightly accelerated, have also been described (Byrd, 1969; Kelleher and others, 1963). Direct comparison between FI and FT schedules of shock presentation has been reported by Malagodi and others (1978). In a multiple schedule FI 6 min FT 6 min, there was no difference between components with regard to response rates or temporal regulation (FT quarter-life values were 0.75 and 0.78 respectively for the two rhesus monkeys). When FI and FT were successively studied over blocks of sessions, response patterning was the same but response rates were lower in the FT schedule (Fig. 5.19). These data are in line with those described in FI and FT schedules of food presentation (the absence of difference in response rates on the multiple schedule may be explained by short-term history or induction from the FI component to the FT component). They also corroborate conclusions drawn from food-presentation studies.

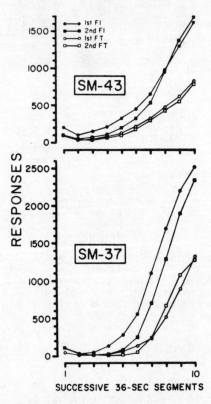

Fig. 5.19. Fixed Interval 6 min (FI 6) and Fixed Time 6 min (FT 6) of shock presentation in two rhesus monkeys. Total number of responses in successive tenths of the 6 min interval during the last ten sessions of each exposure to FI 6 and FT 6. (From Malagodi and others, 1978. Copyright 1978 by the Society for the Experimental Analysis of Behavior, Inc.)

Behaviours such as lever or key-pressing are not the only ones that have been reported in FT-shock schedules. The almost exclusive use of monkey species (Byrd's 1969 study on cats is an exception) has allowed for the recording of various manipulatory and oral behaviours such as biting or drinking, when the subjects have been given the appropriate substrates. These oral behaviours occur just after shock presentation, so that the cumulative records show inversed scallops (see Fig. 5.20). These patterns do not imply any temporal regulation on the part of the subject. Tentative interpretations refer to shock-elicited behaviour (for example Hake and Campbell, 1972; Morse and others, 1967) or adjunctive influences of the electric shock (De Weeze, 1977 for example) (For a more detailed analysis of such behaviours, see Hutchinson, 1977, and Hutchinson and Emley, 1977).

5.5. INTERINDIVIDUAL DIFFERENCES

The experimenter who runs a few animals of the same species under the same set of temporal contingencies will surely come across individual differences. Besides species-specific, developmental, or historical variables, individual specific factors also determine the quality of temporal regulation. Little systematic attention has been given to this point, though Pavlov had already noted that quiet dogs were more easily conditioned to duration than agitated animals with dominant excitatory processes. In the logic of this theory, one would expect temporal conditioning to be linked with typological characters: it should be remembered that conditioning involves, in Pavlov's view, two distinct active mechanisms, excitation and inhibition; that his typology is based on the respective role of these mechanisms, and the balance (or imbalance) between them; and that conditioning to time is essentially controlled by inhibitory mechanisms.

Obviously, Pavlov did not perform the kind of statistical analysis needed to give some validity to his typology. Rather, he derived it intuitively from contrasting individual observations. This sort of clinical approach is also all that can be done in most operant conditioning studies, that typically use very small numbers of subjects. It was, as we know, the deliberate choice of experimenters, on the Pavlovian side as well as among Skinnerians, to concentrate on the analysis of individual behaviour throughout a long period of time, rather than looking at group results in short-term experiments.

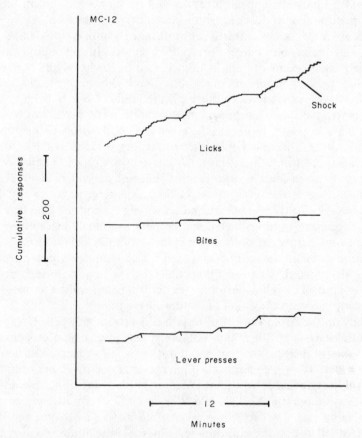

FIG. 5.20. Fixed-Time 4 min schedule of shock presentation in a squirrel monkey. A portion of three simultaneous cumulative records during a single shock session, illustrating the temporal pattern of lever pressing, hose biting and licking. (From Hutchinson and others, 1977. Copyright 1977 by the Society for the Experimental Analysis of Behavior, Inc.)

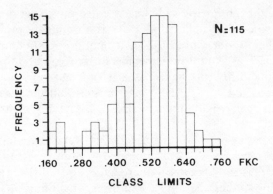

FIG. 5.21. Frequency distribution of the curvature index (FKC) for 115 rats on the thirty-first session of training under FI 2 min. (After Lejeune and Defays, 1979, unpublished data).

As far as temporal regulations are concerned, individual studies are almost a necessity, considering the kind of equipment involved, and the number of sessions needed before stabilized performance is reached. Inter-individual differences observed under such conditions, are nevertheless instructive. For instance, Pouthas (1979) has looked at the details of rats' behaviour under DRL contingencies. In her study, the delay was progressively increased to 10 s and then increased to 20, 30, 40 and 60 s. A new delay was introduced only if 50 per cent of the responses had been correctly spaced through-out the three previous sessions. In this way, the limits of one individual's adjustment could be determined. No more than five subjects were needed to show clear-cut individual differences with respect to the magnitude of the delay that can be mastered: one rat had a poor adjustment even after twenty-six sessions at DRL 10 s, three rats could be transferred to the 20 s delay, and one to 40 s. Collateral behaviour filling the time interval between lever presses seemed different depending upon the individual and the magnitude of the delay. Most of these collateral activities could be classed into three categories: (1) approaching and remaining around the food tray; (2) walking around in the cage; (3) sitting quiet, doing nothing, eventually half-sleeping. Behaviour pertaining to the first category was dominant with short delays (10 s), supplemented by the second category for longer delays (20 s) and immobility became dominant for the largest delay (40 s). Interestingly enough, the rat that performed poorly at 10 s showed a great deal of behaviour of the second category, while the one that adjusted to 40 s mainly showed behaviours of the third category.

In a study on the reinforcement of response duration with cats, Greenwood (1977) employed a criterion of 70 per cent reinforced responses or ten sessions at the same schedule value to increase the required duration in 0.25-s steps. Furthermore, if effi-ciency for a given session undermatched the previous one by 20 per cent, the duration requirement was reduced by 0.25 s. After sixty daily sessions, the four cats had attained response durations of 1.5, 2.25, 2.25 and 4.25 s respectively.

Studies using very few subjects, however informative as to interindividual differ-ences, cannot tell us how the capacities for temporal regulations are distributed within a larger population. Lejeune and Defays (1979) have analysed the results of 115 rats under FI 2 min contingencies. Subjects were first trained under continuous reinforce-ment (each response was reinforced), then transferred to FI 30 s for three sessions, to

FI 60 s for three sessions, to 90 s for three sessions and then, finally, to 120 s. The curvature index (see Chapter 2) was computed for the thirty-first session under the final contingencies for each of the 115 subjects. The distribution obtained is shown in Fig. 5.21. The mathematical analysis indicates that this distribution is normal, with a mean index of 0.5089 and a standard deviation of 0.1114. It would seem that performances under FI, at least when characterized by the index used here, are normally distributed along a continuum. Types of animals in this case would represent extreme or central portions of the distribution. A similar analysis of results in DRL has been done with a population of forty-two rats, who went through a similar training: crf, three sessions in DRL 5, three sessions in DRL 10, three sessions in DRL 15 until the final DRL 20 s contingency was reached. The measure used was the efficiency ratio, that is the ratio between reinforced IRTs and total IRTs (see 2.3.1). The distribution is shown in Fig. 5.22. It is not a normal distribution, but it provides no ground for supporting the hypotheses that there are qualitatively distinct typological classes. What is clear in both cases, is the fact that all individuals do not perform equally well in the same situation and, as far as it can be controlled, with the same history. Cumulative curves shown in Fig. 5.23 provide another illustration of this point.

A number of questions remain to be answered. First, one should ask whether the distribution of individual results would be the same using other indices to characterize the performance. We are faced here with the problem already alluded to in the methods section, concerning the validity of various indices and the correlation between them. One should further ask whether individuals would rank similarly whatever the index used to characterize their performance under given contingencies. Second, one should ask whether the same individual subjects would rank similarly at different times of their exposure to given contingencies. For instance, in Lejeune's study reported above, would the poor performers on the thirty-first session already appear poor on the tenth session and remain so until the fiftieth? This is a trivial problem of reliability of behavioural scores. If an individual's position in the distribution significantly changes from one session to another, it would be hazardous to think of interindividual differences in terms of idiosyncratic levels of capacity to adjust to time. Unfortunately, no data analysis is available to help us answer the second question. Such an analysis could probably be done on records of long-term experiments on the same animals that exist in many laboratories. Third, it should be determined

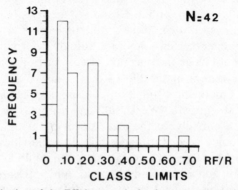

FIG. 5.22. Frequency distribution of the Efficiency ratio for forty-two rats trained on a DRL 20 s. (After Lejeune and Defays, 1979, unpublished data.)

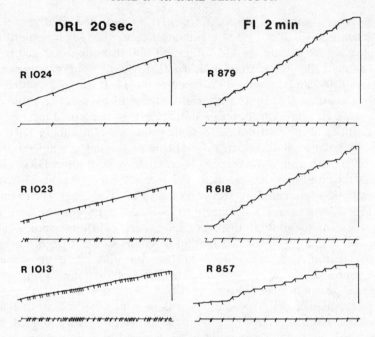

Fig. 5.23. Cumulative records from individual rats on DRL 20 s (left) and FI 2 min (right) after an equivalent number of sessions. From top to bottom: poor to excellent performances. (From Richelle and Lejeune, 1979.)

whether individual subjects rank similarly under different contingencies. This, as we have seen earlier, would provide useful elements concerning a very different question, namely: is there one unique timing mechanism underlying all possible kinds of temporal regulations? But it would also help us in characterizing individual capacity in dealing with time. Fourth, we will still have to trace the origins of observed interindividual differences. Subtle differences in individual history should, of course, not be ruled out, and should be carefully looked for before resorting to some innate capacity. It could be that the way a subject meets the contingencies in the early phase of training is responsible for later performance, and therefore for interindividual differences. For instance, in DRL, the initial rate of responding, possibly quite accidental, or at least without any relation with the capacity for temporal regulation—perhaps due, for example, to the level of emotionality or the level of arousal—could very well result in many or few reinforcements, and accordingly determine later behaviour in a way that will or will not favour the selection of correct IRTs.

Much more remote factors in the life of the animal could also be at work, the capacity to adjust to time being a function of a number of processes that build up through the individual's growth. We should be ready to detect any unsuspected influence going back, eventually, to the early stages of an individual's life. Developmental studies are strongly needed to understand interindividual differences.

Supposing a reasonable degree of validity and of reliability as considered in the first and the second question above, it would be sound to hypothesize some genetic factors, and to approach the problem of interindividual differences with the methods of psychogenetics, for instance by comparing controlled strains, or by proceeding to selective inbreeding of good and poor "timers". As far as we know, this interesting line of research has not been explored and is awaiting researchers of the future, the only available data dealing with sex differences (for example Beatty, 1973, 1977, 1978; Beatty and others, 1975a, 1975b).

5.6. CHRONOBIOLOGICAL FACTORS

Chronobiology has demonstrated the pervasiveness of periodicities in biological systems. Testing for unsuspected effects of rhythms on any dependent variable is becoming current practice in biological research. Variations of body temperature, heart-rate, hormonal activity, drug toxicity or efficiency as a function of circadian cycles are only a few classical examples. However, this now widespread interest is not yet shared by experimental psychologists working with laboratory animals. The neglect of chronobiological variables is most surprising with experiments bearing upon temporal regulations. As suggested in the introduction, the possible relations between biological rhythms and conditioned temporal regulations are of great theoretical interest. In the almost total absence of experimental work, we shall content ourselves with defining the type of research that is strongly needed to fill the gap between contemporary chronobiology and the study of behavioural adjustments to time.

A basic question is: do performances of conditioned animals vary as a function of the moment in a biological cycle at which the experiment takes place? The most easily controlled independent variable, in this respect, would be, of course, the night-day cycle. Looking at them with the view of a chronobiologist, it is amazing that most, not to say all, studies reviewed in this book do not even mention at what time of day subjects were run in the experimental chamber. A fortiori the reader is not told whether the experimental session took place at the same hour each day. Precise details concerning the light-dark periods are also totally lacking in the methods sections and experimenters disregard the natural activity cycles of their subjects. Rats are kept under the same light-dark cycles as their human experimenters who seem to ignore the fact that even laboratory strains might have retained something of their ancestral nocturnal habits.

There are, fortunately, a few exceptions to the rule, especially on the part of experimenters working with unusual laboratory species. Experimenters working on hamsters, for instance, took into account the crepuscular character of that species and ran their experiments late in the afternoon (see Anderson and Shettleworth, 1977; Godefroid and others, 1969). Others took care to invert artificially the day-night cycle by keeping their nocturnal subjects under light condition from evening to morning, and in darkness from morning to evening, so that the experimental session would take place in the dark period during which animals are active in their natural environment. This was done for example by Lejeune (1976), in her study on a prosimian species.

Systematic studies of temporal discrimination as a function of the moment in the day-night cycles have been done on human subjects (see for example Thor, 1962; Thor and Baldwin, 1965) but not on animals, except for a few studies on temporal regulation. Beatty (1973) has shown that DRL performance in rats does not change when sessions take place in the afternoon rather than in the morning. In a more recent paper (Beatty, 1977), he reports that female rats performed more efficiently than male under DRL when they were trained during the daytime part of the day-night cycle.

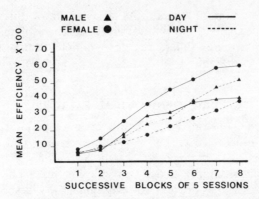

FIG. 5.24. Performance of female and male rats trained on DRL 20 s during day-time or night-time. Each point data is averaged from sixteen subjects. (Drawn after numerical data from Beatty, 1977.)

When training took place during the night, no significant difference was observed. Females trained during the day showed better efficiency than female trained at night. Fig. 5.24 summarizes the results of the four groups of subjects for eight successive blocks of five 30 min sessions in DRL 20 s. Response rate and reinforcement rate do not change consistently as a function of time in the day-night cycle. No details are given, unfortunately, about other aspects of temporal regulation.

In a study of rats' activity as a function of light-dark cycle and food presentation schedule, Evans (1971) compared performance under FI 2 min for rats tested during the dark period and rats tested during the light period. Subjects lived under regular 9-h light-15-h dark cycles, and were tested at a designated time every 48 h. As shown in Fig. 5.25, the operant activity for food was higher for animals tested in the dark, following the widely documented rule that rats are more active at night. The temporal regulation was not impaired by the increase in response rate. On the contrary, as can be seen from the quarter-life figures, it was better in the dark than in the light.

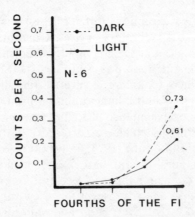

FIG. 5.25. Mean responding per segment on Fixed-Interval 2 min schedule of rats after more than 30 h experience being tested in the dark (dashed line) or in the light (solid line). The number to the right of each line represents quarter-life. Each point data is averaged from six subjects. (Redrawn, after Evans, 1971. Copyright 1971 by Pergamon Press, Ltd.)

Cycles in learning and retention processes and their relation to biological rhythms have been studied by a few authors in situations that do not involve temporal regulations (see Holloway, 1978, for a concise review of this underdeveloped field). In one study involving DRL, Holloway and Jackson (1976) addressed themselves to a different, though not unrelated problem: is an animal able to use the time of the day-night cycle as a discriminative stimulus controlling operant behaviour? If a sound of a given intensity is associated with FR contingencies and a sound of another intensity is associated with DRL contingencies, an animal subject will quite easily learn to respond differentially in the presence of either stimulus. Could it respond differentially if the time of day is used as a discriminative stimulus rather than an exteroceptive stimulus? Supposing no external cue is present that might signal different periods of the day, the subject could only rely on some internal temporal information, that is upon a given state of its own organism linked with the moment in the nycthemeral or some other biological rhythm. What is looked for here is a state-dependent learning effect based on time of day, analogous to state-dependent learning based on levels of deprivation or on drug administration. Though the authors did not focus on the temporal regulation as such in the DRL schedules, their study deserves attention in the present context as an original attempt to investigate an animal's reaction to its own biological cycles. Eight rats were trained on DRL 15 s or FR 15 on alternating days. For half, the sessions for both schedules took place at the same time of the day (morning or afternoon) while for the other half, sessions on FR took place in the morning and sessions on DRL in the afternoon, or vice versa. No reinforcement was delivered during the first 10 min of the fourteen testing sessions. Behavioural measures during these extinction periods would, hypothetically, indicate whether the subjects responded differentially depending upon the time of the day, as rate of responding is distinctively higher in FR than in DRL. Though the conclusions from this experiment are not very clear-cut, some behavioural indices show differential responding in the group that was trained at different times.

A second important, and equally neglected question, concerns the relationship between the temporal parameters used in conditioning experiments and the periods of biological cycles. Would an animal subject adjust better to an FI schedule, for instance, if the interval were made to coincide with the duration of some natural rhythm? The high precision of circadian cycles would suggest that temporal conditioning will be facilitated when the delay to be mastered is not arbitrary with respect to the 24 h cycle, but rather is in phase relation to it (the simplest case being perfect coincidence). Among the numerous experimenters who have contributed to the study of acquired temporal regulations, practically none has had this question in mind when selecting critical delays for the experiment. In fact, the choice of the delays is rarely justified. It seems that in many cases experimenters have just intuitively selected delays in the range that could be mastered by their subjects, or have followed tradition and adopted delays used by previous workers. When the experiment has been aimed at exploring a wide range of delays, and eventually the limits of adaptation to temporal contingencies, the values of the temporal parameters seem to have been dictated by some technical constraints of the equipment rather than by logic. When logic has guided the choice, it has been simple mathematical logic rather than biological logic (see for instance p. 51, the progressive interval schedules designed by Harzem, 1969). Of course, the relationship between circadian cycles and the DRL value, usually within the range of 0 to 100 s, might not be easily put to experimental test. FI contingencies, however, lend themselves to such a test, when long intervals are explored. There are some indications that 24 h is a privileged value in learning situations. Bees, as we have seen in Chapter 4, do not develop "scallops" in FI 2 min schedules. They are known, however, for their remarkable "time sense", often mentioned in the chronobiological literature. As long ago as 1929, Beling trained bees to visit a feeding dish every day between 10 a.m. and noon. After one week of training, he left the dish empty, and observed the number of visits (conditioners would say that he put his bees on experimental extinction, by eliminating the reinforcing event). The great majority of visits took place during the period (10 to 12 h) in which bees had found food during their training. This regularity could be obtained for other times of the day as well and was maintained in constant conditions. It persisted even in the unusual and highly stable environment of a salt-mine 600 ft below the surface of the earth, where possible geophysical cues had been drastically modified compared with a normal bee's environment. It persisted also when bees were transferred from Paris to New York or vice versa, to perfectly identical environments—an experiment that ruled out the influences of unsuspected celestial cues. The interesting point for our purpose is that bees, endowed as they are with such an extraordinary time sense, cannot be trained to a schedule that demands a visit to the feeding stand every 19, 27 or 48 h. This is not because they have been conditioned to a 24 h periodicity during their lifetime, since it is also true of bees hatched in an incubator under constant conditions (Palmer, 1976). Grossman's failure to obtain temporal regulation under a Fixed-Interval schedule of 2 min might be explained by the distance of this arbitrary value from natural periodicities characterizing circadian rhythms. Possibly, the behavioural clock(s) might work more independently from the biological clock(s) in some species than in others.

The privileged status of the 24 h cycle in its relation to operant conditioning has also been demonstrated in rats. Bolles and de Lorge (1962) fed rats at regular intervals of time. For one group, this interval was 24 h, for another group it was 19 h, and for a third group 29 h. All animals were given free access to an activity wheel, and the number of runs was recorded. This procedure is what is termed a Fixed-Time schedule, in which food is delivered at regular intervals of time, and is not contingent upon the animal's behaviour. Rats anticipated the feeding time, with a clear-cut increase of recorded activity in the hours immediately prior to feeding. However, only the animals put on a 24 h schedule showed this anticipation. It was not observed in subjects on 19 h or 29 h cycles. As Bolles and de Lorge's subjects were neither raised nor kept under constant conditions, it could not be concluded whether good anticipation in the 24 h group was due to the effect of environmental cues (or synchronizers, in chronobiological terminology), or to an intrinsic endogenous 24 h cycle.

In another experiment, Bolles and Stokes (1965) compared rats living on a normal 24 h cycle (12 h light–12 h dark) with rats born and reared on a 19 h or 29 h cycle (half light–half dark). Animals were then submitted to a feeding cycle of the same period as characterized their whole life experience. Food was available for a limited period in each cycle, always at the same time after changing the light condition (either after light onset or dark onset). For one group of rats, spontaneous locomotor activity in an activity wheel was recorded as in the previous experiment; for another group, food was contingent upon lever pressing—that is each lever-press during the feeding period

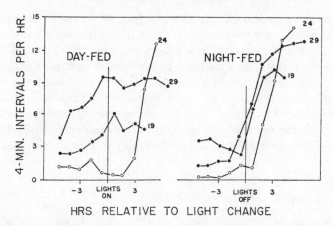

FIG. 5.26. Anticipation of feeding period in rats raised and subsequently fed on 19 h, 24 h, or 29 h cycles. Ordinate: mean number of 4 min intervals per hour in which bar presses occur. Abscissa: time relative to change in illumination preceding feeding (see text). Left: animals fed when light is on; right: animals fed when light is off. (From Bolles and Stokes, 1965. Copyright 1965 by the American Psychological Association. Reprinted by permission.)

(equal to one-tenth of the total cycle) was reinforced by food. Anticipatory spontaneous or operant activity was clearly observed in rats fed on the 24 h cycle, both during light or dark periods, while rats fed (and raised) on the 19 h cycle or the 29 h cycle showed no or little such anticipatory activity. Fig. 5.26 summarizes the data for the bar-pressing situation. Points plotted in this figure are based on non-reinforced responses, emitted *before* the feeding period started. The feeding period started 3.8, 4.8 and 5.8 h after the lights changed for the 19, 24 and 29 h cycle groups respectively (explaining the shift between the three curves in the position of the last points on the abscissa). These can be best compared with the non-reinforced responses emitted during the interval of a Fixed-Interval schedule. The contrast is especially clear-cut for animals fed in the light. Interactions between activity cycles and hunger cycles are complex, and have been discussed by Bolles (1975). It can be concluded from these experiments that it is extremely difficult, if not impossible, for rats to develop temporal regulations for periodic feeding cycles that depart from the 24 h cycle. If we remember that spontaneous temporal regulations develop easily in FI schedules with short intervals (of the magnitude of 2 or 4 min), we might wish to know how temporal regulations would vary as a function of the interval along the whole continuum from 1 min or so to 24 h and beyond.

A third interesting question to which, to our knowledge, no experimenter working on animals has ever addressed himself, is: what happens to temporal regulations under constant conditions? In Bolles and Stokes' experiment, arbitrary cycles were imposed upon the organism, but external synchronizers were still provided, in the form of light-dark regular alternation. It is well known in chronobiology that biological rhythms persist in so-called free-running experiments, though their period somewhat departs from what is observed when they are synchronized on a strictly 24 h basis. This shift has justified the name *circadian*, and is generally considered as characterizing the individual's internal clock. Suppose an animal's temporal regulation remains perfectly unaltered by exposure to constant conditions, while the usual shift in biological rhythm is observed: this would be a suggestive argument for the independence between basic biological clock(s) and behavioural clock(s). A change in behavioural timing performance would indicate either a common mechanism or different mechanisms, with the behavioural clock(s) being subordinated to the biological clock(s). The few data available are from humans (see for instance Fraisse and others, 1968). The techniques used are not quite similar to those used with animals, and data on humans in time-estimation studies raise a number of difficult problems of interpretation. Combining the techniques of modern laboratories in animal behaviour and in chronobiology, a systematic study of this problem is feasible and it would undoubtedly provide us with new insights in our understanding of time in animal behaviour.

PART III
MECHANISMS

VI. *Physiological Mechanisms*

6.1. THE SEARCH IN THE CENTRAL NERVOUS SYSTEM

The function of the central nervous system in timing behaviour has been explored through the three classic methods of lesion, stimulation, and recording. Each method has provided an impressive set of data. The interpretation of such data, however, proves difficult. Many factors require a careful control, as they currently differ from one study to another: technical parameters, characteristics of the experimental paradigm, age, and individual history of the animals. In the lesional method, it must never be forgotten that the observed behavioural changes may result from the unintended interruption of a conduction pathway that crosses the aimed structure rather than from the actual elimination of the structure itself. Secondary chemical changes in other sites of the brain have also been discovered after quite delimited lesions. The main problem with the electric stimulation method is that the spatio-temporal patterning of the biological and the electric impulses strongly differ. The latter simultaneously activates a large number of cells and fibres having normally excitatory as well as inhibitory functions. This has led to the developing utilization of a more refined method of chemical stimulation, involving direct micro-injection of ions, crystals, or presumed neurotransmitters such as acetylcholine, serotonine, and norepinephrine. Finally, problems posed by the recording method primarily involve contamination of the normal electric activity of the brain by toxic reactions due to chronic electrode implantation, electrode polarization, and other possible distortions.

6.1.1. Lesional Studies

DRL schedules (and, above all, DRL 20 s) are most often chosen for studying the effect of brain lesions in animals. FI schedules are also used, but only occasional utilization of other paradigms possessing temporal features is reported. The curious disfavour of temporal discrimination schedules is particularly deplorable inasmuch as they could provide a useful control for the important component of motor inhibition involved in DRL.

A deficit in DRL is revealed by a shift in the response distribution towards short IRTs, and a correlative lack of the usual modal IRT corresponding to the reinforced value of the delay (Fig. 6.1). In FI the animals start responding earlier in the post-reinforcement period (Fig. 6.2). An increase in response rate is generally observed in both paradigms.

A. Role of the Septal and Hippocampal Structures

It appears from a review of the data that the septum and the hippocampus are the most often and most convincingly implicated in the regulation of temporal behaviour. These strongly interconnected regions are generally considered part of the so-called

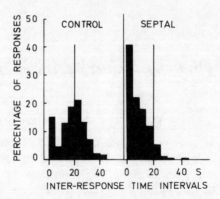

Fig. 6.1. Relative frequency distributions of inter-response times after 34 days of training in DRL 20 s for control and septal rats. (Adapted from Ellen and others, 1964. Copyright 1964 by Academic Press, Inc.)

limbic system classically involved in the affective and motivational aspects of behaviour. The septum is a subcortical set of nuclei lying under the septal area in the anterior medial depths of the forebrain. The hippocampus extends along the floor of the fourth ventricle. Despite some nuances, the data obtained in temporal schedules after destruction of these structures are mostly comparable, and have been given similar interpretations. The DRL schedule is the experimental test which gives rise to the most consistent data. It must be emphasized that adequate timing is a necessary condition for obtaining reinforcement in this schedule, which is not the case in FI. However, interpretation of the DRL deficit is controverted, given the complexity of the performance. Several hypotheses must be confronted before postulating specific dysfunction of timing processes as the cause of the observed changes.

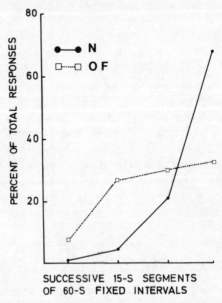

Fig. 6.2. Mean distribution of responses obtained from normal (filled circles, N) and frontal monkeys (open squares, OF) in three 60 s sessions (Adapted from Manning, 1973. Copyright 1973 by Pergamon Press, Ltd.)

The most classic syndrome due to septal injury is the well-known "septal rage" (Brady and Nauta, 1953, 1955) and the less transient hyperactivity evident in spontaneous behaviour as well as in a large variety of experimental situations (see Fried, 1972, for review). It is worth specifying that response output is more affected in interval than in ratio schedules, which suggests that this septal effect depends on some specific contingencies. On the other hand, increased liquid consumption is frequently reported after both anterior and posterior septal lesions (Carey, 1969), and it has been suggested that the septum regulates the reciprocal balance between food and water intake (Coury, 1967). The hippocampus has also been assumed to play a role in this respect (Ehrlich, 1963; Harvey and Hunt, 1965). But more generally, lesions affecting this structure are supposed to produce mnemonic defects, or else hyperactivity, especially in delayed-response tasks. In man, tumour or vascular disturbances localized in the hippocampus can produce temporal disorientations, in addition to troubles in recent memory.

In view of these general data, lesional deficits in temporal schedules might be due firstly to a loss or alteration in the motivational properties of the reinforcement, or a modification of a basic drive itself. For instance, increased appetitive need might generate higher response rates incompatible with the DRL contingencies. It has been shown that the differences between septal and control rats with regard to response output in FI can be reduced (Beatty and Schwartzbaum, 1968) or even suppressed (Harvey and Hunt, 1965) if thirst in septals is diminished by water "preloading". Harvey and Hunt's data, however, indicate that increased response rate is not an invariant consequence of higher drive, but rather depends on the schedule, since it did not occur in DRL. This is an atypical finding. Besides considerations relative to the extent of the lesion, the gradual shaping procedure used in this study may be responsible for the lack of deficit observed, since this factor can largely compensate for septal impairment in the acquisition of DRL behaviour (Caplan and Stamm, 1967). Several data weaken the motivational hypotheses. Increased thirst cannot account for septal deficit in DRL since the IRT distribution of septal rats is quite different from that of severely water-deprived control animals (Carey, 1967a). (Fig. 6.3). Water intake and performance in DRL with water reinforcement can be dissociated on the basis of discrete septal lesions (Carey, 1967b). In FI schedules, increased response output is observed even in non-deprived septal rats (Beatty and Schwartzbaum, 1968) and is not affected by changes in the palatability of reinforcement (Pubols, 1966). In hippocampal rats, overresponding on VI schedules has been proved independent on changes of food intake (Jarrard, 1965).

A second line of interpretation refers to the implication of a septo-hippocampal system in somatomotor inhibition (Burkett and Bunnell, 1966; McCleary, 1966). Connections with the subcallosal cortex are probably also involved (Carey, 1967a; McCleary, 1966). Trouble in motor inhibition is a major factor to be considered as a source of deficit in DRL, since the reinforcement requirements of this schedule can only be met if the animal withholds lever pressing during a given delay. A shift toward short IRTs may thus reflect difficulties in inhibiting motor responses. Such a shift, concomitant with general overresponding, is a primary sign of septal and hippocampal lesions. Hence, impairment of the ability to inhibit goal-directed responses (Burkett and Bunnell, 1966; Ellen and others, 1964) or an increase in the resistance to extinction of responding after short unreinforced IRTs (Carey, 1967a; Clark and

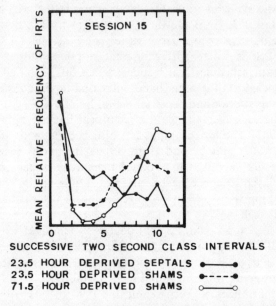

FIG. 6.3. Mean relative frequency distributions of inter-response times in successive 2 s class intervals for the 15th DRL 22 s session, for one group of septal rats (23.5 hour water deprived) and two groups of sham-operated control rats (23.5 and 71.5 water deprived). (Redrawn, after Carey, 1967a. Copyright 1967 by Pergamon Press, Ltd.)

Isaacson, 1965) have been hypothesized. It is worth noting that training on a DRL schedule is most often preceded by shaping in CRF, a schedule that favours rapid responding. Lesioned animals are considered less successful in extinguishing the CRF pattern when transferred to DRL, since their response rate is generally much higher than normal during the first DRL sessions. Accordingly, hippocampal rats can acquire a normal rate of reinforcement in DRL 20 s (though responding more than normals) provided that they have received no preliminary shaping in CRF. After CRF, the high rate of bar-press cannot be properly extinguished (Schmaltz and Isaacson, 1966b). Perseveration would then be the key to the deficit (Schwartzbaum and co-workers, 1964), also explaining why the impairment is most generally observed in passive avoidance behaviour while the acquisition of active avoidance is facilitated.

Fried (1972) later refined the inhibition hypothesis by proposing that one function of the septum is to weaken the habit strength of responses acquired in the presence of incentive contingencies, in order to allow for elaboration of new responses. Moreover, the effect of expected reinforcement has been emphasized as linked to a higher level of activation. For example, hippocampal squirrel monkeys do not continuously press the response bar in FI or DRL as if suffering from a general loss of response inhibition. They pause after reinforcement. In FI and DRL cumulative records present a step-like pattern revealing pause-response alternations; yet, they show an increased tendency to respond as time elapses: they respond more than normals and sooner than normals in anticipation of reinforcement (Jackson and Gergen, 1970). On the other hand, response suppression has been assumed to have aversive properties (Caplan, 1970), as indicated by the high probability of rapid response "bursts" following almost-reinforced IRTs. These bursts can be taken as a "frustrative" index (Stamm, 1964). In

Caplan's study, septal and control rats did not differ in response rate when trained in CRF, but the septals produced more bursts when submitted to a short DRL delay (1.2 s); the bursts diminished when the delay was half-reduced. Following occasional omission of the reinforcement, the response rate increased more in septal than in control groups. These data suggest that the level of activation is enhanced in septals due to increased aversive reaction to response suppression. The septal region would exert an inhibitory control upon reactivity to the reinforcing properties of stimuli. Another argument in favour of the inhibition hypothesis is that in a study by Slonaker and Hothersall (1972) the septal deficit was reduced by placing the rats in a complex environment which favoured "mediating" behaviour. A positive correlation was found between the amount of material chewed and the DRL efficiency in septal rats. At the end of training, their performance did not significantly differ from that of controls trained in standard operant cages. Moreover, a severe loss of efficiency was observed after the material to be chewed was removed (Fig. 6.4). It may be that the chewing behaviour provides a way of derivating excitatory processes so that it facilitates the inhibition of lever pressing (see Chapter 7 and Chapter 9).

Some discrepancies arise when comparing the possible inhibitory functions of the septum and the hippocampus. In the work of Ellen and others (1964), the hippocam-

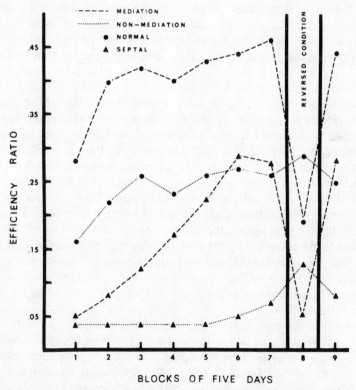

FIG. 6.4. Group mean efficiency ratios (reinforcements/responses) for each block of 5 days of DRL 20 s training in normal and septal rats given mediation or non-mediation conditions. (From Slonaker and Hothersall, 1972. Copyright 1972 by the American Psychological Association. Reprinted by permission.)

pal rats, contrarily to the septals, had no problems in acquiring DRL after CRF training, though showing transient overresponding in the first DRL sessions. This contradicts the data obtained by Schmaltz and Isaacson (1966b), possibly because of differences in the relative extent of the lesions. Ellen and collaborators denied the hippocampus any responsibility in the task-induced inhibitory processes, and postulated that septal inhibition is mediated via the mamillary nuclei and possibly the anterior nuclei of the thalamus, rather than via the hippocampus.

In any case, there are several arguments against the hypotheses of inhibition. Overresponding is not an invariant consequence of septal damage (Braggio and Ellen, 1974; Harvey and Hunt, 1965). The level of stabilized responding after lesion is higher in VI than in DRL even when the density and distribution of reinforcement are equalized in the two schedules (Ellen and others, 1978). Moreover, it depends on the particular value of the DRL delay (Ellen and Braggio, 1973). This argues against an aspecific inability to extinguish conditioned responses. Septal rats learn to space their responses provided that an exteroceptive cue signals the availability of reinforcement, in a so-called "cued DRL" schedule (Ellen and Butter, 1969; Ellen and others 1977a): even if stimulus control may be thought mainly responsible in this case, such animals are obviously capable of motor inhibition. A gradual method of shaping, involving progressive steps up to the final DRL delay, also provides for the acquisition of proper performance after septal damage (Caplan and Stamm, 1967).

A third kind of interpretation emphasizes stimulus control as a major factor in DRL performance. Septal damage would primarily affect the ability to discriminate the stimuli necessary for proper behaviour. Response-produced stimuli are first concerned, as external cues are lacking in DRL. Plausibly, the basis for organizing DRL behaviour is proprioceptive, but external cues such as noises coming from the programming and recording apparatus may become associated with response-produced stimuli and thus provide an anchor for temporal processing. The discrimination of proprioceptive cues may be particularly difficult after septal injury. Ellen and Aitken (1971) found no evidence of temporal patterning in septal rats when all external noises were eliminated from a DRL schedule, whereas some timing subsisted in a previous experiment with the usual noises present (Ellen and others, 1964). No septal deficit appears if external cues are particularly easy to detect, as in the "cued-DRL" of Ellen and Butter (1969). The acquisition of DRL behaviour is facilitated if a gradual transfer of control from exteroceptive to proprioceptive stimuli is provided by progressive fading of the light which signals the availability of reinforcement (Ellen and others, 1977a) (Fig. 6.5). As proposed by the authors, this may be a way of compensating for an attentional type of deficit due to septal dysfunction. The positive effects of gradual shaping (Caplan and Stamm, 1967) possibly sustain this interpretation, especially since the progressive steps employed lose much of their efficiency when lengthened (Van Hoesen and others, 1971).

Another set of arguments comes from studies which indicate the disturbing effect of non-proprioceptive stimuli, such as those resulting from stomach distension. In food-reinforced DRL, the consequences of feeding before the experimental sessions differ in normal and septal rats (Aitken and others, 1975). Prefeeding ameliorated performance, provided that the body weight was not superior to 85 per cent of *ad-libitum* level in the first group, but independently of changes in body weight in the second. The authors argued that stomach distension functions as an additional source of stimulation. In

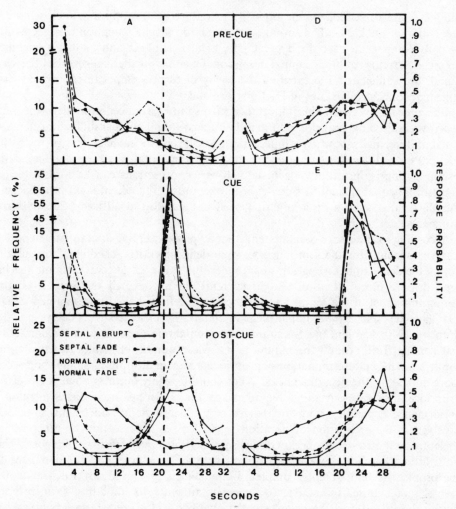

FIG. 6.5. Mean relative frequency of inter-response times (Panels A-C) and mean conditional response probabilities (IRTs/Op: Panels D-F) for each of four groups of rats (with and without septal lesions; with abrupt removal or gradual fading of the cue) during DRL 20 s prior to cue training (top row), during cue training (middle row), and subsequent to cue termination (bottom row). (Redrawn after Ellen and others, 1977a. Copyright 1977 by the Psychonomic Society, Inc.)

normals, it decreases the probability of short IRTs, in agreement with the DRL contingencies, under conditions of deprivation, but acts as a disturbing factor when the activating effect of deprivation is suppressed. Septals, on the contrary, would nevertheless respond to stomach distension as it is the most potent source of stimulation, though irrelevant for efficient DRL behaviour. Hence, their better performance merely results from a generalized suppression of responding rather than from enhanced DRL control. Being unable to discriminate response-produced stimuli, septal animals would meet the DRL requirements only when more clear-cut cues could be used in agreement with the actual contingencies.[1] The involvement of attentional processes is suggested also by the work of Schmaltz and Isaacson (1968a): peripheral blindness due to section of the optic nerve reduced the hippocampal deficit in the acquisition of DRL 20 s. The authors hypothesized that the reduction of sensory input partially compensated for the hyperactivity of diencephalon and midbrain energizing circuits due to loss of hippocampal inhibition. One may further suggest that the visual stimuli are sufficient for preventing the easily disturbed hippocampal animals from utilizing proprioceptive cues in DRL performance.

One objection against the importance of response-produced stimuli in the septal deficit was raised by Braggio and Ellen (1974) who found comparable performances in septal and normal rats when a differential proprioceptive feedback was provided in a DRL 20 s schedule. The authors concluded that the septal rats were not impaired in discriminating proprioceptive feedback. However, objections can be formulated and suggest that either the animals can discriminate only quite accentuated proprioceptive stimuli, or that such a discrimination need not develop at all (see 6.2.2 for further comments).

According to the above data, it seems that a "proprioceptive-attentional" hypothesis can best account for the septo-hippocampal deficit in DRL schedules. But the duration of the delay, and, through this parameter, the existence of central timing mechanisms, deserve more attention than is currently being devoted to them. It is worth noting that most of the septal and hippocampal deficits have been obtained in DRL 20 s schedules. No serious impairment is apparently observed with delays up to 8 s (Caplan, 1970; Caplan and Stamm, 1967): even though response rate increases, the usual modal IRT around the minimum interval required for reinforcement develops with training. The discrimination of proprioceptive stimuli in itself should not become more difficult as the delay increases. What could possibly occur is a memory decay of proprioceptive traces. Another hypotheses is that different mechanisms mediate the performance depending upon the length of the delay. Septal nuclei might be more involved in the estimation of durations of at least a few seconds, or other nervous structures might more properly compensate for the septal deficit when shorter durations are concerned. However, discordant data are available, and this indicates that the problem is undoubtedly complex. For instance, in a study by Ellen and others (1964), septal animals in DRL 20 s showed difficulty in inhibiting their responses during the first 10 s of the delay, but the likelihood of correct responding gradually increased beyond this interval. This suggests that some temporal regulation subsisted at longer delay, though it may be worth recalling that the deficit was more complete when all external cues coming from the apparatus were suppressed (Ellen and Aitken, 1971).

It rests upon future research to determine whether, and to what extent, the proprioceptive index disguises a central deficit. This question will not be answered, however, before data relative to the discrimination of different durations can be compared with the results of DRL studies. The main problem obviously remains: do the lesional effects obtained in temporal schedules reflect a specific deficit, or do they result from a more general impairment? The rate of responses is clearly less critical for efficient

---

[1]Related results were obtained in another schedule which clearly emphasized the proprioceptive stimuli as discriminative cues for efficient performance (Ellen and Kelnhofer, 1971). Rats were trained to make five responses on a "count bar" before responding once on a "food bar" for reinforcement. The septals did not shift from the first to the second bar as appropriately as the normals. Prefeeding the animals before testing tended to induce premature shifting in normals. Both stimulation from stomach distension and septal damage thus seemed to interfere with the discrimination of proprioceptive stimuli.

DRL behaviour than their proper distribution through time; the disappearance of the appropriate modal IRT after septal or hippocampal injury must be considered as the most significant symptom in DRL. This might suggest a specific impairment of timing mechanisms. It is most probable, for instance, that the absence of septal deficit in "cued-DRL" (Ellen and Butter, 1969) primarily depends on the elimination of any timing requirement, since it is sufficient to press the bar when the signal is on. Such arguments in favour of specificity can be gathered in the literature. Needless to say, they still are most questionable and require extensive control.

Other questions remain open to further investigation. The possibly critical septal and hippocampal sites cannot be precisely referred to as a neuroanatomically well-defined system. It has been suggested that the anterodorsal hippocampus, the medial fornix (the most important hippocampus pathway) and the medial septal nucleus compose a functional set, while the posterior hippocampus, the lateral fornix and the lateral septal nuclei form another. This is not consistent, however, with behavioural information since, for example, comparable deficits were found after lesions of antero-dorsal hippocampus, medial fornix and lateral septal nuclei (Johnson and others, 1977). Selective lesions involving various fibre circuits lead Carey (1968) to propose an inhibitory system dependent on reticulo-hippocampal fibres, relayed through the dia-gonal band of Broca and medial septal nucleus, and reaching the hippocampus by the dorsal fornix. On the other hand, the olfactory bulbs have also been implicated in a functional inhibitory system, since lesion of these structures induces changes compar-able to the septal symptoms in animals performing VI and DRL schedules (Thorne and others, 1976). In fact, different functional fibre systems may be involved in various studies, depending on divergences in lesion extent, subjects' age, conditions of training, and other possibly important factors. Even the method employed may be critical: Ellen and Powell (1963) found discordant data when using an angled versus a vertical approach of the lesioning electrode towards a same structure. In addition, species-dependent discrepancies may exist. The extent of the limbic structures, relative to the rest of the brain, decreases as an inverse function of the phylogenetic hierarchy. Hence, their implication in the regulation of behaviour must not be assumed homogeneous. Several data suggest, for instance, that the integrity of the hippocampus is more critical for DRL performance in rats (Clark and Isaacson, 1965; Schmaltz and Isaac-son, 1966b) than in monkeys (Gol and others, 1963; Jackson and Gergen, 1970).

B. Role of other Cerebral Structures

Most of the other structures studied in the context of temporal schedules have more or less intimate connections with the limbic system. We shall make only a rapid survey of the relevant data, as these structures are much less frequently and convinc-ingly implicated than are the septum and hippocampus. For instance, deficits resulting from hypothalamic dysfunction may best be interpreted in terms of motivation and emotion. Emotional changes are also clearly obtained after amygdala destruction, a well-known consequence of which is to produce tameness in animals. However, neither motivational changes nor inhibition deficit seem to account for the impair-ment of DRL acquisition after bilateral damage to the mamillary bodies in rats (Smith and Schmaltz, 1979). In pigeons, lesion of the archistriatum, which is the homologous

of the mammalian amygdaloid complex, leads to improvement of DRL 10 s performance. This was interpreted as due to enhancement of hippocampal activity, to the
extent that those structures have antagonist properties (Zeier, 1969). No deficit has
been noted in cingulectomized rats (Caplan, 1970). Frontal lesions provoked slight
deficits in DRL in rats (Nonneman and others, 1974; Numan and others, 1975), cats
(Numan and Lubar, 1974; Wagman, 1968), and monkeys (Glickstein and others, 1965),
but recovery quickly occurred, indicating participation of other structures in timing
behaviour. Indeed, improvements have also been reported in prefrontal (Stamm, 1963,
1964) and cingulectomized monkeys (Stamm, 1964).

As already mentioned, tasks involving the discrimination of two durations are
unfortunately seldom employed in lesional studies. An example of such a task is given
by Rosenkilde and Divac (1976a, 1976b). Cats had to choose between a right and a left
feeder according to time spent in a restraining cage. Confinement periods were 5 and
(at least) 8 s. A deficit was observed after damage to the prefrontal cortex and to the
anterior part of the caudate nucleus. The authors hypothesized the participation of
both structures, which are functionally related, in timing behaviour. Yet, it must be
noted that the deficit was only transient, as happens in DRL schedules.

Transient deficits, followed by recovery, also affected the retention of DRL behaviour after bilateral caudate lesions in rats (Glick and Cox, 1976; Schmaltz and Isaacson, 1968b). In a study by Johnson and others (1973), however, no difference appeared
between caudate and control groups with regard to the acquisition of the performance. Effects of caudate lesions can thus be thought rather slight, to say the least. As
the caudate nucleus is one of the major integrative centres for the extrapyramidal
motor system, it must be specified in this connection that extrapyramidal structures
have also been implicated in temporal behaviour by the work of Ellen and Powell
(1963), who found impairment in acquisition of FI 60 s after unintended damage to the
zona incerta. Obviously, after lesion of such structures, motor deficits must be carefully ruled out as a possible source of changes in the conditioned behaviour.

Finally, the role of thalamic structures in DRL and FI schedules has been investigated with particular interest given to the group of associative nuclei (essentially
including much of the dorsomedial nuclei, and lateral nuclei), as their dysfunction is
less likely than that of other thalamic structures to reveal only sensory or activation
deficits. Rats lesioned in the medial dorsal thalamus seem slightly inferior to controls
in acquisition of DRL 20 s (Ellen and Butter, 1969). In addition, deficits in acquisition
are found after damage to the lateral nuclei (Schmaltz and Isaacson, 1966b); these
data were interpreted as due to the direct connections of the dorso-medial and lateral
nuclei with the septal area. In FI schedules, data suggesting a possible deficit in the
temporal patterning of the responses are more controverted (Delacour and Alexinsky,
1969; Lejeune, 1971, 1972, 1977a, 1977b; Lejeune and Delacour, 1970). The medial
thalamic structures seem particularly involved in inhibitory control (see Delacour,
1971 for review) and it has been suggested that these inhibition mechanisms mediate
temporal regulation (Lejeune, 1971, 1972; Lejeune and Delacour, 1970). Lejeune
(1977a) recently proposed a more sophisticated explanation of the differential effects of
the lesions in FI, VI and DRL schedules, based on the particular regularity of structure involved in each paradigm (see Chapter 8).

In conclusion, the implication of these various brain regions in timing behaviour is
not consistently reported. Only the septum and the hippocampus are serious candi-

dates. We now have to confront the above data with the results obtained using the stimulation and recording methods.

6.1.2. Cerebral Stimulation

Cerebral stimulation of the brain has been much less extensively used than the lesional method in the study of timing behaviour. Very few indications of its effects are reported. Whether these effects are expected to reproduce the normal functioning of the structure, or to interfere with it, primarily depends on the parameters of the stimulation.

With continuous stimulation applied to the septal area, Kaplan (1965) obtained impairment in retention of DRL 15 s performance in rats. This effect, involving over-responding and decreased reinforcement rate, was similar to that of lesions made in this region. It was interpreted as due to disruption in behavioural inhibition, in connection with works revealing trouble in passive avoidance responding after septal (Kasper, 1964) or amygdala (Pellegrino, 1965) stimulation. However, emotional factors may also intervene, since Kasper (1964) signalled diminished "stress" reactions in his rats. With regard to the "proprioceptive-attentional" hypotheses developed above, it is also not excluded that septal-stimulated animals were unable to use response-produced stimuli properly enough to meet DRL requirements, plausibly through a distracting effect of the cerebral stimulation. It must be said nevertheless that the septal stimulation had specific consequences, since, in Kaplan's study, a control group stimulated in the anterior cingulate cortex did not show any deficit in conditioned behaviour.

The classical conditioning situation has also been used for related studies. In experiments with cats (Flynn and others, 1961) a buzzer or a click served as the CS while a shock to one leg was the US. The CRs were cardiac and respiratory changes in a trace paradigm, and leg flexion in a delayed conditioning procedure. Hippocampal seizures produced by electric application of biphasic square waves disrupted both trace and delayed responses, provided that the discharge spread to the contralateral limbic structures. Along with the few lesional studies implicating the extrapyramidal circuits, the stimulation method indicates a possible role of the globus pallidus in temporal behaviour. Given the rewarding properties of electric brain stimulation first explored by Olds and Milner (1954), stimulation of this and other structures in monkeys was used as reinforcement in DRL 20 s on alternate days with food reinforcement (Brady and Conrad, 1960a). The performance was dramatically altered with pallidal excitation, showing overresponding and a shifting of the response distribution toward the shorter intervals. A peak subsisted around the 10 s interval, however, which demonstrated the persistence of some temporal regulation, though inadequate (Fig. 6.6). In contrast, no change appeared in the usual IRT distribution and response rate with stimulation of the medial forebrain bundle or the anterior medial nucleus of the thalamus. When pallidal stimulation was added to the food reinforcement in a subsequent set of sessions, still more pronounced deficits occurred, relatively few responses being emitted beyond 10 s. This effect was similar to that obtained after administration of amphetamine in food reinforced DRL, but whether the involved mechanisms are comparable remains unknown. This research also indicated that different processes control performance in DRL and VI schedules, since stimulation of the medial forebrain bundle increased response output in VI while being inefficacious in DRL.

It may be of interest to recall in this respect that a number of brain areas, mostly included in the limbic system, have been assumed to possess reward properties. The use of brain stimulation as reinforcement has been extensively studied in interval schedules (Brady and Conrad, 1960a, 1960b; Keesey, 1962; Pliskoff and others, 1965; Sidman and others, 1955; Terman, 1974; Terman and others, 1970, among others) as in other paradigms. Brain stimulation proves to be as potent as food reinforcement in FI or DRL, for instance, provided that stimulation parameters be optimized and that one stimulation not be necessarily considered as equal to one food pellet (Pliskoff and others, 1965). These authors also emphasize the accuracy of the brain reinforcer compared to the long and variable consummatory chain. It must be noted that possibly specific effects dependent on electrode placement were present in this study, since

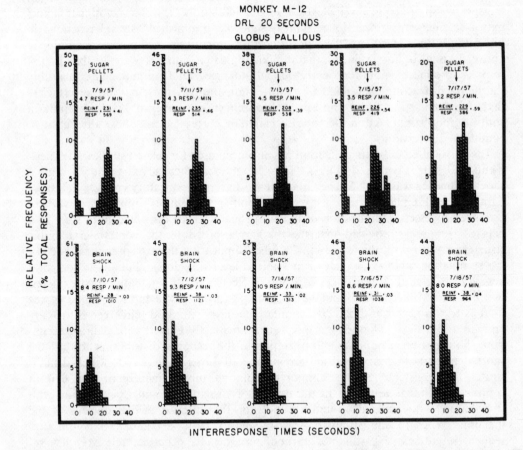

FIG. 6.6. Inter-response time distributions for one monkey on DRL 20 s sessions during an alternate-day procedure of food (top panels) or pallidal stimulation (bottom panels) as reinforcement. Note that the first peak in each distribution only reflects the bursts of rapid responses commonly observed in DRL schedules. (From Brady and Conrad, 1960a. Copyright 1960 by the Society for the Experimental Analysis of Behavior, Inc.)

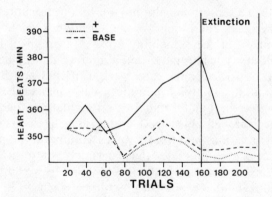

FIG. 6.7. Heart rate of conditioned rats during positive stimuli (+), negative stimuli (−), and immediately preceding stimuli (base) as a function of training. The stimuli are cortical stimulations at 4 or 20 c/s. The data are averaged for blocks of twenty trials. (Redrawn, after Weinberg, 1968. Copyright 1968 by Pergamon Press, Ltd.)

septal relative to posterior hypothalamic stimulation yielded lower response rate, especially in FI. The usual scallops, however, occurred in both cases.

Along another line of research, central stimulation is utilized as a CS or US (see Doty, 1961 for review). For instance, flexion responses may be obtained with repeated excitation of various cortical (and several subcortical) sites preceding electric shock to the leg; or when a sound is consistently followed by stimulation of the motor cortex. Both CS and US may consist of brain stimulation: electric excitation of the occipital area, after association to that of the sigmoid gyrus, induces stereotyped head movements in the dog (Giurgea, 1953, 1955). Thus, a temporal link between two cerebral stimulations may form a basis for conditioned reflexes. Moreover, cortical stimuli differing only in temporal parameters can be discriminated: differential conditioned heart-rate was recorded in rats when pulse trains applied to the striate cortex were or were not associated to a grid shock, depending on their frequency (Fig. 6.7). The existence of a neural mechanism underlying temporal patterns was suggested by this experiment (Weinberg, 1968). The author recalled the speculation of Sokolov (1966) concerning the possible specific sensitivity of defined neuronal sets to certain temporal sequences.

6.1.3. Electrophysiological Correlates

To the extent that it does not interfere with the functioning of the central nervous system, the recording method is a basis for searching normal correlates of timing behaviour. Yet, pertinent indices in this regard have thus far been difficult to assert.

The basic cortical alpha rhythm—or, more precisely, an alpha activity cycle—has been thought to constitute a system for coding and timing the flux of sensory stimuli to which a living organism is continuously submitted (Lindsley, 1952). The alpha rhythm, composed of high-voltage slow waves, of 8–12 Hz frequency, is a classical sign of the resting awake state. Lindsley postulated that alpha activity is "a basic metabolic or respiratory rhythm of the individual brain cell", giving way to the recordable alpha rhythm only when synchronous activity occurs in a great number of brain cells. The basic alpha activity would provide an excitability cycle that would increase or decrease the probability of a sensory impulse provoking a neuronal response, depending upon the phase of the cycle. Thus, timing of the sensory flux would be assured by the frequency and the phase of the excitability cycle in a group of neurons acting in synchronization. Lindsley based his theory upon several data coming from animal and human research, and in particular the early work of Bartley and Bishop (1933), who discovered just such an excitability cycle in the visual system of the rabbit. A correlative hypothesis is that cerebral motor impulses, rather than (or together with) sensory stimuli, may be timed through alpha mechanisms, as it has been indicated that the probability of emitting various movements vary according to the phase of the alpha rhythm. The existence of such processes, thus far largely controverted, would provide a basic mechanism for timing sensorimotor events. How it would be related to temporal estimation and regulation is another unsettled problem.

Given the importance of the temporal relation between the CS and the US in the classical conditioning paradigm, the various electrophysiological changes which have been shown to accompany the establishment of the CRs must be reviewed rapidly, but it is doubtful that they are specifically involved in timing behaviour. It is well known that the occurrence of any novel stimulus induces cortical desynchronization, i.e. disappearance of the alpha rhythm in favour of a fast low-amplitude activation pattern (of 30–60 Hz frequency). Rapid habituation of this "alpha blocking" follows repeated presentation of the stimulus provided that it has no functional significance. The desynchronization pattern, however, reappears if the neutral stimulus is given the role of a CS by temporal association with a US. With inter-stimulus intervals (ISI) of at least a few seconds duration, the alpha blocking may progressively be confined to a short response to the CS, followed by hypersynchronous slow waves which may take up nearly the whole ISI (Fig. 6.8). The significance of this hypersynchronous activity remains unknown, though it has been proposed as an index of the Pavlovian "internal inhibition" process. It is probably not of cortical origin, but rather has some relation with the hippocampal theta activity, which is also recorded during conditioning, and consists of rhythmic slow waves, of approximately 4–7 Hz frequency. Theta activity has been implicated in the timing of motor behaviour since it appears to be synchron-

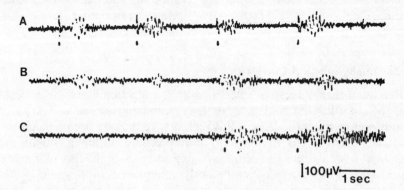

Fig. 6.8. Hypersynchronous waves recorded from the visual cortex of a cat. A: after monotonous presentation of a great number of flashes of light at regular intervals. B: persistence at approximately the same intervals for a few seconds immediately after interruption of the photic stimulation. C: reappearance of the spindle bursts when the photic stimuli are again presented. (From Hernandez-Péon, 1960. Copyright 1960 by Elsevier/North-Holland Biomedical Press.)

ized with bar-pressing for self-stimulation in the rat (Buño and Velluti, 1977). It is probably determined by septal efferents, as it is abolished after anteromedial septal lesion (Donovick, 1968).

At any rate, it must be noted that activities like hypersynchronous or other electro-physiological waves, even though they seem to depend in some way on temporal factors involved in conditioning, can be experimentally dissociated from learning pro-cesses. For instance, the administration of a drug can alter EEG activity without inducing any change in conditioned performance. Ross and others (1962), in their search for EEG correlates which would specifically distinguish between the behaviours obtained from monkeys in a Sidman avoidance, and in a DRL schedule, only found that the EEG pattern reflected an aspecific state of alertness in both tasks, when compared to that of an intercalated Time-Out period. One could think that differen-tial EEG correlates of timing would have been easier to detect in this study if the avoidance shock-shock interval (20 s) and the DRL delay (21 s) had more clearly diverged so as to involve possibly different timing mechanisms. In any case, adminis-tration of pentobarbital or dl-amphetamine had clear differential effects upon the conditioned performances (only DRL was markedly impaired) whereas the frequency of the EEG increased equally in both situations. Thus, we have no alternative but to conclude with the authors that EEG rhythmic activity does not seem to reflect neural processes involved in a learning, or, more precisely for our concern, in timing behav-iour (however, some relationships have been evidenced in man between EEG sleep stages and time perception—see for example Carlson and Feinberg, 1976 and Carlson and others, 1978).

From a number of studies investigating different mechanisms of the living organism, it seems possible to assume that every physiological response is potentially time-con-ditionable. The alpha blocking is an example of such an assumption. Since the casual discovery of its conditioning by Durup and Fessard (1935) in man, many experiments have followed, involving various conditioning paradigms. Delayed, trace, and cyclical alpha-blocking responses, for instance, can be obtained (Jasper and Shagass, 1941), and even seem to be more accurate than subjects' overt judgements about the inter-vals involved. Many animal studies have also been done (Morell and others, 1956; Rowland and Gluck, 1960, among others). It is not certain, however, that conditioning rather than sensitization processes have been obtained in this context (Milstein, 1965). Photic driving, another well-known electrophysiological phenomenon, is also suscep-tible to conditioning (Morell and Jasper, 1956). Let us recall the basic contingencies required for its appearance: when intermittent visual (or, though less readily, acoustic) stimulation is presented, with a frequency which is not too different from the range of spontaneous brain rhythm frequency, the latter may become synchronized to the former. Photic driving occurs in various cortical and subcortical sites (Fig. 6.9). It has been presumed to constitute a mnemonic trace of the temporal sequence of the sen-sory stimuli presented, thus providing a model for comparison with subsequent stimu-lations (John and Killam, 1960), but this hypothesis has raised many objections, firstly because this phenomenon, being limited to the range of the normal EEG rhythmic activity, does not constitute a general process. Furthermore, hippocampal slow waves are probably connected with it (Liberson and Ellen, 1960).

The study of photic driving has given way to the interesting experiments of Popov (1944, 1950) about "cyclochrony", which demonstrate that the central nervous system

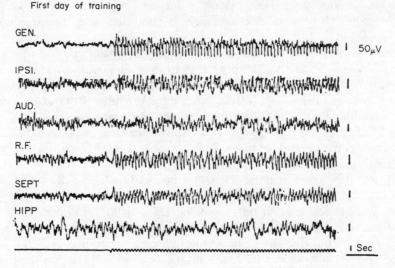

FIG. 6.9. Photic driving in the lateral geniculate nucleus (GEN), visual cortex (IPSI), auditory cortex (AUD), midbrain tegmentum (R.F.), septum (SEPT), and hippocampus (HIPP), on the first day of avoidance conditioning. The signal consists of 10 c/s photic stimulation. (From John and Killam, 1959.)

possesses the ability to reproduce a preceding cyclic activity with its temporal order. Intermittent photic stimuli, of 1–2 Hz frequency, drive the rhythmic activity of striate cortex in rabbits; in most cases, such activity persists at the same frequency for several cycles after cessation of the visual stimulation. Popov compared this phenomenon to the time-conditioned salivary reflexes studied in dogs by Pavlov and his followers. More recently, similar properties have been discovered at the level of the cortical evoked potentials, which can still occur some time after cessation of the stimuli presentation, precisely at the moment when the stimuli ought to be delivered. These so-called "emitted potentials" have generally been interpreted as linked with information-processing and retrieval processes (see, for example, Weinberg and others, 1974).

If all these phenomena may to some extent be put in relation with temporal factors, it seems questionable to further assume that they represent indices of timing as such. A possibly more promising temporal correlate has been recently found in the "contingent negative variation" (CNV) of the steady cortical potential. This CNV, discovered by Walter and his co-workers (1964), is a negative polarized slow potential shift that appears at the surface of the cortex when a subject attends to some significant event. A reaction time task with preparatory period has been typically used in related studies: the CNV develops during the interval between the warning stimulus (S 1) and the imperative stimulus (S 2). Its latency ranges between 200 to 500 ms, i.e. it follows (or adds to) the late components of the cortical potential evoked by S 1. Its maximal duration can reach a few tens of seconds. The CNV is generally not detected in a single EEG record, but rather appears after summation of several sampled periods, which neutralizes the EEG background. It then presents itself as composed of a more or less abrupt initial rise toward a negative peak that is reached before or at the moment of S 2, and then followed by a sudden release of negativity (called the "resolution") around S 2 occurrence (Fig. 6.10).

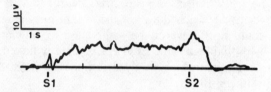

FIG. 6.10. CNV recorded on the anterior supra-sylvian gyrus of a cat during the inter-stimulus interval of a reaction time task. Note the evoked potentials at S 1 and S 2, both auditory stimuli. ISI duration = 4 s. Negativity upward.

Extensively studied in man, this phenomenon has thus far been explored in a few experiments in monkeys (Borda, 1970; Donchin and co-workers, 1971; Hablitz, 1973; Low and others, 1966; Rebert, 1972) and cats (Chiorini, 1969; Irwin and Rebert, 1970; Macar and Vitton, 1978; Macar and others, 1976; Rowland and Goldstone, 1963). Evidence for similar slow potential shifts has been reported in the dog (Kamp and others, 1969). The CNV has been recorded from the reticular formation, the hypothalamus and various thalamic nuclei in monkeys (Rebert, 1972); in man, it has been found in the brain stem (McCallum and co-workers, 1973) and on every cortical area. Though usually studied in tasks which require motor responses, this phenomenon can be dissociated from the cortical motor potentials since it precedes a mental response as well (Low and co-workers, 1966; Walter, 1966). It is generally taken as an index of several psychophysiological processes and in particular, attention and activation (Tecce, 1972). Yet, a recently developed hypotheses suggests that the CNV would be an index of timing processes when periods of a few seconds are involved, since temporal parameters are significant components of every situation in which it is recordable (Macar, 1977). In cats reinforced for pressing a lever precisely 5 to 7 s after the onset of a visual signal, the amplitude and time course of cortical CNVs were found to vary as a function of response accuracy; that is, the largest CNVs were linked with correct responses (Macar and Vitton, 1978) (Fig. 6.11). CNVs tended to habituate with overtraining, suggesting a possible transfer of the involved timing processes from cortical to subcortical circuits when performance automatization was obtained. CNV data open to a temporal interpretation were also reported in man (McAdam, 1966; Ruchkin and others, 1977). A critical question concerns the relation of this gross EEG index to the underlying brain unit activities. From a speculative point of view, it might

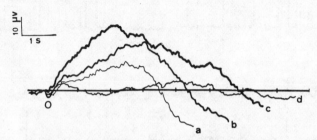

FIG. 6.11. CNVs recorded on the anterior supra-sylvian gyrus of a cat during the delay of a temporal schedule. In order to obtain a reinforcement, the animal had to press a lever 5 to 7 s after the beginning of a visual signal (0). The CNVs precede responses made: (a) 4 to 5 s after 0; (b) 5 to 6 s after 0; (c) 6 to 7 s after 0; (d) 7 to 8 s after 0. Negativity upward.

be proposed that the CNV reflects synchronized neuronal activities, combined in particular patterns relative to each duration the organism is confronted with. Such neuronal activities could not, however, be assimilated to cellular spike discharges, since it seems actually probable that the EEG is less directly linked to the spikes than to the post-synaptic potentials which partly determine them (Buchwald and others, 1966; Creutzfeld and others, 1966).

6.1.4. Single Unit Studies

Intra- and extra-cellular recordings through microelectrodes in the central nervous system have furnished interesting results in the classical and instrumental conditioning frameworks. In the non-anaesthetized rabbit, neurons of the dorsal hippocampus, amygdaloid complex, and hypothalamic perifornical nucleus exhibited clear CRs to time during a classical procedure of elicitation of motor CRs by repeated presentation of paired acoustic and electric skin stimulation (Kopytova and Kulikova, 1970; Kopytova and Mednikova, 1972; Mednikova, 1975). The unit CRs consisted of an increased (excitation type) or decreased (inhibition type) spike activity, consistently obtained after a few tens of pairings. These CRs could persist throughout several trials after omission of the US. Comparable phenomena were recorded in cells of the visual cortex following the occurrence of repeated flashes (Bagdonas and others, 1966); the authors postulated the existence of timing neuronal mechanisms. Moreover, intercellular recording of sensorimotor cortex showed periodic oscillations of the cellular membrane potential, which, as spike inhibition, followed the interruption of rhythmic electric stimulation applied to the forelimb or the cortical area (Voronin, 1971). The periodicity of such phenomena approached that of the external stimuli. Thus, apparent CRs to time are observable even at the membrane processes level (Fig. 6.12).

On the other hand, neuronal correlates of timing have been presumed to appear in the dorsolateral prefrontal and anterior cingulate cortex of monkeys submitted to an instrumental schedule (Watanabe, 1976). The animals had to depress a lever for 3 or 5 s in order to provoke the illumination of a button on the test board; the lighted button could be struck for reinforcement after another period of at least 2 s during

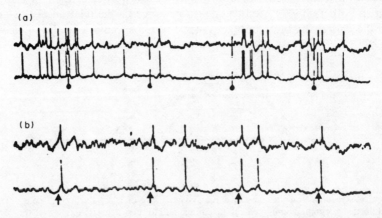

FIG. 6.12. Depolarization waves and unitary periodical discharges. A: during electrical skin stimulation; B: 20 s after interruption of stimulation. (Adapted from Voronin, 1971.)

which the lever had to be kept depressed. Among the three types of spike patternings recorded, one seemed to be linked with the subjective estimation of the delay: sustained firing enhancement began after the button illumination onset and stopped just before the lever release, whatever its latency. Thus, the duration of this activity was short when the lever was released early and long when it was released later. Watanabe concluded that neurons of this sort were involved in time estimation processes. However, this interesting experiment would require further control for possible interference of preparatory motor components.

These several indications may once again evoke the hypothesis of Sokolov (1966) concerning the sensitivity of certain neuronal sets to particular temporal patterns. On the other hand, loci specificity of such sets seems thus far difficult to assume. Rather the available data would lend support to Pavlov's view that temporal counting must be a basic property of every neuron.

6.2. PERIPHERAL MECHANISMS

6.2.1. Visceral Pacemakers

Assuming that the central nervous system plays a major role in timing processes does not preclude a possible participation of visceral pacemakers. Many biological clocks are found in the organism. One may imagine that one or several of them provide periodic indices which form the basis for temporal regulations. The respiratory and, above all, cardiac cycles have received particular attention in this respect.

An important line of investigation has been concerned with the relation which may exist between measures of temporal performance and indices of physiological arousal, in particular pulse rate, blood pressure, and respiration rate. This type of research, usually done using human subjects, has yielded rather meagre results (see Wallace and Rabin, 1960, for review) and presently seems to be unpopular. It is suggested, however, that the results differ according to the method employed, the production of time intervals giving generally more consistent data than the reproduction or estimation methods (Hawkes and others, 1962). The method of production seems, more probably than others, mediated by involuntary processes (Jones and Narver, 1966). There is no ground for developing these topics here, as they essentially belong to human research. Yet, a particular aspect of the relation between vegetative pacemakers and the temporal dimension of behaviour must be mentioned: that is, the modulating function of cardiovascular processes in sensorimotor activity. Even though this phenomenon has been mainly investigated in man, evidence for possibly responsible mechanisms is to be found mostly in acute animal preparations. Many sensorimotor events have been shown to fluctuate according to the moment of the cardiac cycle in which they occur. A periodic evolution of the medullar reflexes has been shown in animals for almost half a century (Emery, 1929). In man, tendinous and Hoffmann reflexes are depressed 350 ms after the R wave of the ECG (Coquery and Requin, 1967). Reaction time is shorter when the imperative signal is delivered synchronously with the P wave (Birren and others, 1963; Callaway, 1964; Requin, 1965). Subjects who are asked to reproduce the duration of a standard signal tend to make relative overestimations when the onset and termination of the stimuli are positioned immediately after the R wave, and relative underestimations 300 to 400 ms after the cardiac systole (Requin and Granjon, 1968); however, a more recent experiment yielded less significant data (Granjon and

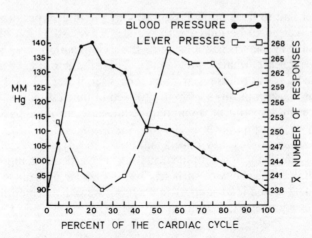

FIG. 6.13. Running averages of lever-presses, 30 per cent adjusted, superimposed on the mean blood pressure wave. Measures taken in a monkey responding on a 10 s Sidman avoidance schedule. (Redrawn and modified from Forsyth, 1966. Copyright 1966 by the Society for Psychophysiological Research.)

Requin, 1970). As for the periodic fluctuations of spontaneous motor activity, controverted data appeared in man for responses on a key (Callaway, 1965; Granjon and Requin, 1970; Requin and Bonnet, 1968), but conclusive results were found in one monkey working on a 10 s Sidman avoidance schedule (Forsyth, 1966). The animal's blood pressure and heart-rate were recorded throughout the experimental sessions. A significant negative correlation was observed between the relative frequency of the lever presses and the height of blood pressure, since lever presses were the least frequent approximately 150 ms after the beginning of the blood pressure elevation that follows the systole (Fig. 6.13).

The distribution of spontaneous lever presses in the cardiac phase were also studied in cats submitted to a DRL 5 s LH 1 s schedule with two levers available (Macar, 1970). A delay of 5 to 6 s was required for reinforcement between the responses made consecutively on the first (A) and the second (B) lever. No cyclic fluctuation appeared in the distribution of the responses made on either lever, but another interesting phenomenon was observed. The cardiac frequency showed a characteristic evolution during the A–B interval: a strong tachycardia developed at response A, persisted until response B, and was then followed by a return to baseline level (specific tachycardias like the one observed here have been interpreted as reflecting attention focused on internal events by Lacey and Lacey, 1970). Differences were noted with regard to the accuracy of the temporal regulation: the tachycardia was more pronounced when the response B occurred too early, and less when it occurred too late, relative to the correct period of 5 to 6 s (Fig. 6.14). Given that the number of cardiac beats per time unit increases as a function of cardiac frequency, this phenomenon could lead to equalize, in each trial, the number of beats after which the response B was made. Such an equalization, which, however, the data proved very approximate, would suggest that the differences in cardiac rhythm affect the temporal regulation of the animals, thus providing some evidence for the possible influence of cardiac mechanisms on timing processes. Yet, it is clear that the trends observed both in cardiac rhythm and

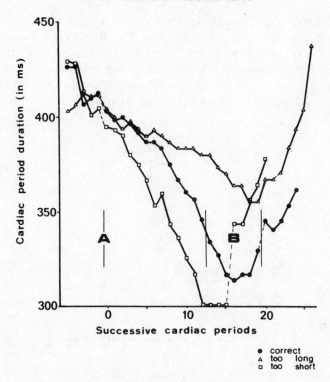

FIG. 6.14. Recording of heart rate in one cat during the delay of a DRL 5 s LH 1 s schedule with two levers available (average of about forty trials). Duration of the cardiac period before the A response (five periods), during the A-B sequence, and after the B response (five periods). The vertical lines show the occurrence of the A response, and the interval during which the B response may have occurred in the case of reinforced trials.

in DRL performance might possibly result from the influence of some other common factor, rather than reflect a causal relation.

Three main interpretations of the periodic fluctuations of the excitability of nervous structures during the cardiac cycle have been proposed (Granjon and Requin, 1970). The boldest states that a specific index, repeated in each cardiac cycle (e.g. one of the ECG waves), would provide a periodic anchor for temporal perception. The information given by such an index, which every sensorimotor event would refer to, would then be processed at a central level. A less demanding viewpoint holds that the cardiac pacemaker might regulate sensorimotor activity by delaying stimulus perception or response execution up to a suitable moment in the cardiac cycle, or by filtering these sensory and motor events, thus constituting a permanent sampling mechanism. This hypothesis is similar to the one concerning the possible function of an alpha excitability cycle (Lindsley, 1952). Finally, the cardiovascular processes might just cyclicly perturb sensorimotor activity by introducing an irrelevant variance into its functioning.

Lacey and Lacey (1977, 1978) recently added to the body of knowledge bearing on the question of modulation of sensorimotor activity by cardiovascular processes by showing that the extent of the decelerative cardiac response typically obtained in the fixed fore-period of a reaction time task depends on the moment when the imperative stimulus is presented within the cardiac cycle. The earlier the stimulus occurred in one cycle, the more prolonged this cycle became (Fig. 6.15). These data seem open to generalization, since they appeared in various other tasks and in particular, tasks involving self-initiated responses. In addition, it was found that sensorimotor events occurring late in the cardiac cycle produced a lengthening of the subsequent period, instead of the heart period in which the events occurred.

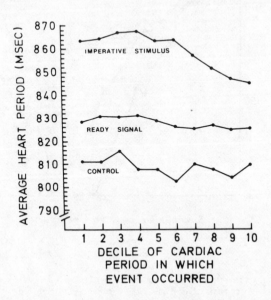

FIG. 6.15. Relationship of concurrent heart period to relative cycle time of event occurrence, in a reaction time experiment with fixed foreperiod in man. Ordinate is average of individual median heart periods ($N = 66$). Note that cardiac deceleration (i.e., increase in heart-rate period) occurs between measures taken at a control point (in the intertrial interval) and the warning stimulus, but still more often between the warning and the imperative stimulus. A dependency is found between the position of the latter in the cardiac cycle and the degree of heart period increase. (Redrawn after Lacey and Lacey, 1977. Copyright 1977 by the Psychonomic Society, Inc.)

Though cortical mediation is presumable in such behavioural studies, these phenomena are comparable to data coming from acute animal preparations concerning vagal control of the heart. A cardiac cycle is prolonged if the vagus is electrically stimulated early in the cycle; the effect is transferred to the subsequent cycle if the stimulation occurs later (Dong and Reitz, 1970; Levy and Zieske, 1972; and others). The release of acetylcholine upon vagal stimulation and its action on the sinoatrial node have been shown responsible for such effects (Levy and others, 1970). It is to be noted that stimulation of the carotid sinus baroreceptors—and, more extensively, of vago-aortic afferents (Gahery and Vigier, 1974)—results in inhibitory effects on the central nervous system (Heymans and Neil, 1958). Such stimulation is ensured by an increase in blood pressure at each systole; as known for a long time, this leads to inhibition of various reflexes (Koch, 1932; Pinotti and Granata, 1954) and spontaneous movements (Koch, 1932) and to progressive disactivation of the EEG (Bonvallet and others, 1954). Inhibition of pyramidal neurons of the motor cortex has been reported more recently (Coleridge and others, 1976).

From the inhibitory effects due to stimulation of the baroreceptors, Lacey and Lacey (1978) inferred that reduced baroreceptor activity may facilitate sensorimotor processes. Thus, heart-rate changes found in various studies would have some instrumental function in the regulation of behaviour. This assumption is suggested by data obtained in human subjects working on a DRL 15 s LH 4 s schedule. When the cardiac cycle occurring immediately before the key press was measured, as an index for the heart-rate level, in those trials which were correctly timed, a systematic relation was discovered between this measure and the position of the key press: the latter was postponed to later and later portions of the cardiac cycle as the immediately prevailing heart-rate increased. As concluded by the authors, such data are "in search of a theory". At any rate, they strongly suggest the existence of a significant relationship between sensorimotor and vegetative processes, and emphasize temporal factors as a major determinant of the effects thus far reported.

Finally, the possible modulating properties of the respiratory cycle have also been studied, though less frequently. As most studies have once again used human subjects, we shall only cite as an example the observations of Poole (1961) who found periodic distributions in the probability of occurrence of various events, from sleep spindles to spontaneous tapping movements. Influences possibly mediated by the reticular activating system were presumed to affect respiration or be secondary to it. A modulating effect of the respiratory cycle on sensorimotor activity, similar to that reported in cardiovascular processes and perhaps linked to it, is thus suggested by this and other studies.

In conclusion, it may be of interest to emphasize a particular observation common to many studies in the field of vegetative pacemakers: less intra- than inter-individual variance is generally reported. In fact, very consistent data are often found in only a few subjects of a group. This would suggest that the construction of a particular time base may differ from one subject to another. Moreover, as noted by Requin and Granjon (1968), the choice of a proper time base might be more limited in certain situations than in others: for instance, well-defined conditioning procedures would be more demanding than tasks involving spontaneous activity; the experimenter would thus have a greater chance of obtaining correlates of a definite time base in the first case. If so, one could draw the conclusion that multiple time bases are continuously constructed in response to the particular requirements of each situation, and replaced by others when they become useless. This would fit in with the fact that a living organism never stops adapting itself to numerous dissimilar temporal parameters.

6.2.2. Proprioceptive Feedback and Timing

If the "internal clock" cannot be located in the central nervous system or the viscera, perhaps temporal regulation and estimation are derived functions of one of the sensorial systems such as vision, audition, or proprioception. The idea of a proprioceptive time-keeping mechanism is not new in psychology. For example, Münsterberg (1889) long ago suggested that the sensation of duration is related to variations in muscular tonus that occur with breathing and human subjects rely on these variations in order to measure time. However, if variants of this hypotheses have often been discussed and explored in relation to temporal aspects of human behaviour (see below), this possibility has received little attention in studies of animal timing behaviour.

The term *proprioception* refers to the sense of body position in space and relative position of body segments, movement force and extent, muscular tension, and physical pressure (Sherrington, 1906). It includes *kinesthesis* or the sense of movement. The primary sensory receptors or *proprioceptors* consist of the kinesthetic receptors located in the muscles, tendons and joints and the vestibular receptors located in the inner ear. Impaired discrimination of information arising from these receptors has occasionally been evoked in animal studies as a possible explanation of the effects of certain lesions (see 6.1.1), drugs and environmental manipulations on timing behaviour.

As described under 6.1.1, septal-lesioned rats and cats show poorer performance on DRL (Burkett and Bunnel, 1966; Caplan, 1970; Ellen and Aitken, 1971, 1977b; Ellen and others, 1964, 1977a, 1977b; Numan and Lubar, 1974) and FI (Beatty and Schwartzbaum, 1968; Ellen and others, 1977b; Ellen and Powell, 1962a; Schwartzbaum and Gay, 1966) schedules of reinforcement than normal subjects.[1] Typically, septal-lesioned animals have increased response rates and altered temporal distributions. However, if an exteroceptive stimulus signals the end of the required delay on a DRL schedule, lesioned rats show performances comparable to those of normal subjects (Braggio and Ellen, 1976; Ellen and Butter, 1969; Ellen and others, 1977a; Ellen and Powell, 1966). This suggests that septal lesions may not simply produce response disinhibition (McCleary, 1966, or facilitation (Schnelle and others, 1971). Ellen and Butter (1969) have suggested that in DRL studies using exteroceptive stimuli, exteroceptive-cue discrimination comes to replace the normal response-produced proprioceptive-cue discrimination that takes place in non-lesioned subjects exposed to the standard DRL situation. According to their hypotheses, response-produced proprioceptive feedback actually serves two functions in the standard DRL situation. First, proprioceptive feedback stimuli provide discriminative cues for bar-pressing. Secondly, these feedback stimuli acquire reinforcing properties through association with the primary reinforcer (see Grice, 1948; Skinner, 1953; Spence, 1947 and Mowrer, 1960, with regard to proprioceptive stimuli as secondary reinforcers) and thus help maintain collateral responding (behaviour other than bar-pressing) that mediates the DRL delay interval (see Chapter 7). The DRL deficiency of septal-lesioned rats is therefore due to their having lost the ability to utilize such feedback stimuli. However, as already mentioned, the results of a more recent study by Braggio and Ellen (1974) seem to show that septal-lesioned rats do discriminate proprioceptive feedback. In this study, septal-lesioned rats and normal rats were divided into three groups and exposed to a DRL 20 s schedule of food reinforcement. For two of the groups, differential response feedback was provided by systematically varying the force required to depress the response-lever. For one group, the force required to press the lever decreased linearly from 49 to 21 g as a function of time since the last lever-response; at $t = 0$ s, the force required to press the lever was 49 g and at $t = 20$ s or more, 21 g. For a second group of rats, this relation was inverted; the force required to operate the lever increased from 21 to 49 g as a function of time since the last lever-response. For the third group, the force requirement remained constant at 21 g throughout the

[1]In addition to the timing hypotheses discussed here, proprioceptive stimuli have also been attributed secondary, non-temporal roles in timing situations (e.g. Kintsch and Witte, 1962; Rosenkilde and Lawicka, 1977). However, space does not permit discussion of these studies.

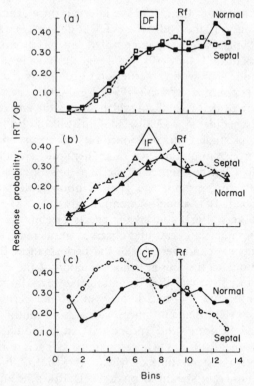

FIG. 6.16. Conditional probability of each inter-response time (IRT/OP) for normal and septal rats run under varying conditions of force for bar-pressing in DRL 20 s. Each IRT bin has an interval size of 2.22 s. Reinforcement occurred after the ninth bin. DF: decreasing force; IF: increasing force; CF: constant force. (From Braggio and Ellen; 1974. Copyright by the American Psychological Association. Reprinted by permission.)

inter-response interval. Results show that in the two conditions of differential feedback, normal and septal-lesioned rats do not differ significantly with regard to number of reinforcements obtained or efficiency ratio. In contrast, in the constant force requirement condition, septal-lesioned rats obtain fewer reinforcements and have lower efficiency ratios than normal subjects (Fig. 6.16). Similarly, IRT /Op analysis shows temporal regulation in normal and lesioned rats in the two conditions of differential feedback but only in normal rats in the constant force requirement condition. Finally, modal IRT is the same for non-lesioned rats in all three force conditions. Braggio and Ellen conclude that the septal-lesioned rats were able to discriminate response-produced proprioceptive stimulation since they responded efficiently in both conditions of differential feedback, but that differential proprioceptive feedback probably was not crucial to normal DRL performance as modal IRT was the same for all three groups of non-lesioned subjects. It should be noted, however, that the septal-lesioned rats possibly did not really learn to discriminate proprioceptive feedback. In the decreasing force condition, the animals could have continually responded with a response-force just slightly superior to 21 g; responding would not have been recorded before the end of the DRL delay. In the increasing force condition, they could have been responding with minimal force at the beginning of the delay and then increasing their response force throughout the rest of the interval; once again, responding would have been recorded only at the end of the DRL delay. This kind of increase in response-force with the passing of a temporal delay has been described on a FI schedule of reinforcement by Haney (1972) and pigeons have been observed "collaterally" pecking the response-key with sub-criterion force during inter-response intervals on a DRL schedule (Topping and others, 1971). In addition, in a recent study, Ellen and others (1977a) showed that if the light which signals the end of the required delay on a cued-DRL 20 s schedule is gradually faded instead of abruptly removed as has been the case in previous studies, septal-lesioned rats show post-cue performances that are indistinguishable from those of normal rats. Therefore, it would seem that exteroceptive stimulus control can be transferred, perhaps to response-produced feedback, and the septal deficit may be more one of inattention to less salient but relevant stimuli, as has been pointed out under section 6.1.1.

Richelle and co-workers (1962) studied the effects of chlordiazepoxide on FI and DRL responding in both cats and rats and observed an increase in response rate and disruption of the temporal characteristics of the performances. They suggested that these effects might be due to the muscle relaxant action of the drug and an eventual impairment of proprioceptive cue discrimination. However, a study by Cook and Kelleher (1961) seems to show that muscle relaxants do not perturb temporally regulated behaviour to the extent that substances with a more central action do.

Finally, Beasley and Seldeen (1965) proposed that a modification of the gravitational environment might affect DRL responding as such a manipulation would affect the proprioceptive system. Indeed, when exposed to 5 h acceleration at 5 g preceding daily sessions, rats stabilized on a DRL 10 s schedule showed flattened relative frequency distributions and an increased proportion of long inter-response

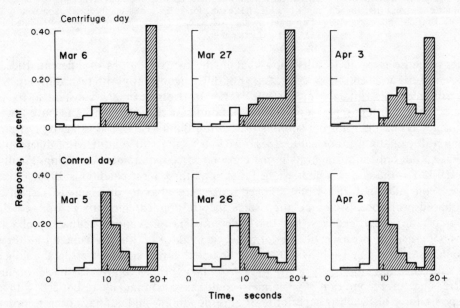

FIG. 6.17. IRT distributions for three centrifuge and three control days for rat C 12 on a DRL 10 s schedule. (From Beasley and Seldeen, 1965. Copyright 1965 by the Society for the Experimental Analysis of Behavior, Inc.)

times close to 30 or 40 s (Fig. 6.17). Sequential analysis of IRTs showed that this resulted exclusively from long IRTs produced in succession early in the session. After this phase, the subjects came back abruptly to their usual IRT, close to the critical 10 s value. This sequential organization of IRTs after acceleration contrasts with the organization usually observed in these subjects, showing alternating groups of long IRTs and of IRTs adjusted to the delay. However, rats stabilized on a FR 5 schedule showed a large overall decrease in response rate, due to an increase in post-reinforcement pausing after acceleration and not to a decrease in response rate when the subjects were responding. Moreover, free-feeding animals exposed to 5 g acceleration showed significantly reduced weight gain. Post-acceleration food intake was decreased, as compared with baseline measures. Taken together, these results seem to point to a more general motivational effect of acceleration.

Two major shortcomings can be found in the studies described above. First, in two of the studies it is assumed from the start that a proprioceptive-based discrimination process constitutes the timing mechanism involved and any disruption in the temporal characteristics of responding is the result of a disruption of this process. Secondly, in all of the studies the manipulations possibly touch more than just the proprioceptive system and the effects on the temporal characteristics of responding may actually be due to a more general effect on motivation, attention, etc. One may ask how, precisely, a proprioceptive timing mechanism might work and, therefore, how one could act more specifically on such a mechanism. Some interesting suggestions, both theoretical and methodological, can be found in the research done on human motor skills and in particular, in studies of "perceptual anticipatory timing".

A. Anticipatory Timing

In the study of human motor skills, the term *anticipatory timing* is used to refer to situations where the subject must time his motor response to occur in coincidence with an environmental stimulus event and must, therefore, get his response underway before the stimulus event actually occurs (Adams, 1966). An example of this kind of task is trying to hit a pitched baseball with a bat. *Perceptual anticipatory timing* refers to situations where, contrary to that of the above example, the subject cannot directly preview the upcoming stimulus event and must learn the temporal characteristics of its occurrence in order to anticipate correctly (Poulton, 1955, 1957). An example of this kind of situation is trying to step on the accelerator of a car at precisely the moment the red traffic light changes to green. Typically, the experimental situation is one of "discrete visual tracking" (Fig. 6.18). For example, the subject stands or is

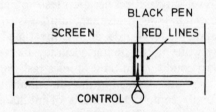

FIG. 6.18. A simplified representation of apparatus used in discrete visual tracking experiments. (Adapted from a description by Adams and Creamer, 1962b.)

seated in front of a screen where a pair of vertical red lines appears alternately on the left and the right, changing from side to side every x seconds. Between these discrete, regular appearances the lines are not visible on the screen. The subject grasps the knob of a horizontal sliding control that directs the movements of a black pen on the same screen and his task consists of trying to keep the black pen writing between the two red lines as they change from side to side. If the subject waits until the stimulus appears and then starts to move the control, his tracking response will show a reaction-time lag. However, if he correctly anticipates the stimulus and initiates his response in advance, his response will closely coincide with its arrival. Such responses are termed "beneficial anticipations", as a response for which absolute error is less than 133 ms (Adams and Xhignesse, 1960).

Adams and co-workers (Adams and Creamer, 1962a, 1962b; Adams and Xhignesse, 1960) and Schmidt and Christina (1969) have hypothesized that in situations requiring temporal perceptual anticipation, the subject relies on proprioceptive time-keeping mechanisms in order to time the interval between successive stimulus occurrences and thus temporally regulate his successive responses. According to the Adams and Creamer "Proprioceptive Trace Hypothesis", decaying proprioceptive after-effects in short-term memory provide the necessary cues; "... a response at time t is assumed to generate a time-varying stimulus trace, and the stimulus characteristics of the trace at time $t + \Delta t$ is the cue set to which the subject learns the response that is anticipatory to the environmental event..." (Adams and Creamer, 1962b, p. 218). In a recent article, Adams (1977) argued that slowly adapting joint receptors are the anatomical basis of this mechanism (see Greer and Harvey, 1978 and Kelso, 1978, for critical discussion of this interpretation). Schmidt and Christina's "Proprioceptive Input Hypothesis" differs from the Trace Hypotheses in that incoming proprioceptive feedback produced by movements made *during* the interval furnishes the necessary cues; "... the subject uses the response-produced proprioceptive feedback from earlier portions of a response series as cues for initiating later responses..." (Schmidt, 1971, p. 385).

Both the Trace Hypotheses and the Input Hypotheses predict that the quality of anticipatory timing will in part depend on task variables (tension on the control, amplitude of the required movement, etc.) that influence proprioceptive level (Bahrick, 1957) and in a general way, *augmented feedback* should lead to enhanced performance. In agreement with this prediction, in a study by Adams and Creamer (1962b), subjects showed better timing in the experimental situation described above (Fig. 6.18) when the control was spring loaded (with 1 or 4 lb) than when it was unloaded. Ellis and others (1968) had subjects estimate a 2 s interval by moving a control along a track and found that estimates were more accurate when the required movement amplitude was 65 cm than when it was 2.5 cm. Using a similar task, Ellis (1969) found that resistance proportional to velocity and acceleration lead to more accurate and consistent estimates of a 2 s interval in fewer trials than an unresisted movement. Also, when subjects spelled 2-, 3- and 4-letter words during the timed interval and auditory feedback was masked by white noise, timing accuracy increased with word length. However, except in this last condition, in these studies task variables and, supposedly, proprioceptive feedback were manipulated in the same limb that executed the timing response, therefore eventually confounding effects of mechanical and proprioceptive factors; that is, the augmented feedback conditions may have simply created a mechanically more favourable situation for responding (Schmidt and Christina, 1969). In order to avoid this kind of confusion, Quesada and Schmidt (1970), Schmidt and Christina (1969) and Christina (1970, 1971) manipulated movement variables in a limb different from the limb being used to perform the timing response. In the Quesada and Schmidt study, subjects held on to a control with their left hand while a motor moved the control vertically down a 26.5 inch track to a stop. With the index finger of their right hand, the subjects depressed a response-key. Two s after the end of the passive left-arm movement, an exteroceptive stimulus appeared and the subjects' task consisted of trying to release the response-key at precisely the same moment the stimulus appeared. Results showed superior timing performance in these conditions when compared to a no left-arm movement condition. Using a voluntary left-arm task Schmidt and Christina (1969) had subjects make a small 0.5 in. movement, a 3.5 in. rotary movement, or a 11.5 in. rotary movement while making a timed right-finger response. Results showed significantly more beneficial anticipations with the 3.5 in. rotary movement than with the small 0.5 in. movement. The 11.5 in. rotary movement did not lead to significantly greater accuracy than the 3.5 in. rotary movement; however, the larger movements showed more between-trials inconsistency and, therefore, perhaps were not providing reliable proprioceptive cues. Finally, for half of the subjects in the rotary movement conditions, left-arm movement extent was correlated with moment of response emission and Schmidt and Christina hypothesized that these subjects were using proprioceptive positional cues to time their right-finger responses. In a similar study, Christina (1970) found that 43 per cent of the subjects that moved a left-arm control while timing a right-finger response tended to co-vary the velocity of their left-arm movement with the moment of emission of their right-finger response. In addition, subjects with left-arm movement showed greater anticipatory response consistency than no left-arm movement control subjects. In a follow-up study, Christina (1971) actually instructed subjects to use their left-arm movement (rotating a crank handle counter-clockwise) to time their right-hand response (lifting their right index and middle fingers from a key when a moving pointer aligned with a stationary one; interval = 1.5 s). Results showed that subjects in this condition temporally anticipated with less absolute error than subjects without left-arm movement and 94 per cent of the subjects co-varied extent of left-arm movement with right-hand response time.

B. Related Animal Studies

Though the results of the studies described above seem to lend support to a proprioceptive feedback interpretation, other interpretations are possible (see below). Before discussing the shortcomings of the approach used in these studies, two recent studies of animal timing behaviour (Fowler and others, 1972; Greenwood, 1977) in which a similar approach was employed will be described.

In an experiment reported by Notterman and Mintz (1965), rats were trained on a 1.6 s response-duration schedule of reinforcement and their response forces were recorded. The minimum response force requirement was 2.5 g. However, subjects emitted high forces in the order of 50 g and individual responses showed striking force oscillations. This suggested that the rats may have been using neuromuscular feedback to time their responses and the oscillations may represent some kind of kinesthetic "scanning". In order to explore this hypotheses, Fowler and others (1972) trained six rats on the same schedule and then submitted two responses (those following the

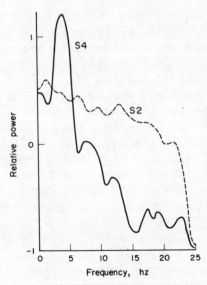

FIG. 6.19. Spectral density functions estimated from single responses for the best (subject 4) and worst (subject 2) subjects responding on a 1.6 s response-duration schedule. (From Fowler and others, 1972. Copyright 1972 by the American Association for the Advancement of Science.)

twentieth and fortieth reinforcements) from each subject for the same session to spectral analysis. They found first, that subjects showing good temporal regulation also showed highly regular intra-response force oscillations and secondly, relative spectral power value (in bandwidth 2–5 Hz) was positively correlated with response efficiency; that is, the extent and regularity of response-force variations were related to quality of temporal regulation. These results are summarized in Fig. 6.19.

Greenwood (1977) recently completed a study similar to one that had been done in relation to non-temporal determinants of temporally regulated behaviour by Topping and others (1971). These authors trained six pigeons on a DRL schedule using two different response-force requirements. For half of the subjects, the minimum force required to operate the response key was 15 g. For the other half, the force criterion was 45 g.

For both groups, DRL delay was progressively increased from 4 s to a final 20 s. Eventually in support of a proprioceptive interpretation, stabilized performance showed a significantly higher efficiency and lower response rate for the 45 g group. However, observation of these subjects revealed that they frequently emitted light ("collateral") responses of less than 45 g force during inter-response times. Such responding was not recorded and this, therefore, could account for the superior performance of the 45 g group. In order to avoid this kind of difficulty and like Fowler and others (1972), Greenwood (1977) employed a response duration schedule as the "filled" character of the response intervals would preclude infra-criterion responding. Four male adult cats were trained on a two-lever response duration schedule; a response on one lever (B) was reinforced (a fishy-smelling milk reinforcer was used) providing it followed a response on another lever (A) of sufficient duration. During this first phase, phase "35 g A", the response-force requirement for both levers was 35 g. Response-A duration was progressively increased from 0.25 s to a different final

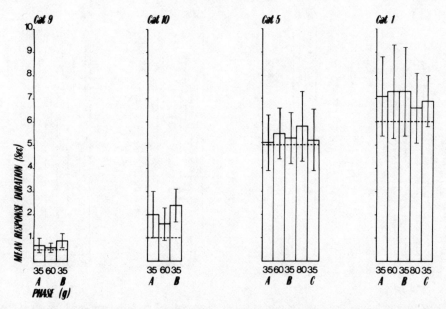

FIG. 6.20. Mean and standard deviation of response A duration, averaged over last ten sessions at different force requirements for cats responding on a two-lever response-duration schedule. Broken lines indicate duration criteria. (From Greenwood, 1977.)

delay for each subject: 0.50 s for cat 9, 1.00 s for cat 10, 5.00 s for cat 5 and 6.00 s for cat 1 (see 5.5 for progression criteria). After a month's stabilization at these delays, all the subjects' performances showed good temporal regulation (Fig. 6.20 and 6.21); mean response-A duration averaged over the last ten sessions was greater than the delay criterion and an average of 57 to 88 per cent of the response-A–response-B sequences were reinforced. During the subsequent phases, phases "60 g", "35 g B", "80 g" and "35 g C" (these last phases involved cats 1 and 5 only), while the response-A duration criterion and the response-B force criterion remained at their previous values, response-A force criterion was alternately increased and decreased, each force requirement being maintained for a dozen sessions or so. The results of the experiment are summarized in Fig. 6.20 and 6.21 and it can be seen that only one subject had results in support of a proprioceptive feedback interpretation (subject 5). For this cat, each increase in response-force criterion produced an increase in average mean response duration and, consequently, an increase in the average proportion of reinforced A-B sequences. For the other three subjects, although generally good performances were maintained, no systematic effects of response-force requirement were observed; average mean response A duration and proportion of reinforced response sequences were sometimes greater, sometimes smaller during phases "60 g" and "80 g" than during preceding and subsequent "35 g" phases. For all subjects, the mean number of A-B response sequences showed no systematic variations over the different force phases. It should be noted that the results of cat 5 are subject to the same criticism that was addressed to the early studies done by Adams and Creamer (1962b) and by Ellis and others (1968); experimental manipulations designed to influence proprioceptive feedback level were made in the same limb that was used for the timing response (response A) and, therefore, the improved performance may have resulted

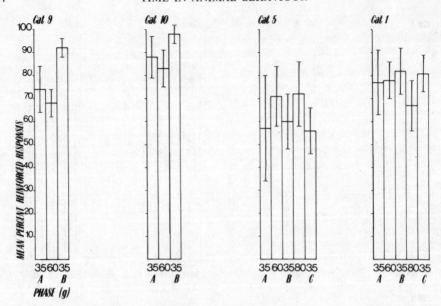

F IG. 6.21. Mean and standard deviation of proportion of reinforced A-B response sequences, averaged over last ten sessions at different force requirements for cats responding on a two-lever response-duration schedule. (From Greenwood, 1977.)

from purely mechanical factors. Finally, it is of interest that all throughout the experiment, cats 1 and 5 presented stereotyped collateral behaviour during lever-A responding. For the most part, this behaviour consisted of a rhythmic "reaching movement" in the direction of the milk-cup (Fig. 6.22). The significance of this kind of behaviour for proprioceptive interpretations will now be considered.

C. Proprioceptive Feedback and Collateral Behaviour

The response-force variations described by Notterman and Mintz (1965) and in the Fowler and others (1972) study may have resulted from the presence of rhythmic movements of the type that accompanied lever-A responding in the Greenwood (1977)

F IG. 6.22. Rhythmic "reaching" movement in direction of milk-cup shown by cats during lever-A responding on a two-lever response-duration schedule. (From Greenwood, 1977.)

study. Indeed, that collateral behaviour showed up on response-force recordings as small variations in response-force (Greenwood, unpublished data). In any case, whether the "collateral" force variations in the other two studies were actually due to the presence of other collateral behaviour or not, that such behaviour is commonly observed on temporal schedules (see Chapter 7) may be an important fact for proprioceptive interpretations.

It will be recalled that in the Schmidt and Christina (1969) "Input Hypotheses", incoming proprioceptive feedback produced by movements made during the timed interval furnishes the necessary stimulus cues. According to Christina, this hypotheses further holds that the subject ".... acquires a movement that occupies a given time (initiated at time t) and learns to execute a timed motor response at time $t + \Delta t$ (Δt a constant) via PFB [proprioceptive feedback] generated by the movement made from time t to $t + \Delta t$" (Christina, 1971, p. 99). Evidence for the hypotheses has been drawn from the studies by Ellis (1969), Ellis and others (1968), Schmidt and Christina (1969) and Christina (1970, 1971) described above as well as from other less recent studies (Goldfarb and Goldstone, 1963; Goldstone and others, 1958) all indicating that the accuracy of a temporal estimation or regulation of an interval is enhanced by the presence of movement during the interval. However, a relation between superior performance and the presence of stereotyped collateral behaviour during inter-response intervals has also been observed on temporal schedules in animal studies and is the source of another, yet similar explanation of temporal regulation and estimation: the "mediating behaviour" interpretation. According to this hypotheses, a behaviour that precedes a reinforced operant response is also, though to a lesser extent, reinforced and as the presence of such behaviour increases IRT length and, therefore, the probability of reinforcement on spaced-responding schedules, the likelihood of the collateral-operant link again appearing also increases. Eventually a whole sequence or "chain" of collateral behaviours, long enough to fill the required delay, is established, each behaviour serving both as secondary reinforcement for the preceding behaviour and discriminative stimulus for the emission of the following collateral (or operant) response. Hearst and others (1964) have suggested that the discriminative stimuli provided by such collateral responding may be primarily proprioceptive in nature. In a slightly different context, Berryman and others (1960) and Hurwitz (1963) have suggested that discrimination of runs of repetitive behaviour— or "counting" behaviour—in rats is based on proprioceptive response-produced stimuli. It can be seen that studies describing collateral behaviour in animals on temporal schedules may actually provide additional support for the proprioceptive interpretations, just as these interpretations may provide a theoretical complement to the hypotheses elaborated in the context of these studies.

D. Feedback and Voluntary Movement

The "Input", "Decay" and "Mediating Behaviour" hypotheses are not the only possible explanations of why movement would enhance timing performance. For example, Richelle (1972; see Chapter 7) has suggested that while movement made during an interval to be timed may provide temporal cues for responding, such movement may primarily provide a mechanism for compensating *inhibition* accumulated during the interval.

One important criticism of the proprioceptive interpretation has come from Jones (1973, 1974) who points out that the anticipatory timing studies involve voluntary movements; that is, they manipulate *outflow* as well as inflow or input, and, therefore, enhanced performance may rely on efference instead of reafference. For instance, efferent signals going from the central nervous system to the muscles used in the responses may have provided the temporal cues. Jones suggests using passive left-arm movements as this would involve afference but eliminate efferent signals.

Actually, the question of whether or not proprioceptive feedback is important for the timing of motor responses is only part of a much larger question: whether or not proprioceptive feedback is important for voluntary movement in general. This "inflow/outflow", "feedback/motor programme", "peripheral/central mechanisms" debate can be traced back to the original experiments of Mott and Sherrington (1895) and Lashley (1917). When Mott and Sherrington deafferented limbs in monkeys and observed that the subjects no longer used their limbs in purposive movement, they concluded that somatic sensation is essential to the performance of voluntary movement. Lashley observed that a patient who had lost all sensation from his inferior limbs as a result of a gun wound could still correctly position his legs, even when blindfolded. He concluded that ". . . the chief mechanism for the control of movement is located in some other body segment than that of the moving organ" (Lashley, 1917, p. 185). Today's proponents of the "Motor Programme Hypotheses", that is the idea that movements are centrally represented and can be executed in absence of peripheral feedback (Keele, 1968), cite more recent evidence from nerve block and deafferentation studies (see Adams, 1968; Glencross, 1977; Hinde, 1969; Jones, 1974; Taub, 1977 for reviews). For example, in a series of studies (Knapp and others, 1958, 1963; Taub and others, 1965, 1966; Taub and Berman, 1963, 1964), Taub, Berman and co-workers demonstrated that monkeys with one or both forelimbs deafferented, conditioned prior to surgery or naïve, could learn to avoid shock when the avoidance-response consisted of flexing the elbow or grasping with the fingers of the deafferented limb in the presence of a buzzer or brief click. Monkeys sustaining total spinal cord deafferentation also retained the conditioned response. In a further study, Taub and others (1972) succeeded in shaping thumb-forefinger prehension in monkeys deafferented at birth.

Schmidt (1971, 1973) has argued that while these studies show that proprioception is not essential to learning or performance of *spatial* aspects of movements, they do not show that proprioception is not important in temporal aspects of motor responses. However, as Jones (1974) points out, timing is an essential part of even the simplest voluntary movements as they involve the co-ordination of two antagonistic muscle groups. Therefore, it may be that while the motor system incorporates a timing mechanism, this mechanism is not based on proprioceptive feedback. Or, as Adams (1976) has suggested, proprioception may be only one of the possible sources of feedback that can be used in skilled motor behaviour.

6.3. DRUGS AND TIMING

A review of studies concerned with the effects of psychotropic drugs on temporal regulations would have to cover several thousand references, due to the explosive development of psychopharmacological research since the early fifties and to the use made of operant techniques in a vast majority of studies carried out in that area (see, for instance, Iversen and Iversen, 1978; Blackman and Sanger, 1978). This would lead us far beyond the limits of the present monography, and, interesting as it might be for the behavioural pharmacologist, it would not bring very meaningful answers to the kind of questions to which we are addressing ourselves here. Since temporal regulations of behaviour are still very poorly understood, and involve a number of intricated factors, the interpretation of the effects of drugs upon them is by no means straightforward. Many drugs affect many different kinds of behaviours, and among those many kinds of behaviours adjusted to time. And these effects can be attained through very different ways. On the whole, drug studies add to the complexity of the picture rather than simplify it. What the student of biological and psychological time would most eagerly wish to discover, of course, is a drug that would affect selectively one or all types of temporal regulations of behaviour. Elucidating the mechanisms of action of such a miracle drug would bring us closer to the site and mechanisms of the underlying clock(s). Unfortunately, this drug has not been found yet. While there are hundreds of demonstrated changes in temporal regulation due to drugs, in no case can this effect be considered as a primary effect. It is always secondary to some other less specific effect. We shall review a few illustrative examples.

Amphetamine, a psychostimulant, has been very widely studied in behaviour research. Though one would expect a general increasing effect on rate of responding, it is not always observed. Instead, it seems that the effect of amphetamine is, to some extent, a function of the initial rate of responding. Roughly, when the initial rate is low, it is increased by the drug, but when the initial rate is high, a decrease is observed after drug administration. Dews suggested that many aspects of drug action on behaviour could be interpreted in terms of initial rate dependency (*rate dependency hypothesis*) (Dews, 1955; Dews and Wenger, 1977). Whether or not this relationship is universally confirmed need not detain us here, since most temporal schedules induce fairly low or medium rates of responding and therefore the effect of amphetamine is usually an increase of rate. It is accompanied by a disruption of temporal regulation, as can be seen from the cumulative records shown in Fig. 6.23. It must be noted that this effect on temporal regulation is not a necessary correlate of rate increase. Fig. 6.24 shows a typical example where the number of responses is multiplied by a factor of 3.4 while the temporal pattern is fairly well preserved. In the case of atropine, time keeps controlling behaviour.

This is not enough, however, to consider amphetamine as a drug acting on temporal mechanisms proper. For this conclusion to be validated, one would have to prove that the disruption of temporal regulation cannot be accounted for by some less specific effect also found in non-temporal schedules. Segal (1962b) has compared performance of rats under DRL and under Variable-Interval contingencies in a concurrent schedule: responses on one lever were reinforced according to a DRL 16 s, while simultaneously responses on another lever were reinforced according to a VI 1 min schedule. As classically observed, amphetamine shifted the peak of IRT distributions toward lower values, though something was retained of the temporal regulation—IRT distributions on the DRL lever were still clearly distinct from IRT distribution on the VI lever, which showed no evidence of temporal control. The number of responses on the VI lever between two consecutive responses on the DRL lever was unaffected. The increase in rate was more or less proportional in both components. Segal concludes

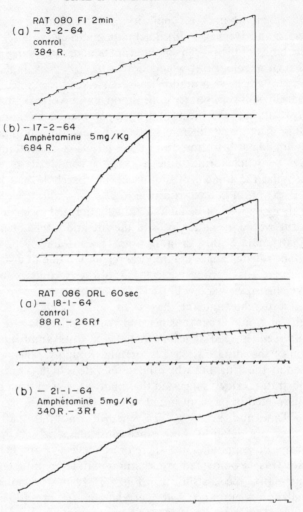

Fig. 6.23. Effect of amphetamine on FI (top) and on DRL (bottom) in rats.

that "the drug's disruption of timing behaviour was not due to a derangement of internal timing mechanisms, nor to interference with the topography of pattern of behaviour. Rather, it might be a secondary result of the accelerated emission of overt behaviour patterns mediating the temporal spacing of DRL bar presses". In other words, the drug would affect temporal regulation via its aspecific action on collateral behaviour (see Chapter 7).

Many other drugs similarly disrupt temporal patterns of behaviour, only as a by-product of some more general aspecific effect. Minor tranquillizers, especially those belonging to the class of benzodiazepines (chlordiazepoxide, diazepam) alter temporal regulation in FI and DRL schedules in rats and, more drastically, in cats (Richelle, 1962; Richelle and others, 1962; Richelle 1969, 1978). Though these drugs are quite different from amphetamines, and other psychostimulants, they produce unexpectedly

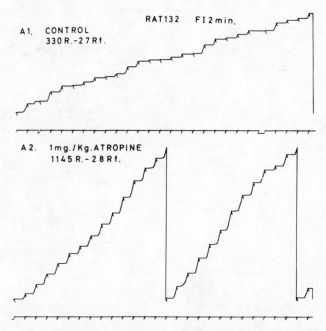

FIG. 6.24. Effect of atropine on FI performance in rats, showing increased overall rate without impairment of temporal regulation (After Fontaine and Richelle, 1967.)

an increase in rate in these schedules. Fig. 6.25 illustrates the latter effect on FI 2 min in one rat. The correlative disruption of temporal regulation is shown in Fig. 6.26 for the same subject plus two other animals. The effect on timing can be thought of as secondary to the effect on rate, which is usually attributed to a general disinhibitory effect of benzodiazepines. In chronic treatment with chlordiazepoxide, the rate-increasing effect is maintained while the alteration of temporal regulation is progressively attenuated; this shows that it is but a transitory by-product of rate increase (Richelle and Djahanguiri, 1964).

Alteration of temporal regulation may also accompany a rate decreasing effect, which is commonly observed, for instance, with major tranquillizers or neuroleptics (chlorpromazine and other phenothiazines). But this effect is no more specific than it is with benzodiazepines: neuroleptics produce deterioration of so many aspects of behaviour that there is no ground to suppose some elective action on temporal mechanisms.

Less attention has been given to possible improving effects of drugs on temporal regulation, which might reveal more interesting to elucidate the problem of internal clocks. Here again, a specific effect is not easy to assert, and an apparently improved temporal regulation may turn out to be but an accidental by-product of some aspecific effect. For example, tremorine, a Parkinsonlike agent, produces a decrease in response rate under FI and DRL schedules, eventually leading to complete suppression of conditioned behaviour. On the way to this extreme effect, there are stages where temporal regulation is better than before drug administration, but no direct action upon timing mechanisms themselves is needed to account for this effect (see Fig. 6.27).

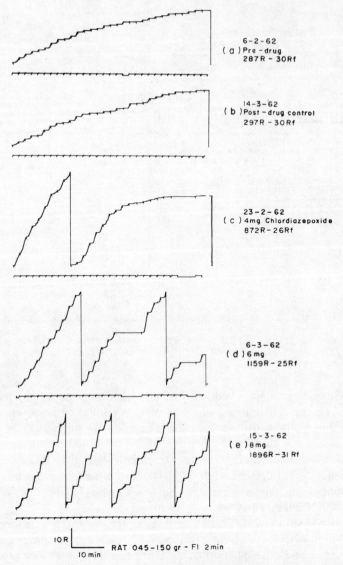

FIG. 6.25. Cumulative record showing the effect of Chlordiazepoxide on FI 2 min performance in one individual rat. Doses are absolute values, that should be multiplied by five to six to obtain an approximation of doses kg of body weight. Number of responses per session (R) and of reinforcements (Rf) are indicated at the right of each curve. (From Richelle and others, 1962. Copyright 1962 by Pergamon Press, Ltd.)

Sulpiride, a compound pertaining to the class of benzamides, produces an improvement of temporal regulation in DRL, both in terms of efficiency ratio and of IRT distribution, as shown in Fig. 6.28. But it also has enhancing effects on performance in situations which do not involve temporal regulation (Fontaine and others, 1975; Richelle, 1979). Thus it seems that improvements as well as disruption of temporal regulation as described until now do not reflect any direct action of drugs on timing mechanisms.

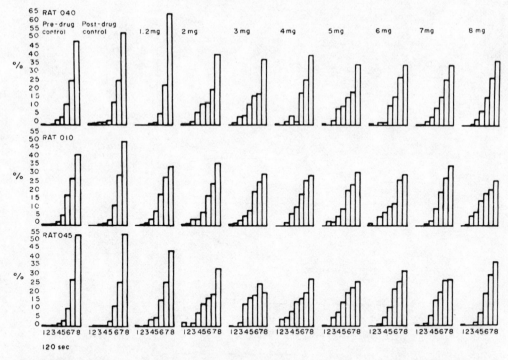

FIG. 6.26. Distribution of responses in eight consecutive segments of the 2 min interval in FI for three rats after administration of increasing doses of Chlordiazepoxide. (From Richelle and others, 1962. Copyright 1962 by Pergamon Press, Ltd.)

This state of affairs is to be compared with what is known of the effect of drugs on biological rhythms. Though many drugs affect the amplitude of biological rhythms, almost none has been found to alter their period—an effect that would suggest a specific action on the internal clock. Deuterium oxide (D_2O) or heavy water might be an exception. It has been shown to lengthen the period of various biological rhythms in plants and animals, unicellular and multicellular. But D_2O has very general and pervasive effects on living organisms, which make the interpretation of its action on rhythms a difficult one (see Pittendrigh and others, 1973). To our knowledge, the action of D_2O on conditioned temporal regulation has never been explored. This would be an interesting step for future research.

6.4. MISCELLANEOUS

BODY TEMPERATURE. The relation between body temperature and time estimation has been studied almost exclusively in man. A short summary of hypotheses developed in that context seems appropriate. In 1922, Bard showed a relationship between the evolution of organic tissues and time estimation. Piéron (1923) suggested that modifications of physiological organic processes induced by changes in temperature were responsible for changes in "mental time".

As noted by Bell (1975, 1977) two kinds of hypotheses have been proposed. One hypothesis states, in a very general and simple way, that changes in body temperature

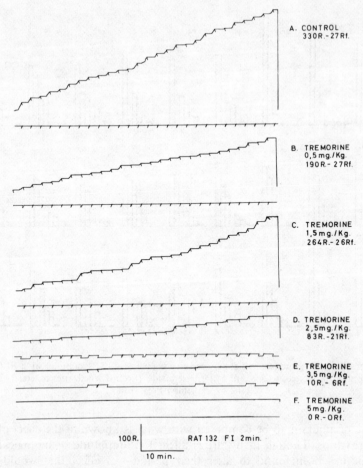

Fig. 6.27. Cumulative records of one rat showing the effect of tremorine on FI 2 min schedule. Number of responses (R) and Reinforcement (Rf) are indicated at the right side of each curve. (From Fontaine and Richelle, 1967. Copyright 1967 by Springer Verlag.)

will influence time estimation. A number of studies have been carried out in this perspective, starting with those by Francois (1927, 1928). A second hypothesis more specifically states that the effects of modifications in body temperature on time estimation are mediated through some central "metabolic pacemaker" or "biochemical internal clock". This hypothesis was formulated by Hoagland (1933, 1935, 1943 and 1951) and further developed by Treisman (1963, 1965) and by Cohen (1964, 1967). It must, of course, take into account the constraints inherent to the biochemical processes resorted to, and it requires, for its confirmation, a high degree of homogeneity of effects in all subjects (see on this point the discussion by Bell, 1975, 1977). An increase in body temperature, associated with a speeding up of central metabolic processes would lead to an acceleration of subjective time. Conversely, a reduction of temperature slows down the metabolic processes and lead to a slowing down of subjective time as well. Studies carried out along these lines, which are of interest to chronobiologists (Palmer, 1976), make use of high temperatures found in pathological states

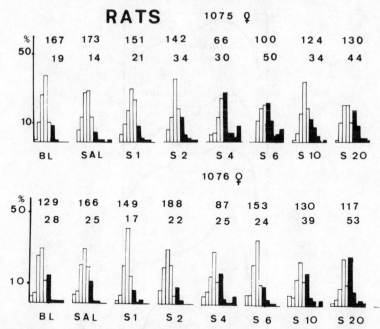

FIG. 6.28. IRT distribution under DRL 20 s in two rats under various doses of Sulpiride (S), compared with control baseline without drug (BL) and with saline (SAL). The figures above each histogram indicate the total number of responses (first line) and of reinforcements (second line) for the session or group of sessions (in case of Baseline, averaged from ten sessions.)

(fever) (Hoagland, 1933), of normal circadian variations of body temperature (Pfaff, 1968), of various techniques for heating or refrigerating, such as diathermy (Francois, 1927; O'Hanlon and others, 1974), of specially designed clothes such as vapour barrier suits (Fox and others, 1967), of cooling or heating ambient air (Bell and Provins, 1962, 1963; Fox and others, 1967; Lockhart, 1967) or of immersion in hot or cool water (Baddeley, 1966; Bell, 1965, 1975).

On the whole, results tend to support the general hypothesis, but do not clearly validate the specific biochemical hypothesis (Bell, 1975; O'Hanlon and others, 1968). Readers interested in this problem will find in Bell (1977) and in Green and Simpson (1977) an illustrative controversy, showing the methodological traps involved in these kinds of studies. Unquestionable evidence for the existence of a metabolic clock is still to be produced.

Out of the range of disruptive heat stress (see for example Pepler, 1971), acceleration in time estimation or in manipulatory tasks have been observed in man when the body temperature was increased (Fig. 6.29). In animals, the rate of operant responding, of food consumption, and of spontaneous activity may be reduced as a result of temperature increase obtained by overheating of the environment (Barofsky, 1969; Hamilton, 1963; Hamilton and Brobeck, 1964; Stevenson and Rixon, 1957). Under DRL contingencies in rats (Barofsky, 1969), both the rate of responding and the number of reinforcements obtained decrease when the ambient temperature passes from 25°C to 35°C (of course, this increase in ambient T° produces an increase in body T°). This effect depends, however, upon the value of the critical delay. It is very

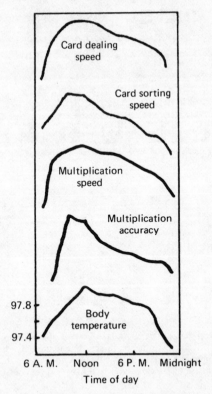

FIG. 6.29. Average daily rhythms in speed and accuracy of performance of simple tasks by one human subject over 20 days. Temperatures taken orally are recorded in F°. (Data of Kleitman, 1933, as plotted by Palmer, 1976. Copyright 1976 by Academic Press, Inc.)

pronounced for DRL 15 to 30 s, but almost negligible for DRL 100 s. Moreover, the IRT distributions at 35°C show an increase in the relative frequency of very long IRTs, together with a reduction in the proportion of IRTs just above criterion. This accounts for the paradoxical deterioration of efficiency (Fig. 6.30). On the other hand, collateral behaviours are changed and a variety of new behaviours (that is not usually observed in operant situations) appear, such as saliva spreading, immobile postures, or attempts to escape. All these are, obviously, potentially useful reactions, in an over-heated environment (Hainsworth, 1967; Robinson and others, 1968). It can be concluded that systematically induced variations of thermic state result in behaviours which are reinforced by their thermoregulating effects. Lengthening of IRTs might be a mere by-product of such behaviours. Therefore, speaking of a specific effect of increased temperature on temporal regulation might not be appropriate.

Manipulating body temperature as the independent variable raises a problem which is familiar in other areas, especially in psychopharmacology, namely the specificity of the effect observed. Does a change in body temperature affect the internal clock proper (whatever and wherever it is) or alter other functions of the organism, which in turn are responsible for the behavioural effects observed?

The particular method used is also a crucial factor in studying such variables as body temperature. It would seem advisable to use discrimination tasks equally feasible

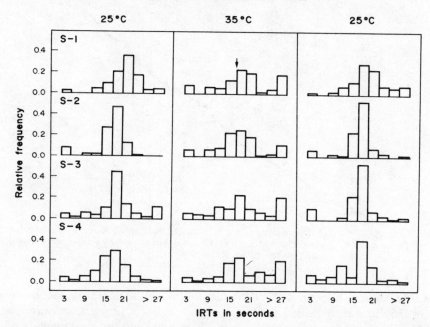

FIG. 6.30. Relative frequency of IRTs in 3 s intervals, plotted for four rats for the session-before, session-of, and session-after high ambient temperature exposure on a DRL 15 s schedule. The arrow marks the contingent IRT. (From Barofsky, 1969. Copyright 1969 by the Society for the Experimental Analysis of Behavior, Inc.)

with humans and with animal subjects, since these tasks require minimal motor energy and do not impose any temporal regulation upon the response itself. This would discard possible interferences between thermo-regulating mechanisms and the temporal regulations of responses as required by the contingencies, as was the case in Barofsky's rats. The studies of Lockhart (1971) and by O'Hanlon and others (1974) for example are along these lines. Human subjects were tested for threshold for the perception of the succession of dichoptic flashes and for critical fusion frequency threshold (CFF). Temporal accuracy was improved under hyperthermia. This would suggest a relationship not only between central parameter and perception of time passing but also between critical parameter and perception of simultaneity.

As far as we know, there seems to be only one recent study with animals that satisfy to the methodological requirements stated above. It has used FI contingencies, which do not require temporal regulation of responding though they do eventually induce such regulation— and in which the number of reinforcements is not related to the rate of responding (except for extreme cases of erratically low responding). The species used by Rozin (1965) was the goldfish, trained under FI 60 s in water maintained at 30°C (Fig. 6.31). A so-called ABA experimental design made it possible to dissociate the effects of temperature on the rate of responding and on the temporal regulation. After control sessions at 30°C, the water temperature was changed to 20°C, and then again brought to 30°C. The rate of responding decreased in lower temperature (phase B, 20°C). The three subjects emitted respectively 1.86, 1.87 and 2.27 times more responses at 30°C than at 20°C. The rate of responding seemed to be a direct function of the metabolic rate in these poikilothermic organisms. However, the quality of the

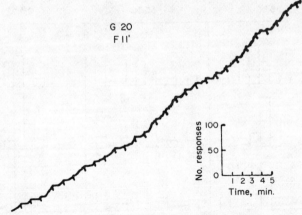

FIG. 6.31. Cumulative record of one session from a well-trained goldfish (G 20) on a FI 1 min schedule. Downward pips indicate reinforcements. (From Rozin, 1965. Copyright 1965 by the American Association for the Advancement of Science.)

temporal regulation, as measured by the distribution of responses in successive segments of the interval, remained unaltered when the water temperature was lowered (Fig. 6.32). These results suggest that in goldfish spontaneous temporal regulation does not depend upon metabolic processes affected by temperature changes. An extension of a similar procedure to homeothermic organisms recommends itself.

ACCELERATION. One study has been devoted to the effects of acceleration on DRL performance. As discussed under 6.2.2, the authors (Beasley and Seldeen, 1965) suggested that acceleration would affect DRL performance through a modification of proprioceptive feedback of the responses. However, as it has been pointed out, it seems that the effect of acceleration on DRL performance is only a by-product of a more general, non-specific effect. Conclusions from one single rat study on acceleration would therefore seem premature.

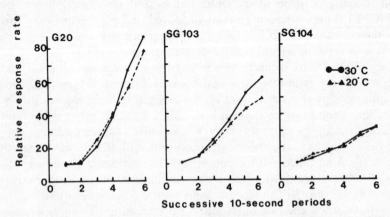

FIG. 6.32. Relative response rates in successive 10 s segments of a FI 1 min schedule for three goldfish at 30°C and 20°C. The number of responses in the first 10 s segment at each temperature was arbitrarily set at ten, and the other values were adjusted. (From Rozin, 1965. Copyright 1965 by the American Association for the Advancement of Science.)

HYPOXIA. One study by Beard and Wertheim (1966) reports significant reductions in operant response rates on Fixed-Interval (FI) and DRL schedules in white rats. The subjects were exposed to hemic hypoxia (CO increased to 100, 250, 500, 750 or 1000 ppm). The reduction in response rate was directly related to carbon monoxide concentration. These results preclude, however, any conclusion in terms of an impairment of temporal regulation proper. Indeed, carbon monoxide depresses response rates in other schedules as well (VI) and induces motivational changes (Beard and Wertheim, 1966; Seitz and Keller, 1940).

GONADS. Female rats have been shown to acquire more rapidly DRL behaviour than do male subjects. This superiority in acquisition speed can be abolished with ovariectomy. On the contrary, gonadectomy at the first day of life remains without effect on DRL acquisition in male rats. Endogenous androgens seem therefore to be relatively unimportant in the development of sex differences in DRL schedule learning (Beatty, 1973; Beatty and others, 1975a, 1975b).

BLOOD CHEMISTRY, LEUCOCYTES AND TISSUES. Persinger and others (1978) have evaluated with rats the incidence of abrupt shift from DRL 6 s to DRL 12 s on blood chemistry, leucocytes, and tissues. Significant changes ($P \leqslant 0.05$) are induced by schedule shift in alkaline phosphatase, SGOT and relative blood lymphocyte numbers, as compared to control rats maintained on DRL 6 s. However, the main differences are described between conditioned animals maintained at 80 per cent of free feeding weight and *ad-lib*. controls. This small effect of DRL schedule shift contrasts with heavier alterations formerly described after FR schedule shift (Valliant, 1978).

VII. *Collateral Behaviour*

7.1. EXPERIMENTAL DATA

Wilson and Keller (1953) were the first to note the presence of stereotyped *collateral behaviour*, that is, behaviour other than the explicitly reinforced operant response, in rats responding on a DRL schedule. Since then, collateral behaviour has been observed on most temporal schedules and in a variety of laboratory animal species. In a typical example, rats were observed "nibbling" their tails on a grid-bar of the cage floor on a DRL schedule. Laties and others (1965), who reported their systematic observation of these behaviours, have noted their frequency and duration. Fig. 7.1 shows the percentage of reinforced lever-presses for one individual subject on DRL 22 s as a function of the presence and duration of tail nibbling in the inter-response time. In this particular case, the longer the tail-nibbling activity, the more likely the following lever-press was reinforced.

Rats have also been observed running in activity wheels on FI (Roper, 1978; Skinner and Morse, 1957), FT (King, 1974), VI (Levitsky and Collier, 1968), and DRL (Lince, 1976) schedules, lever-pressing on stimulus duration discrimination schedules (Snapper and others, 1969), and even consistently mounting and copulating with female rats on periodic shock schedules (Barfield and Sachs, 1968; Caggiula and Eibergen, 1969). Rats develop *polydipsia*—or excessive water consumption—on a variety of schedules such as FI (Fig. 7.2: Falk, 1969; Keehn and Riusech, 1979; Roper, 1978; Segal, 1969), DRL (Deadwyler and Segal, 1965; Pouthas, 1974; Pouthas and Cave, 1972; Segal and Deadwyler, 1964, 1965a, 1965b; Segal and Holloway, 1963), and fixed-time (Hsiao and Lloyd, 1969; Segal, 1969; Segal and others, 1965; Singer and others, 1974). They develop also eating on a fixed-time schedule of water presentation (Wetherington and Brownstein, 1979). Consistent grooming or exploratory-like behaviour has been displayed by rats on FMI (Mechner and Latranyi, 1963) and double-avoidance (Blancheteau, 1967a) schedules.

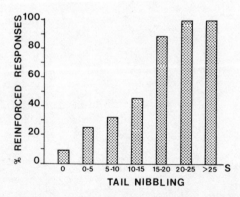

FIG. 7.1. Percentage of reinforced responses as a function of the amount (duration) of tail-nibbling in rats under DRL 22 s. (After numerical data from Laties and others, 1965.)

Pigeons have been known to pace or circle around the experimental chamber (Hearst and others, 1864; Holz and others, 1963; Kramer, 1968), to peck extra keys (Lydersen and Perkins, 1972; Nevin and Berryman, 1963; Schwartz and Williams, 1971), or to peck a lone response key with subcriterion force (Topping and others, 1971) during inter-response intervals on Fixed-Minimum-Interval (FMI) and DRL schedules. They have shown *polydipsia* on Fixed- and Variable-Interval schedules (Shanab and Peterson, 1969). Even attack behaviour has been displayed by pigeons on Fixed-Interval (Cherek and others, 1973; Richards and Rilling, 1972), DRL, (Knutson and Kleinknecht, 1970) and Fixed-Time (FT) schedules (Flory, 1969).

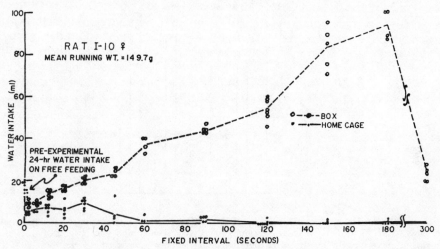

FIG. 7.2. Polydipsia as a function of length of food Fixed-Interval schedule. Food limited to 180 pellets (45 mg each) for all sessions. (From Falk, 1966. Copyright 1966 by the Society for the Experimental Analysis of Behavior, Inc.)

Monkeys have also shown stereotyped collateral behaviour on FT (Villarreal, 1967), DRL and Sidman avoidance (Hodos and others, 1962) as have cats responding on DRL and response-duration schedules (Greenwood, 1978) and in discrete-trial, temporal discrimination situations (Rosenkilde and Divac, 1976). Not only do laboratory animals consistently show collateral behaviour on temporal schedules, human subjects also display reliable patterns of collateral behaviour while responding on DRL (Bruner and Revusky, 1961; Carter and MacGrady, 1966; Kapostins, 1963; Laties and Weiss, 1962; Randolph, 1965; Stein, 1977; Stein and Flanagan, 1974; Stein and Landis, 1973) and VI (Catania and Cutts, 1963) schedules. Although Wilson and Keller's rats displayed heterogeneous sequences of collateral behaviour, most studies describe homogeneous, repetitive behaviour. Most importantly, the presence of collateral behaviour has been associated with superior temporal performance in a number of studies:

(a) Where consistent collateral behaviour is observed, temporal responding is often more efficient. This was the case, as we have seen above (Fig. 7.1), for rats in the study by Laties and others (1965). Other examples can be found in Blancheteau (1967a), Holz and others (1963), Kramer, (1968), Shwartz and Williams (1971), Slonaker and Hothersall (1972), Topping and others (1971). In an unpublished study, Lincé (1976) trained rats on a two-lever DRL schedule: a response on lever A initiated the critical 8 s delay, after which a response on lever B was reinforced if it occurred within a 5 s limited hold. The subjects had free access to an activity wheel. For those subjects who used this facility, the presence and the number of runs in each Response-A-Response-B Time Interval were recorded. A-B sequences with and without wheel activity were treated separately, and it was shown that efficiency was better for sequences in which the animals had been running in the wheel (Fig. 7.3).

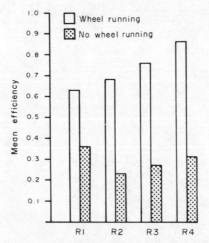

FIG. 7.3. Collateral wheel-running in rat on a two-lever DRL 8 s LH 5 s. Ordinate: mean efficiency ratio. White blocks: data for IRTs in which wheel running occurred; shaded blocks: data for IRTs in which no wheel-running occurred. Results are from four individual animals (numbered R 1, R 2, R 3, R 4). (After Lincé, 1976, unpublished data.)

(b) When collateral behaviour is disrupted as a result of drug-administration or alterations in the physical support of the behaviour, the temporal characteristics of responding are also often disrupted. Laties and others (1969) prevented rats from nibbling the grid-bars by placing a panel over the grid-floor. Consequently, the performance deteriorated as illustrated in Fig. 7.4. In the study already mentioned, Lince (1976) maintained free access to the activity wheel but prevented its rotation. This resulted in a decrease in efficiency. Other examples are described in Deadwyler and Segal (1965), Hodos and others (1962), Laties and others, (1964), Segal (1962b), Segal and Deadwyler (1964, 1965a, 1965b), Slonaker and Hothersall (1972), and Stein and Landis (1973).

(c) Extinction of collateral behaviour usually accompanies extinction of temporal responding (Cherek, and others, 1973; Laties and others, 1965, 1969; Levitsky and Collier, 1968; Segal and Deadwyler, 1965a, 1965b; Villarreal, 1967; Wilson and Keller, 1953).

It seems then, that collateral behaviour may play a functional role in temporal regulation and discrimination, eventually constituting some sort of "behavioural clock". In order to further explore this possibility, much research has been concentrated on gaining better control over collateral behaviour, thus making recording and manipulating them less difficult.

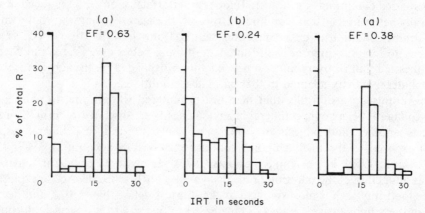

FIG. 7.4. IRT distributions and efficiency (EF) in a rat under DRL 18 s when collateral behaviour (nibbling grid-bar) is prevented (b) as compared with normal condition (a) before and after (b). (Redrawn, after Laties and others, 1969. Copyright 1969 by the Society for the Experimental Analysis of Behavior, Inc.)

In many cases, the development of a specific behaviour has been *favoured* by providing the subject with supplementary response levers or keys for pressing (Bruner and Revusky, 1961; Davis and Wheeler, 1967; Zuriff, 1969), blocks of wood for nibbling (Laties and others, 1969; Roper, 1978; Slonaker and Hothersall, 1972), water bottles for licking and drinking (Roper, 1978; Smith and Clark, 1974, 1975), or wheels for running (Lince, 1976; Roper, 1978; Smith and Clark, 1974, 1975). The study by Laties and co-workers (1969) already cited is an interesting example of inducing collateral behaviour by providing material support for it. Rats which did not exhibit consistent collateral behaviour "spontaneously" were given a piece of pine wood. They eventually began nibbling it actively, and their performances became more efficient: the number of reinforcements obtained increased, and the IRT distributions shifted towards the reinforced value. A co-variation was observed between the number of reinforcements and the amount of wood nibbled.

However, as Sidman (1960) pointed out, the best way to control collateral behaviour would be to make it an operant, that is to explicitly reinforce it, and this has been done in a number of studies (Lince, 1976; McMillan, 1969; Segal-Rechtschaffen, 1963).

Another approach, as Kramer and Rilling (1970) have suggested, is to try to *eliminate* collateral behaviour altogether and then see if good temporal performance can still be obtained. While several studies have attempted to reduce collateral behaviour by employing a locomotor-limiting schedule like the response-duration schedule (Greenwood, 1977; LeCao and Lince, 1974; Platt and others, 1973) or a smaller-than-standard-size conditioning chamber (Skuban and Richardson, 1975; Staddon, 1977), other studies have taken more drastic measures and employed physical-restraint apparatuses (Frank and Staddon, 1974; Glazer and Singh, 1971; Richardson and Loughead, 1974a). In one of these, Frank and Staddon (1974) exposed pigeons to a FI 2 min and a DRL 5 s or 10 s schedule under two spatial conditions. In the "restrained" condition, the pigeons were physically confined in a small plastic tube that allowed for only the head and neck movements necessary for key-peck operant responding and very small body movements. In the "free" condition, the pigeons were free to move about in the normal-sized conditioning chamber. Temporal performance on all three schedules was generally better in the "free" condition. However, the major finding was that a change in restraint condition in either direction, that is, from "free" to "restrained" or "restrained" to "free", temporarily disrupted the timing behaviour. Frank and Staddon hypothesized that recovery under a new condition depended on the subjects' developing collateral activity adapted to that particular condition.

This points up a difficulty that may be encountered when trying to gain control over collateral behaviour: collateral behaviour has been observed to "drift" from one form to another, and subjects may abandon even a pre-trained collateral response when exposed to the final temporal contingencies (Laties and others, 1969; Lince, 1967; Zuriff, 1969). For example, Pouthas (1979) has observed collateral activities in rats as a function of the duration of the delay under DRL schedule. Active exploration of the food tray area is most frequent at the shortest delay, that is 10 s, and becomes less and less frequent as the delay is increased to 30 or 40 s. Conversely, locomotor activity along the cage walls is infrequent at 10 s and amounts to 80 per cent of the total number of collateral behaviours at 30 s. Sitting quietly ("half-sleeping") becomes the main collateral (in)activity at the longest (40 s) delay for those rats who adjust to it (Fig. 7.5).

A related problem is the possibility that *covert* collateral behaviour develops at less observable levels of the organism (Glazer and Singh, 1971; Hodos and others, 1962; Saslow, 1968). This might explain why collateral behaviour was not observed in a number of studies that employed temporal schedules of reinforcement (Anger, 1963; Belleville and others, 1963; Catania, 1970; Kelleher and others, 1959; Reynolds and Catania, 1962; Richardson and Loughead, 1974b; Saslow, 1968; Stubbs, 1968). We shall further discuss these problems in the next section.

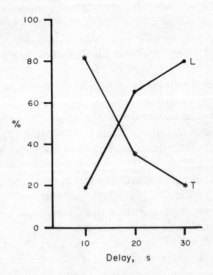

FIG. 7.5. Relative frequency of two types of collateral behaviour as a function of the DRL value. T = activity around food-tray; L = locomotor activity. (Redrawn, after Pouthas, 1979. Copyright 1979 by Presses Universitaires de France.)

7.2. ORIGIN AND FUNCTION OF COLLATERAL
BEHAVIOUR

We shall now turn to the questions of the origin of collateral behaviours and of their function in temporal regulation. When collateral activities were first observed in the fifties and early sixties, experimenters, concentrating on operant responses controlled by the experimenter-defined contingencies, were not ready to consider classes of behaviours that they had not explicitly defined as targets for reinforcement. When some of them, who had not completely abandoned direct observation in spite of increasing automatization of conditioning experiments, came across these unforeseen phenomena, they tended to force them into the familiar framework of operant behaviour theory. The classical interpretation is that collateral behaviours appear for any one of several phylogenetic and ontogenetic "reasons", such as conditioning history and motivational factors (see Segal, 1972, for a more complete list), and then are maintained "superstitiously" through *adventitious reinforcement* (Hernnstein, 1966; Skinner, 1948). Skinner (1948) seems to have originally observed the phenomenon in an experiment in which eight hungry pigeons were given food every 15 s independently of their behaviour at that moment (FT 15 s schedule). Six of the pigeons developed stereotyped repetitive behaviour; cage circling, pecking towards the cage floor, etc. According to Skinner's interpretation, these behaviours came to be strengthened through their occasionally being followed by food and therefore accidentally reinforced. Like a superstitious person who insists on carrying a particular "good luck charm" because something good once happened to him when he had it with him and nothing too bad has happened since, the pigeons' repeated behaviours that had fortuitously met with reinforcement and as the behaviours were emitted more frequently, they were reinforced more frequently and further strengthened. In spaced-responding situations, the interpretation is usually applied in the following way: a behaviour that occurs during an IRT and precedes a reinforced operant response is also superstitiously reinforced. As the presence of such behaviour increases IRT length and therefore the probability of reinforcement, the collateral-operant link is further strengthened (e.g. Deadwyler and Segal, 1965; Holz and others, 1963; Wilson and Keller, 1953). Thus, a collateral response, like an operant response, is controlled by a response-reinforcer relation and the law of effect maintains collateral behaviour in the same way it maintains operant behaviour. Often associated with this view was the idea that collateral behaviours provide the subject with all the necessary cues for timing. All an animal would have to do in order to space its lever responses correctly in a DRL schedule would be to discriminate a given point in a chain of activities. Collateral behaviours, according to this hypothesis, are *mediating* behaviours dispensing with timing mechanisms altogether.

A number of experimental facts and theoretical analyses make this hypothesis—in both its aspects—no longer tenable. Objections to the interpretation of collateral behaviour in terms of adventitious reinforcement come from the new look at the "superstition" phenomenon initiated by Staddon and Simmelhag (1971) in a seminal paper, and more generally from the current trend to reconsider the status of the explicit operant in its relation with other classes of experimentally induced or "natural" behaviours. Employing the same type of schedule Skinner employed in his superstition experiment, that is fixed- or variable-time free food schedule, Staddon and Simmelhag, and a number of authors that followed (Anderson and Shettleworth, 1977;

Reberg and others, 1977, 1978, for example) have shown that behaviours appear and are maintained for reasons other than response-reinforcer linking. Some behaviours, such as pecking in pigeons and exploration of a food area in rats, seem to occur more frequently as the moment of reinforcement approaches and are undoubtedly related to anticipation of food. Because of their tendency to occur toward the end of the inter-reinforcement interval, these behaviours have been called *terminal* activities. As can be seen from the example in Fig. 7.6, they are clearly distinct from the kinds of behaviour that are observed after food reinforcement and during the first part of the interval. These have been called *interim* activities. Running, drinking, attacking, preening, and wing-flapping are now classical examples. Of course, the probability that one or another behaviour will develop depends upon the material support offered by the situation: wheel-running supposes a wheel to run in, drinking the availability of water, and so on. Much attention has been given to excessive drinking since Falk (1961) first described *polydipsia* in rats under intermittent food schedules and extensively explored the variables involved in this particular kind of schedule-induced *adjunctive* behaviour (Falk, 1971, 1977).

If not all classes of behaviour are likely to occur towards the time when food is delivered, it could hardly be argued that terminal activities are strengthened by a process of adventitious reinforcement *sensu stricto*. Interim activities are still less likely to be maintained by adventitious reinforcement, since they never occur in close tem-

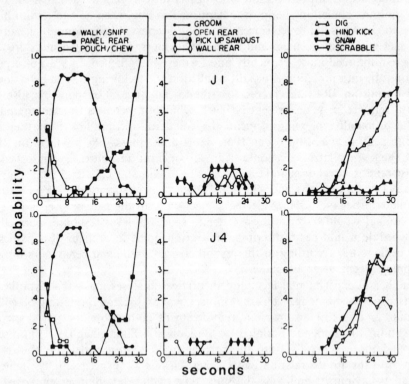

FIG. 7.6. Probability of a given Activity Pattern in 2 s units of interfood intervals (Fixed Time 30 s) for two individual hamsters. (From Anderson and Shettleworth, 1977. Copyright 1977 by the Society for the Experimental Analysis of Behavior, Inc.)

poral contiguity with the reinforcer. Moreover, as we have seen above, a drift from one kind of interim activity to another kind after a number of sessions is often observed. Even if one resorts to some chaining process to account for reinforcement of a given behaviour by a distant reinforcing event, how can one explain that a given behaviour maintained for hours suddenly gives place to some other activity that was not previously associated with the reinforcer?

Another strong, though indirect argument against the superstition hypothesis is offered by studies of *autoshaping* (Brown and Jenkins, 1968). The discovery that key-pecking in pigeons is not, as was previously believed, a pure operant response shaped and maintained by its pairing with a reinforcer, but is elicited by food-related visual stimuli (and eventually maintained even when it results in the omission of food), has challenged Skinner's interpretation. If the presence of assumedly operant key-pecks is better explained by resorting to some other process than reinforcement contingencies, it would appear rather illogical to insist that other key-pecks not explicitly controlled by such contingencies be accounted for by the mechanism of operant conditioning (for a more detailed discussion of autoshaping and its bearing on the superstition hypotheses, see Staddon, 1977 and Schwartz and Gamzu, 1977).

Terminal and interim activities are of course also observed when the reinforcement is contingent upon a specified operant response, under Fixed-Interval rather than Fixed-Time schedules (Fig. 7.7). In this case, terminal activities parallel the operant responses, and indeed the latter present themselves as a particular class of terminal behaviour (Anderson and Shettleworth, 1977; Roper, 1978). Interim activities fill the pause, as they eventually fill the interval between responses in a DRL schedule. As interim and terminal activities are less clearly distinguishable in situations such as DRL, response latency, response duration, and the like, we shall retain the general expression *collateral behaviours*.

Having disposed of the superstition hypothesis with regard to the origin of collateral behaviours, let us turn now to their possible mediating function in temporal regulation. In its strongest version, the hypothesis states that subjects discriminate or

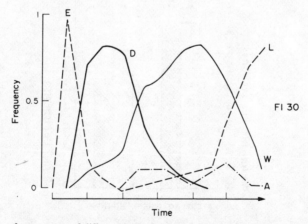

FIG. 7.7. Frequency of occurrence of different behaviours as a function of time since last reinforcement for one rat on FI 30 s schedule. E = eating food pellet, D = drinking from the water bottle, W = in running wheel, L = lever contact, A = general activity (sniffing, etc.). (Redrawn after Roper, 1978. Copyright 1978 by the Society for the Experimental Analysis of Behavior, Inc.)

react to their own collateral behaviour in such a way that they need not really estimate time. A number of factual and theoretical arguments, however, indicate that collateral behaviours are not substitutes for timing mechanisms.

(a) If an animal reacts to successive links of a chain of collateral behaviours, each being the stimulus for the next one and the last being the stimulus for the operant response(s), one would predict that preventing the chain, or part of it, from occurring would result in non-occurrence of operant behaviour. As the examples cited above have shown, this is not what happens. When collateral behaviour is prevented, operant responding simply occurs earlier, and the temporal regulation is impaired.

(b) Discriminating a terminal link in a chain of behaviour is admittedly a possible task for animals. However, in many cases, collateral behaviours do not consist of a chain of activities, but of the repetition of one stereotyped activity. If the subject need not be able to estimate time in that situation, it must nevertheless be able to count, which may not pose simpler problems as noted by Richelle (1967, 1968) and by Staddon (1977). Discrimination of spent muscular energy or of fatigue state could possibly be involved, but this has not been demonstrated.

(c) Collateral behaviours, though very frequently observed, are by no means *always* observed. Some animals are seen to sit quietly in the cage during the pause of a FI schedule or the IRT of a DRL. Some animals which eventually exhibit the best adjustment to long delays replace locomotor activity with apparently half-sleeping rest as the delay is increased (Pouthas, 1979; Richardson and Loughead, 1974b). Pavlov (1927) noted that animals easily conditioned to time were of the quiet type, and are often observed sleeping during intervals between unconditioned stimuli. If, as a rule, animals which produce collateral behaviour show better performance with than without such behaviour, animals who do not produce it are not necessarily poorer performers. An example is shown in Fig. 7.8, based on data from Lincé (1976): rats who

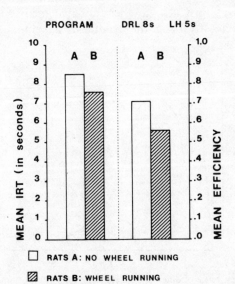

□ RATS A: NO WHEEL RUNNING

▨ RATS B: WHEEL RUNNING

FIG. 7.8. Compared performances of rats exhibiting and rats *not* exhibiting collateral behaviour in a two-lever DRL 8 s LH 5 schedule. Left: mean IRT; right: mean efficiency ratio. (After Lince, 1976, unpublished data.)

run in the activity wheel do not show better IRTs nor better efficiency ratios than rats who do not run in the wheel. It would seem that collateral behaviour helps those subjects who make use of them, but it is not necessary for all individuals.

(d) Making a specified collateral behaviour an additional condition for reinforcement may or may not improve performance. For example, in one manipulation, Lincé (1976) made the behaviour "running in the wheel" part of the required operant: the sequence Response A–Response B was reinforced only when the critical delay was met and, in addition, at least one complete rotation of the wheel had been recorded. The number of runs remained unchanged for two (out of four) subjects that had previously exhibited wheel-running as collateral behaviour. The efficiency ratio remained unchanged for one of them and paradoxically decreased for the other. One subject that had previously used the wheel, employed it more often under the new contingencies, but without benefit with regard to efficiency. The same was true for the fourth subject, which had not previously employed the wheel.

(e) Collateral behaviours do not occur only when they would be expected to occur if they were really mediating behaviour, that is when temporal regulation proper takes place. In the two-lever situation used by Lincé, the interval between Response A and Response B must be estimated, but the animal is free to emit a response on A, starting the delay whenever it chooses to do so. Wheel running occurs during this neutral Response-B–Response-A interval as well as during the critical Response-A–Response-B delay.

(f) Another difficulty with assigning the role of mediating behaviour to collateral activities is their variability both in terms of sequential order and in terms of duration (Staddon, 1977). How can an animal estimate time with the help of so unreliable a clock?

(g) Restricting collateral behaviour through physical constraint generally results, as we have seen, in impaired temporal regulation. If collateral behaviours are necessary mediators for timing, it should be impossible for an animal to learn a temporal discrimination while it is submitted to severe physical constraint which drastically limits, if not eliminates, overt collateral activities. Extremely suggestive data reported by Glazer and Singh (1971) indicate that this is not so. In one experiment, they exposed rats to DRL 10 s contingencies under three different conditions of body restriction: complete restriction, partial restriction, no restriction (plus a stress-control condition). Subjects under complete restriction failed to learn to space their responses. In a second experiment, they trained rats on DRL 10 s either with complete restriction or without restriction over twenty sessions and then switched them to the opposite condition for the next sessions. Rats initially trained under complete restriction, and who showed no evidence of temporal regulation in that condition, exhibited very rapid adjustment when transferred to the non-restricted condition (Fig. 7.9). Within three sessions they had caught up with control animals exposed to the contingencies without physical constraint throughout the entire series of sessions. This is actually a case of latent learning. It suggests that some behavioural clock can be operating while the performance gives no sign of temporal adjustment. There might be more here than a far-fetched analogy with a classical observation in chronobiology: the measured expression of a biological rhythm can be suppressed and still the basic mechanism— the work of the clock—may continue unaltered.

(h) In the absence of overt collateral behaviours, some authors have resorted to

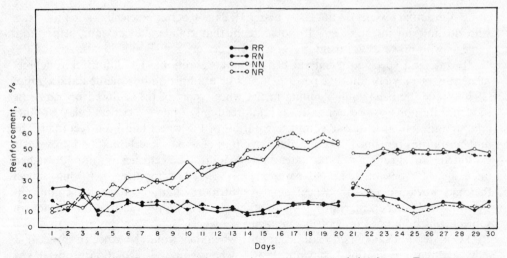

FIG. 7.9. Latent learning of temporal regulation under severe space restriction in rats. Two groups were trained under restriction (R) until day 20, and two groups under normal no-restriction (N); the condition was reversed for the two groups from day 21. (From Glazer and Singh, 1971. Copyright 1971 by the American Psychological Association. Reprinted by permission.)

hypothetical covert mediating behaviour. Glazer and Singh (1971) made this suggestion in order to account for their results, only to fall into inextricable difficulties. In the only study where the problem of covert cues has been attacked with the tool of physiology, Saslow (1968) concluded to the central control of timing. It might be important to note, however, that he was working on very short delays (not exceeding 600 msec).

(i) Other arguments against a mediating function of collateral behaviours in temporal regulation can be drawn from the analysis of the chaining hypothesis as applied to operant responses in Fixed-Interval schedule (see 3.1.1., and Dews, 1962, 1965b, 1970; Hienz and Eckerman 1974; Wall, 1965). Suppressing operant responses at a given point in the scallop of a Fixed-Interval does not alter the temporal pattern; time, not the previous operant response, controls at any given moment the rate of responding during the active phase of FI. If timing is not based on chaining of fairly regular responses, it would be surprising that it would rely upon the mediation of the far less regular interim activities.

It can be concluded that collateral behaviours do not, in any strict sense, *mediate* time estimation. They play at best an auxiliary function in temporal regulation. How can we understand this function? Current interpretation of adjunctive or interim activities resort either to the notion of *displacement* familiar to ethologist (Falk, 1971) or to more or less sophisticated models where various motivational states, incentives, or stimulations alternate to produce typical periodic sequences of behaviours (Staddon, 1977). Inherent in these interpretations, though usually not explicitly stated, is the notion of active inhibitory mechanisms. Displacement activities as defined by ethologists develop when an oriented behaviour is for some reason prevented. In Staddon's or McFarland's version (McFarland, 1974; McFarland and Silby, 1975), interim activities *inhibit* operant responding.

We suggest that collateral behaviours are induced by temporal schedules as activities compensating for the inhibition involved at various degrees in all kinds of temporal regulations. According to this view, it is not correct to say that collateral behaviour inhibits operant responding. The terms must be inverted: operant responding is inhibited by the periodic structure or the temporal requirements of the situation, and collateral behaviours develop as by-products of this inhibitory state. We shall further discuss the interpretation of collateral behaviour in the frame of an inhibitory hypothesis in the final chapter.

VIII. *Temporal Information: Temporal Regulation and External Cues*

In the preceding chapters, we have been looking for the possible mechanisms of temporal regulation of behaviour in the central nervous system, in peripheral visceral or proprioceptive cues, and in overt collateral behaviour. In the present chapter, we shall address ourselves to the role of external cues. As is well known in chronobiology, endogenous mechanisms for time adaptation must be admitted only after external determinants have been ruled out. The debate is not definitively resolved, and some experimenters such as Brown (1976) continue searching for unsuspected external events that might account for the control of biological rhythms. We are facing a similar problem in the study of behavioural regulations. Before assigning the underlying machinery to the organism, we must carefully examine all possible external cues, try to characterize their part in observed performance, and evaluate what is left to some time-structuring device within the subject. We shall successively review the experiments in which external stimuli are explicitly introduced as time-indicators—or *external clocks*—the role of the periodicity of the reinforcing event, or in other words, the periodic structure of the contingencies, and finally, the function of stimuli associated with primary reinforcement or with operant responses.

8.1. EXTERNAL CLOCKS

External stimuli varying as a function of time have been introduced in Fixed-Interval Schedules (Auge, 1977; Caplan and others, 1973; Farmer and Schoenfeld, 1966; Ferster and Zimmerman, 1963; Kelleher and Fry, 1962; Kendall, 1972; Laties and Weiss, 1966; Segal, 1962a; Skinner, 1938; Squires and others, 1975), in DRL schedules (Macar, 1969, 1971b; Reynolds and Limpo, 1968), in avoidance schedules (Findley, 1963; Grabowski and Thompson, 1972; Thomas, 1965), in reinforcement of response latencies (Uramoto, 1973), and in autoshaping situations (Ricci, 1973). These external clocks are made either independent of or dependent upon the subject's responses. In the latter case, a further distinction must be drawn between those clocks that are contingent upon the operant response proper, or *feedback clocks*, and those that are contingent upon some other response, usually an *observing response* (*optional clocks*). Feedback clocks will be discussed under heading 8.3, at the end of this chapter.

8.1.1. Types of External Clocks

A variety of external clocks have been used by researchers. In some cases, the stimulus changes *continuously* through time in a systematic manner: for example, a pointer moves regularly across the dial of a voltmeter, a luminous patch moves progressively in a slot, the intensity of a light or a sound increases progressively (Ferster and Skinner, 1957; Ferster and Zimmerman, 1963). In other cases, the clock is *discontinuous.* For example, different stimuli such as geometric patterns are presented in orderly succession, or one stimulus dimension, such as colour or intensity, is changed by discrete steps at regular intervals (Auge, 1977; Caplan and others, 1973; Hendry and Dillow, 1966; Kendall, 1972; Laties and Weiss, 1966; Segal, 1962a; Thomas, 1965) or identical stimuli are repeated, separated by empty intervals of defined duration (Macar, 1969; Uramoto, 1973). Another possibly important distinction is between *moving* and *non-moving* stimuli. Changes of colour or luminous intensity may take place in a fixed spot in space, while in displacement of a luminous slot or of a pointer, time is in some way converted into movement. Stimuli are either *localized* (as visual clocks, moving or not-moving) or *diffused* (as auditory clocks). Of course, a distinction can also be made with regard to the sensory modality concerned.

8.1.2. Control Exerted by External Clocks

As might be expected, external clocks improve the precision of temporal regulation, and the efficiency of performance. In Fixed-Interval schedules, the duration of the pause is increased, the relative frequency of responses emitted early in the interval drops drastically and the frequency of responses produced in the last part of the interval increases abruptly (Boren and Gollub, 1972; Kendall, 1972; Segal, 1962a). An example of these effects is shown in Fig. 8.1.

In a DRL 35 s schedule, pigeons' performances improve considerably with an added clock (Reynolds and Limpo, 1968). In DRL schedules with two levers, cats show better adjustment after the addition of repeated discrete auditory signals (Macar, 1969) (Fig. 8.2). Uramoto (1973) obtained a similar improvement in the same species in a reinforcement of response latencies situation.

In Sidman avoidance schedules, stimuli associated with the response-shock interval produce maximal avoidance in the end of this interval (Findley, 1963; Grabowski and Thompson, 1972; Thomas, 1965). Suppressing external clocks sometimes results in extinction of avoidance behaviour (Grabowski, and Thompson, 1972). The effect of external clocks has also been reported in complex schedules (Kelleher and Fry, 1962, Shull and others, 1978; Silverman, 1971; Squires and others, 1975).

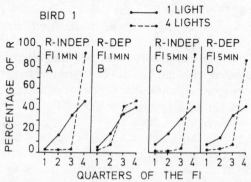

FIG. 8.1. Average percentage of total responses in each quarter of the Fixed Interval. The external clock (key-lights) is response-independent (R-INDEP) or response-dependent (R-DEP). The single light in the control condition is response-dependent. FI 1 and 5 min data are group means (two pigeons) over the final sessions of an experimental procedure. (Redrawn, after Segal, 1962a. Copyright 1962 by the Society for the Experimental Analysis of Behavior, Inc.)

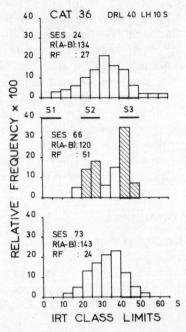

FIG. 8.2. Distributions by 5 s classes of inter-response times on a two-lever DRL 40 LH 10 s schedule. Data are from three sample sessions for cat 36. Middle panel shows the effect of adding three discrete auditory signals (S 1, S 2, S 3 striped columns), the last of which coincides with the limited hold period. Session numbers, number of response on lever A-response on lever B sequences-R (AB)- and numbers of reinforcements (RF) are given for each distribution. (After Macar, 1969, unpublished data.)

The control exerted by external clocks is further evidenced by results of experiments in which the time-correlated stimulus is inverted or falsified. Inverting the clock in a Fixed-Interval schedule may produce fully inverted scallops as shown by Skinner (1938) and Caplan and others (1973). Clearly, the stimuli (or the state of the stimulus) associated with the beginning of the interval control low rates of responding, and stimuli associated with the end of the interval control high rates (Auge, 1977; Farmer and Schoenfeld, 1966; Ferster and Skinner, 1957). In Macar's study (1969), cats were trained on a DRL two-lever schedule. In order to be reinforced, a response on lever B had to follow a response on lever A by a delay of at least 40 s and not more than 50 s. Adding a repeated auditory signal of 10 s duration and separated by 10 s interval (so that one signal coincided with the limited hold) resulted in much better timing of responses, the majority of responses on B being controlled by the third stimulus. When the duration of the auditory signals and of the intervals between them was reduced to half of the original values, a shift in responding was observed which showed the control by the third auditory signal (Fig. 8.3). Similar observations were reported on the same species by Uramoto (1973) after the number of flashes in a 5 s period that defined the requirement in a schedule of reinforcement of latencies of responses was changed. Inverting the external clock superimposed on a Sidman avoidance schedule also resulted in an inversion in the distribution of IRTs in rhesus monkeys (Grabowski and Thompson, 1972).

One can question whether an animal using external clocks provided by the experi-

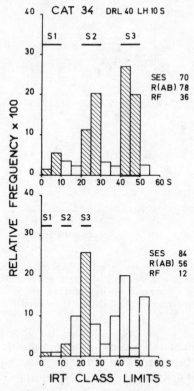

FIG. 8.3. Distributions per 5 s classes of inter-response times on a two-lever DRL 40 LH 10 s schedule. Data presented are two sample sessions for cat 34. Top: inter-response time distribution with three discrete 10 s auditory signals (S 1, S 2, S 3-striped columns), the last of which coincides with the limited hold. Bottom: effect of the reduction of duration (10 s → 5 s) and displacement of the signals. Other details as in Fig. 8.2. (After Macar, 1969, unpublished data.)

menter still shows time estimation. All it has to do in order to be reinforced is to respond under the control of the specific stimulus or aspect of the stimulus that is associated with the availability of reinforcement. This is not basically different from the control exerted by any external discriminative stimulus. The fact that the controlling stimulus is time correlated might be irrelevant.

Several observations indicate that external clocks are merely auxiliary and no full substitutes for timing processes.

8.1.3. Limits of the External Clock Control

The degree of control exerted by external clocks varies from one individual to another and as a function of the quality of temporal adjustment before the introduction of the external clock. An excellent performance is little improved, if at all, by the addition of an external clock. Marcucella and others (1977) have shown that in a DRL 25 s LH 5 s schedule, the degree of control exerted by an exteroceptive stimulus added to the limited hold is inversely related to the quality of the performance prior to its introduction (Fig. 8.4). Rats whose efficiency is high do not change their behaviour,

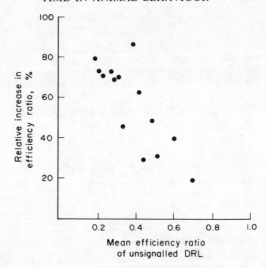

FIG. 8.4. Relative increase in efficiency ratio on a DRL schedule produced by the introduction of a discriminative stimulus, as a function of the efficiency ratio obtained during the last five sessions of the unsignalled condition. Each point represents one individual rat. (From Marcucella and others, 1977. Copyright 1977 by the Psychonomic Society, Inc.)

and do not emit short latency responses to the stimulus. In contrast, rats who perform poorly greatly benefit from the added stimulus. In some cases, adding the stimulus even disturbs the temporal adjustment in well-performing animals.

The control exerted by external clocks also depends upon the duration of exposure to the time-related stimulus. In pigeons, Segal (1962a) has observed the development of accelerated rates of responding during the last of four coloured stimuli covering a 5 min Fixed Interval. This means that some other variable besides the external clock is assuming part of the control of behaviour. Results from Macar (1969) reproduced in Fig. 8.3 also show that after falsification of the clock in that study, a number of responses still occurred during the limited hold, demonstrating temporal adjustment proper. In a study by Caplan and others (1973), two groups of pigeons were trained for six and sixty sessions respectively on a Fixed-Interval 90 s. The external clock, present from the first session, was a light increasing in intensity in six discrete steps during the interval (every 15 s). When the clock was reversed on a final test session, inverted scallops were observed, as noted earlier. However, subjects exposed to extended training showed an increase in rate during the last part of the interval even in the inverted condition (Fig. 8.5).

Finally, the control by external clocks depends upon the value of the delay, and upon the type of contingencies present. In DRL schedules with short delays, adding a signal to the limited hold does not eliminate unreinforced responses. It would seem that time remains the controlling variable in DRL when the delay is small (up to 10 s for rats) while the added stimulus becomes the main variable for long delays (Marcucella, 1974; Shimp, 1968). Experiments on FI schedules tend to reveal an opposite relation. Perfect control by external clocks has been obtained in short FI, while for longer delays (for example 3 to 6 min) temporal discrimination is demonstrated by changes in rate during the last stimulus (Boren and Gollub, 1972; Segal, 1962a).

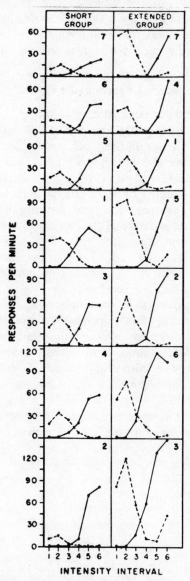

FIG. 8.5. Response rates for individual pigeons per 15 s segments before (solid lines) and after (dashed lines) reversal of a visual external clock (six light intensity steps) on a FI 90 s schedule. Left: six-session training before clock reversal (short group). Right: sixty-session training before clock reversal (extended group). All points are means of twenty trials within a daily session. (From Caplan and others, 1973. Copyright 1973 by the Psychonomic Society, Inc.)

The control exerted by external clocks thus depends upon several variables. Appropriate discriminative stimuli, systematically related to time, can take control of the behaviour in temporal schedules, and free the subject, totally or to a large extent, from the task of estimating time. It does not follow, however, that such an estimation does not take place even in the presence of external clocks. That it actually does take place is another argument in favour of some temporal regulating mechanisms. It is most interesting to note that stimuli that are offered to the organism as a possible substitute for time-keeping eventually induce temporal regulation of behaviour precisely because they are presented in a very orderly and regular fashion with respect to time.

8.1.4. Discriminative or Reinforcing Function of External Clocks

Studies on external clocks, independent of or contingent upon the operant response, suggest that stimuli correlated with the first part of a Fixed Interval acquire the value of negative discriminative stimuli and, as such, reduce the rate of responding. Data obtained by Segal (1962a), Kendall (1972) and Farmer and Schoenfeld (1966) confirm this interpretation formulated by Dews (1970) and further supported by Auge (1977). This last experimenter associated different stimuli with successive thirds of a 32 s Fixed-Interval. If responses emitted during the last third of the interval were followed by a brief stimulus of the same nature as the one associated with the first third of the interval, the rate of responding decreased in the last third. Conversely, short stimuli similar to the one associated with the third part of the interval produced an increase in rate when made contingent upon responses emitted during the first part. The function of such stimuli can be interpreted as reinforcing as well as discriminative. It is difficult to decide between these two possibilities in situations where the stimuli are presented as feedback for the operant response that also produces, at other times, the primary reinforcement. In order to dissociate the effects of contingent stimuli from those of the primary reinforcer, some experimenters have used a special procedure, involving what has been termed *observing responses* (Wycoff, 1952). While submitted to a schedule of primary reinforcement using a specified operant response, subjects may emit some other response, usually on another manipulandum, which produces stimuli associated with various conditions of reinforcement. Such stimuli, when they actually control the observing response, can be considered conditioned reinforcers (Fig. 8.6).

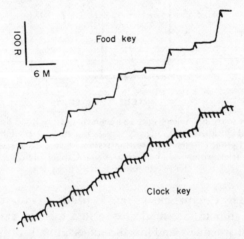

Fig. 8.6. Concurrent performances of one pigeon on FI 6 min (food key) and optional clock (clock key). Pips on food-key record mark reinforcements. Those on clock-key record indicate minutes since reinforcement (elongated pips). (Adapted from Hendry and Dillow, 1966. Copyright 1966 by the Society of the Experimental Analysis of Behavior, Inc.)

Two hypotheses have been proposed to explain the maintenance of observing responses (for a more detailed discussion, see Fantino, 1977). In the first of these hypotheses, the stimuli control behaviour because of their association with the primary reinforcement; in other words, they become *conditioned positive reinforcers* (Dinsmoor and others, 1969, 1972; Jenkins and Boakes, 1973; Jwaideh and Mulvaney, 1976; Katz, 1976; Mulvaney and others, 1974). In the second interpretation, stimuli contingent upon observing responses are endowed with positive or negative value, but maintain observing behaviour because in either case they reduce uncertainty as to the occurrence of primary reinforcement; in other words, they control behaviour because of their *information value* (Berlyne, 1957, 1960; Blanchard, 1977; Hendry, 1969; Hendry and Coulbourn, 1967; McMillan, 1974; Schaub, 1969). This last hypothesis was supported for the FI schedule by results reported by Hendry and Dillow (1966); but an experiment by Kendall (1972) did not confirm the earlier conclusion. It would seem that only those stimuli that are temporally close to or contiguous with the primary reinforcement do in fact take control of observing responses in Fixed-Interval schedules. This is in line with the data from experiments on response-independent clocks and on response contingent clocks reviewed above.

It must be emphasized that in the context of contingencies inducing temporal regulation of behaviour, studies on observing responses reinforced by external clock stimuli have been limited so far to Fixed-Interval schedules. As shall be seen in the next section, discriminative stimuli precisely correlated with time are not particularly useful in these schedules which already involve regular recurrent events. The less temporal organization there is in the contingencies, the more useful any added temporal information such as external clock stimuli would appear, and also the more observing responses a subject could produce in order to receive such information. This hypothesis has not yet been tested in animals, since, as already noted, studies comparing FI schedules with other temporally defined schedules are lacking. One study on human adults by Baron and Galizio (1976) furnishes some interesting findings, however. These authors compared the reinforcing value of an external clock stimulus in a Fixed-Interval schedule and in a Time-Out avoidance schedule. This value was significantly higher in the latter situation, possibly due to the lack of *temporal cues* that are inherent in free-operant avoidance schedules. In the next section, we shall turn to an analysis of the role of periodic events in schedules of reinforcement, and their effect on temporal adjustment of behaviour.

8.2. THE STRUCTURING POWER OF CONTINGENCIES

8.2.1. Definitions and Hypotheses

We concluded the preceding section by suggesting that the fewer temporal cues provided by the situation, the more active the subject will be in obtaining additional cues if it is given the possibility to do so. To explore this hypothesis further, we need to clarify what sort of temporal cues are inherent in different schedules of reinforcement, and what degree of temporal organization is actually present. A main aspect of contingencies, with respect to the temporal control on behaviour, is the distribution through time of primary reinforcers and associated events. It is clear that Fixed-Interval schedules provide for regular delivery of food reinforcers at repeated identical intervals of time if the subject produces appropriate responses. By contrast, a Variable-Interval schedule, where reinforcements are delivered according to a random distribution through time, does not offer such regularity.

Theoretically, a DRL schedule is no less "regular" than a Fixed-Interval schedule: it similarly provides for the possibility of successive reinforcements at identical intervals if the subject spaces its responses by the required delay. However, although the structure of the contingencies is theoretically comparable, the schedules do not generate comparable patterns of behaviour. It is important to distinguish the structure of a schedule, as it is defined by the experimenter, from the way it is in fact experienced by the subject as it interacts with the situation. If we compare the actual distribution of reinforcement through time in a FI 30 s situation and in a DRL 30 s, it will appear very different. Several authors have, in various contexts and under various terms, emphasized this basic distinction. For instance, Zeiler (1977) contrasts *direct* variables specified in the contingencies with *indirect* variables that result from the interaction between the subject and the contingencies; Morse (1966) contrasts *constant* influences deriving from the specified contingencies—such as the minimal inter-reinforcement time in a FI schedule—with *fluctuating* influences deriving from the subject-schedule interaction; Zeiler (1976) makes a distinction between *nominal* (that is maximal given the contingencies) and *obtained* rate of reinforcement.

Not only is the theoretical or the actual degree of periodic structures different in different kinds of contingencies (for instance FI versus DRL), it can also be varied within a given kind of contingency. For instance, the degree of regularity of reinforcement in a FI schedule can be changed by using a discrete interval procedure with varying inter-interval times, and in a DRL schedule by modifying the relationship between reinforcement duration and schedule parameters. Lejeune and Mantanus (1977) have compared three pairs of pigeons in DRL acquisition. The relationship between reinforcement duration (either one single grain of vetch or 3 s access to the grain magazine) and DRL delay is shown in Fig. 8.7. After training on CRF, the pigeons were exposed to the DRL contingencies starting with a 2 s delay that was increased to a 10 s delay in 1 s steps, according to efficiency criteria. Group 1 pigeons presented a steady increase of mean modal IRT value from the start (see Fig. 8.8) and their IRT distributions displayed a sharper peak at 10 s. The trend to a better temporal regulation in terms of matching of modal IRT to criterion in the short reinforcement group may be accounted for by reference to an analysis in terms of schedule structure. Group 1 procedure was the only one where, from the start, there was no overlap between reinforcement duration and DRL delay (group 2) and no double requirement for timing (group 3 where criterion IRTs after reinforced and unreinforced IRTs are different). It is worth noting that a regular increase in modal IRT (see Fig. 8.8) occurred in group 2 from an 8 s delay and beyond and in group 3 from a 6 s delay and beyond. This suggests that a necessary condition for DRL schedule control is a clear-cut difference between schedule requirement and duration of the reinforcing event, a condition that was met from a 2 s delay and beyond in group 1.

This concept of the periodic structure of the contingencies might prove to be useful in understanding a number of intriguing facts in the study of behavioural adjustment to time. One of these, mentioned on several occasions in previous chapters, is the contrast between FI and DRL with respect to the order of magnitude of the delay to which temporal adjustment develops. While performance remains fairly well adapted to increasing delays in FI (up to 100,000 s for example in Dews' study, 1965b) limita-

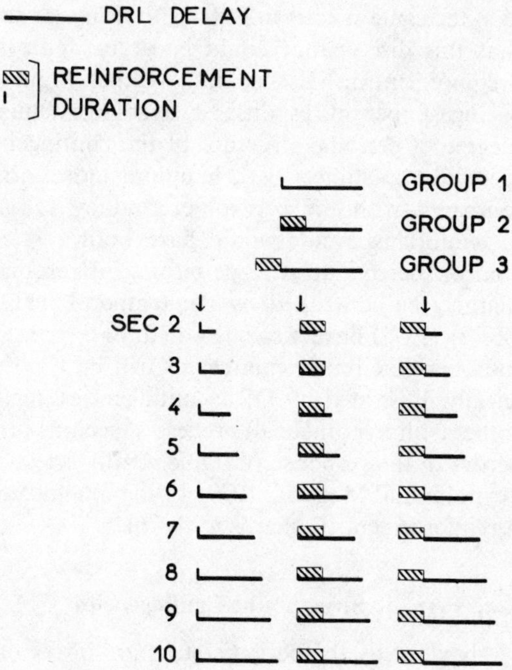

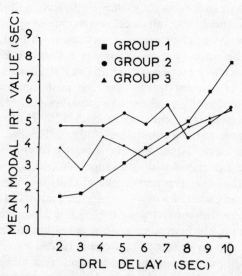

FIG. 8.7. Schematic representation of three different relationships between reinforcement duration (short-bar versus long-striped block) and DRL delay (2 → 10 s). (After Lejeune and Mantanus, 1977, unpublished data.)

FIG. 8.8. Mean modal IRT value in three groups of two pigeons according to the schedules schematized in Fig. 8.7 and to progressive increase in delay of the DRL (2 → 10 s). See text. (After Lejeune and Mantanus, 1977, unpublished data.)

tions are observed in the same species in DRL with delays as short as 20 or 30 s. It has been suggested that this discrepancy could be accounted for by the absence of a response-spacing requirement in FI schedules. We shall argue here that the temporal patterning and spacing of operant responses is, to some extent, a function of the actual (not theoretical) degree of periodic structure of the contingencies. A highly periodic structure would provide the subject with temporal indicators, the control of which might be best compared with the control exerted by *synchronizers* on biological rhythms. Periodic reinforcing events would have both a *selective reinforcing* and a *synchronizing* action on behaviour. The patterning of responses observed under FI schedules would largely be controlled by the temporal regularity of reinforcers, as Lund (1976) and Zeiler (1977) have also suggested. In contrast, when the actual structure is irregular, non-periodic reinforcing events can have no synchronizing function. In such cases, typically illustrated by DRL contingencies, temporal regulation would develop through other (still hypothetical) processes, such as progressive differentiation of temporal properties of the response (Catania, 1970), selection of collateral mediating behaviours (Reynolds and McLeod, 1970), or the modification of response rate as a result of a rate of reinforcement (Richardson, 1976).

8.2.2. Control by the Periodic Structure of Contingencies

The control of behaviour by the periodic structure of contingencies is documented by a wide range of observations, under negative as well as under positive reinforcement schedules. It will be recalled that under negative schedules, such as shock deletion procedures or Fixed-Interval procedures of shock presentation, the temporal patterning of responses develops only when shocks occur with regular periodicity, that is to say when there is an actual periodic structure (Dunn and others, 1971; Kadden and others, 1974; Sidman, 1962, 1966). Other striking examples are found in the inverted scallops that result from superimposing electric shock at fixed intervals of time on a baseline of positively reinforced behaviour (Appel, 1968; Azrin, 1956; La Barbera and Church, 1974), or in various cases described by Davis and others (1975) under the questionable term *autocontingencies* (see Staddon, 1975).

The power of the control exerted by the periodic structure of FI contingencies has been demonstrated in different ways. First, the post-reinforcement pause has been shown to be little affected by food deprivation (Powell, 1972b) or by changing the amount of reinforcement (Harzem and others, 1975; Hatten, 1974). Secondly, the pause is also resistant to various conditions of physical constraint (experimental space reduced to a volume hardly larger than the subject's volume, for instance). The shortening of the pause associated with a change of condition is only transitory, and occurs independently of the direction of the change, as Franck and Staddon (1974) have shown in pigeons under FI 5 min. Thirdly, the pause is little affected by changing the rate of responding without changing the inter-reinforcement time. This has been shown by presenting a blackout after each non-reinforced response (Neuringer and Schneider, 1968), by adding a pacing requirement (paced FI) (Elsmore, 1971b; Farmer and Schoenfeld, 1964b) or by defining the operant as a number of responses rather than as a single response (Shull and others, 1972). Fourth, scheduling response-contingent food during the interval affects the pause in a cage with a lone response lever (Harzem and others, 1978). The same happens with free-food scheduling (according to

Fixed Time or Variable Time) because of occasional contiguity between food and responses (Lattal and Bryan, 1976). On the other hand, pauses remain unaffected if food is made contingent upon non-responding during the interval, or upon responding on a second manipulandum (Shull and Guilkey, 1976). Fifth, the pause is also resistant, to some extent (Barrett, 1975; Lowe and Harzem, 1977), to manipulations of the response-reinforcement dependency, as is the case in conjunctive schedules FR1FT (Morgan, 1970; Shull, 1970b; Shull and Brownstein, 1975) or in Fixed-Time schedules (Appel and Hiss, 1962; Shull, 1971b; Lattal, 1972; Marr and Zeiler, 1974; Zeiler, 1968, 1977). As already discussed in Chapter 5, in contrast to the resistance of the pause to change, the rate of responding is more readily varied. As a rule, it decreases, and the accelerated pattern may fade out and give place to a bitonic pattern of pause-activity-pause (Shull, 1970b, 1971b; Staddon and Franck, 1975).

If, on the other hand, in some kind of discrete trial FI schedules the relationship between *pause duration* and *relative reinforcement proximity* is suppressed, the FI pattern disappears. This is the case in response-initiated FI schedules such as Tandem FR1FI (Chung and Neuringer, 1967; Shull, 1970a) or Chain FR1FI (Shull and Guilkey, 1976; Shull and others, 1978) where both components are associated with discriminative stimuli. In these schedules, reinforcement is followed by a pause that is a function of the duration of the FI component (Fig. 8.9). This pause is followed by sustained VI-like responding until reinforcement. The relationship between FI parameter and post-reinforcement pause duration has been described for FI values from 1 to 15 s (Chung and Neuringer, 1967), 3.75 to 60 s (Shull, 1970a) or 40 and 80 s (Shull

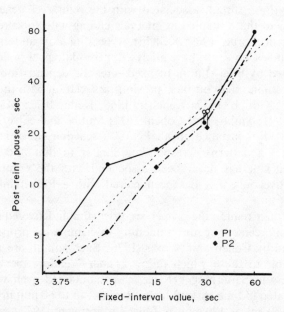

FIG. 8.9. Relationship between post-reinforcement pause duration (ordinate, log scale) and Fixed-interval value (abscissa, log scale) in a tandem FR 1 FI schedule. Data are from two pigeons (P 1 and P 2): each point represents median values based on the last five sessions devoted to condition. Closed symbols indicate medians from initial exposure, open symbols indicate redetermined medians. Dashed diagonal will be matched if pause equates FI duration. (From Shull, 1970a. Copyright 1970 by the Society for the Experimental Analysis of Behavior, Inc.)

and others, 1978). The importance of this schedule for the study of temporal regulation, if any, is in the inter-trial interval (that is the post-reinforcement pause), not in
the FI component. The dissociation between post-reinforcement pause and relative
reinforcement proximity also holds for FR schedules where post-reinforcement pause
grows in direct relation with ratio requirement (Felton and Lyon, 1966; Powell, 1968).
Temporal properties of both schedules are similar and, within certain limits (Cross-
mann and others, 1974), post-reinforcement pauses in FR schedules seem essentially
governed by temporal factors (Barowski and Mintz, 1975, 1978; Killeen, 1969; Zeiler
1972b).

8.2.3. The Periodic Reinforcing Event: a Further Analysis

In order to further refine our analysis of the effects of the periodic structure on
learned behaviour, we shall now turn to situations in which periodic events are not
always primary reinforcers, but brief exteroceptive stimuli that have or have not been
associated with the primary reinforcer. These situations are known as *second-order
schedules*, as labelled by Kelleher (1966). In the type of second-order schedule considered here, a primary reinforcer is delivered at the end of a Fixed Interval only after
a number of similar intervals reinforced by a brief auditory or visual stimulus have
been completed. For instance, in a FR 5 (FI 2 min) second-order schedule, the subject
goes through five 2 min. intervals before it gets food. At the end of the first four
intervals it is presented with a brief sound or light. This brief stimulus may or may not
have been previously associated with the primary reinforcer. Early experiments suggested that only those stimuli associated with the reinforcer generated a pattern of
behaviour similar to that obtained in intervals ending with food reinforcement (Byrd
and Marr, 1969; de Lorge, 1967; Kelleher, 1966; Staddon and Innis, 1969; Stubbs,
1969). Further studies, however, have analysed methodological pitfalls and shown that
the control exerted by brief stimuli in fixed intervals depends more on the physical
properties of the stimuli than on their previous association with the reinforcer (Cohen
and others, 1973; Gollub, 1977; Kelleher, 1966; Kello, 1972; Malagodi and others,
1973a; Stubbs, 1971; Stubbs and Cohen, 1972; Stubbs and Silverman, 1972; Stubbs
and others, 1978). For instance, in an FI 100 s schedule, a visual stimulus of 2.5 s
duration generates a pattern of responding similar to that obtained with food reinforcement, even if it has never been associated with the reinforcer, while a 0.5 s
stimulus is effective only if it has been paired with a food reinforcer (Stubbs and
others, 1978).

In addition to their reinforcing properties, brief stimuli delivered at fixed intervals in
place of food reinforcement acquire inhibitory discriminative properties, as shown by
Stubbs (1971) and by Cohen and Stubbs (1976). Such stimuli are followed by pauses
strictly comparable to those which follow primary reinforcements in FI schedules.
Their effectiveness in patterning behaviour is maintained even when primary reinforcers are separated in time by very long intervals (up to 60 min in Kelleher's study in
1966; averaging 30, 60 or 120 min in Zeiler's experiment, 1972a, where 7 per cent of
the thirty intervals of a session in FI 2, 4 or 8 min respectively were ended by food
reinforcement) (Fig. 8.10).

Another critical variable is second-order programming of the FI. The control by
periodicity, that is the patterning of typical FI behaviour, is maximal when a variable

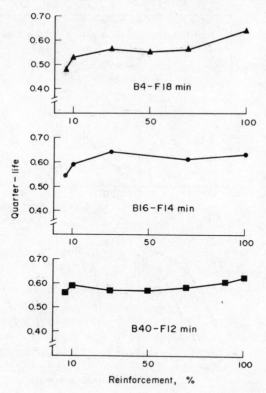

FIG. 8.10. Quarter-life values for three pigeons on discrete-trial FI 2, 4 or 8 min, according to different percentages of reinforcement (7 → 100 per cent). Omitted reinforcements are replaced by a key colour change. Each point represents the median quarter-life for the last five sessions of each condition. (From Zeiler, 1972a. Copyright 1972 by the Society for the Experimental Analysis of Behavior, Inc.)

number of stimuli-reinforced intervals are required—VR (FI) as opposed to a fixed number FR (FI). In the first case, no sequential effect develops across the successive links, and the periodic control is maximized, as shown in Fig. 8.11 (Corfield-Sumner and Blackman, 1976).

Conversely, omitting the brief stimulus (making the schedule a *tandem-schedule*) results in uniform rate of responding, the schedule of primary reinforcement taking over the control of behaviour (Kelleher, 1966—see Fig. 8.12; Stubbs, 1969; Stubbs and Silverman, 1972 among others).

However, the maintenance of the control exerted by second-order tandem schedules may depend essentially upon previous training. The FR 15 (FI 4 min) performances of Kelleher's pigeons are probably a unique case, the author signalling that "the birds had long experimental histories on various schedules of reinforcement" (pp. 476–7). With naïve animals, the absence of responding after primary reinforcement often disrupts its periodic distribution and responding progressively fades out in tandem schedules exceeding three or four FI components. The same remarks apply to the chained second-order schedules. This emphasizes anew the importance of the periodic distribution of the discrete reinforcing events in the control of behaviour in second-order schedules with FI components. As already mentioned in Chapter 5, second-

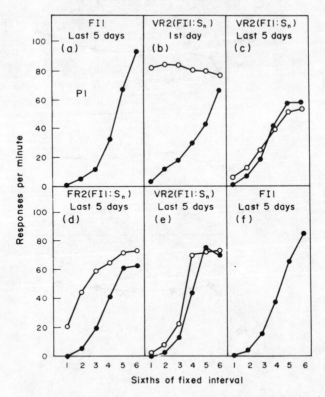

FIG. 8.11. Mean response rates (R/min) in successive sixths of Fixed Interval following food (filled circles) and following a brief stimulus (open circles) for pigeon 1. Data summarize five-session blocks of the schedules (see text), except in panel b which includes only the first session (From Corfield-Sumner and Blackman, 1976. Copyright 1976 by the Society for the Experimental Analysis of Behavior, Inc.)

order schedules have also been employed with monkeys, using electric shocks (Byrd, 1972) or cocaine injections (Goldberg and others, 1975; Kelleher and Goldberg, 1977) as the primary reinforcer.

The main point here for our purpose is that periodic events that need not be paired with the primary reinforcer induce temporal regulation of behaviour and maintain regular patterns of responding. The kind of control exerted by such periodic stimuli cannot be fully accounted for by resorting to the classical concept of conditioned reinforcement. Data summarized here suggest another hypothesis that requires further exploration: namely, that periodic events of the sort discussed here have a synchronizing effect on behaviour.

8.3. REINFORCEMENT AND RESPONSE-RELATED EVENTS

8.3.1. Stimuli Associated with the Reinforcement

Fixed-Interval schedules are, as we have seen, highly structured contingencies with regard to periodicity of reinforcement. Second-order schedules show, however, that periodic exteroceptive stimuli never paired with food can be used as a reinforcer, so

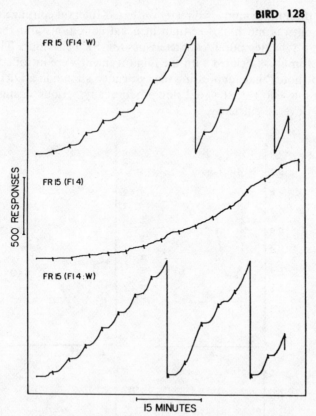

BIRD 128

FR 15 (FI 4 W)

FR 15 (FI 4)

FR 15 (FI 4 : W)

500 RESPONSES

|5 MINUTES

FIG. 8.12. Cumulative records of one pigeon on a second-order FR 15 (FI 4 W) schedule with a white light (W) presented at the end of each component between primary reinforcement (top and bottom records). Middle record shows the effect of white light suppression, the schedule being then a tandem FR 15 (FI 4). Pips indicate the end of each FI component (either a white light, food, or the end of an interval without any exteroceptive signal). (From Kelleher, 1966. Copyright 1966 by the Society for the Experimental Analysis of Behavior, Inc.)

that the periodicity need not be the periodicity of *primary* reinforcers. One could further ask whether in the usual FI schedules the crucial periodic event is the primary reinforcer as such or the stimuli associated with it. In experiments on Fixed-Interval schedules, the delivery of the primary reinforcer (generally food) is accompanied by a number of external stimuli produced by the food dispenser. In fact, these stimuli free the subject engaged in post-pause responding from the task of estimating the moment at which the reinforcer is likely to be delivered.

Using a modified Fixed-Interval 2 min procedure already described above (Chapter 2), Deliège (1975) eliminated auditory and visual stimuli normally associated with the delivery of food. Rats initially trained under a usual Fixed Interval showed a quite different pattern of responding after being transferred to this modified schedule. Typically, the regular pause and run alternation was drastically changed, the rate of responding decreased and eventually complete extinction developed, despite the fact that positive contingencies continued to be in vigour. These effects were reversible: the usual pattern of behaviour was reinstated when the stimuli were reintroduced. It was also possible to train naïve animals directly on Fixed-Interval contingencies without a stimulus. Training was much slower than in usual conditions and the performance exhibited none of the temporal characteristics of FI behaviour. These, however, appeared when stimuli associated with the reinforcement were introduced (Fig. 8.13).

It must also be noted that suppressing the associated stimuli in a DRL 20 s schedule had no comparable effect, that it did not produce any serious change in temporal adjustment or in overall efficiency.

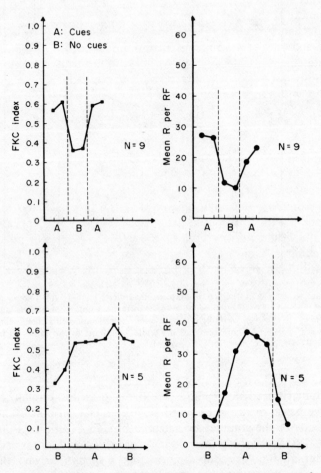

FIG. 8.13. Curvature indices and mean responses per reinforcement for two groups of rats ($N = 9$ and $N = 5$) on a classical FI 2 min schedule (A) and a FI 2 min schedule without external cues (B). Each group is submitted to a different sequence of conditions (ABA or BAB). Each point represents the average data of five sessions. (After Deliège, 1975, unpublished data.)

Obviously, stimuli associated with the delivery of food became secondary reinforcers, and they also exerted an activating action on operant responding. However, as Deliège has shown in a series of control experiments, their function in FI schedules cannot be fully accounted for by these notions. Part of their effect of temporal patterning can be assigned to a synchronizing function, to which we have resorted above in the context of second-order schedules. In this respect, associated stimuli would be analogous to *synchronizers* as defined by chronobiologists. It is well known that biological events come into phase with periodic external events, and also that the measured manifestation of rhythm often fades out in the absence of such synchronizers in *free-running* conditions. This is strikingly similar to the decrease in operant responding and to the impairment of temporal patterning observed when associated stimuli are eliminated in Fixed Interval. This hypothesis is also supported by studies on second-order schedules, as already mentioned, and by studies on effects of cerebral lesions (Lejeune, 1974, 1977a). The hypothesis focuses attention on the periodic alternation of pauses and activity in Fixed-Interval schedules, rather than on the nature of the response activity itself. Temporal patterning would be mainly evidenced by the regular reinstatement of activity after a pause of fairly stable duration. Scalloping or break-and-run patterns would appear as secondary phenomena.

The analogy with synchronized rhythms in chronobiology may be pushed one step further. The precise regularity of a biological rhythm under the control of external synchronizers does not allow one to predict the regularity in the absence of such events. Similarly, in Fixed-Interval schedules, those subjects which exhibit the best temporal patterning in the presence of associated cues are not necessarily those which maintain the best performance when the stimuli are eliminated. In this respect, FI contingencies would not be favourable conditions for revealing an animal's actual capacity for estimating duration since in a sense, they give too much help to the subject in coping with time. Schedules such as DRL would appear to call for this capacity in a quite different way, because these schedules provide no substitute for the organism's timing mechanisms. This interpretation, which requires further experimental support, would provide some explanation for the discrepancy between performance under FI and under DRL contingencies, or more generally between what we have called spontaneous and required temporal regulation.

8.3.2. Feedback from the Operant Response

In this context, it would seem that feedback from operant responses would be of particular interest in those schedules with a weak periodic structure, that is the schedules where the synchronizing function of the reinforcing event is lacking. We shall deal only with the feedback from operant responses which are instrumental for the delivery of primary reinforcers, as we have already discussed *observing behaviour* in the section on external clocks.

Two main hypotheses may be formulated with regard to stimulus feedback from operant responses. The first states that responses followed by a specified stimulus are better discriminated, and that better discriminated operants are more amenable to the control exerted by the contingencies (Bolles and Grossen, 1969; Kuch and Platt, 1976). The second is based on the reinforcing properties of sensory events as such. Kish (1966) and many other authors have shown that sensory feedback of operant responses can have positive or negative reinforcing effects, depending upon the nature and intensity of the sensory feedback itself, and upon the species. Stimuli systematically produced by the operant response may have an intrinsic reinforcing effect, distinct from their conditioned reinforcing effect acquired through their association with the primary reinforcer. The data available concerning the role of response feedback in temporal behaviour are scarce, and have been obtained in very heterogeneous contexts. It is therefore not easy to draw a coherent view from them.

It is convenient, at this point, to make a distinction between non-differential feedback and differential feedback. Non-differential feedback does not provide any information as to the position of the response in time, nor as to the efficiency of the response. The noise produced by pecking a key or pressing a lever is a trivial example. It follows every response. This is also the case for proprioceptive feedback, as discussed in Chapter 6. Each response can also be made to produce some other stimulus that has been previously paired with reinforcement. This produces, as would be expected, an increase in rate of responding under FI contingencies (Deliege, 1975). In DRL schedules, Kelleher, Fry and Cook (1959) suggest that modifying non-differential feedback results in disinhibition of responding. This hypothesis is akin to the one proposed in several studies on the disinhibitory effect of external stimuli in FI and DRL, to be reviewed in Chapter 9 (Contrucci and others, 1971; Singh and Wickens, 1968). However, suppressing the auditory feedback of responses in a DRL 20 LH 5 s has no effect whatsoever (Fig. 8.14) and naïve rats can be trained on a DRL 20 s without auditory feedback (Fig. 8.15).

In Sidman avoidance schedules, Bolles and Popp (1964) describe a deterioration of rats' performances when the response-produced click is eliminated and when proprioceptive feedback is to some extent suppressed by the use of a special lever. These data are in line with the hypothesis that avoidance behaviour is maintained because of the reduction of the conditioned aversiveness of stimuli following the response (Anger, 1963).

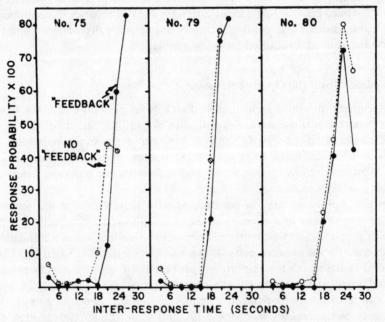

FIG. 8.14. Response probability distributions from one session on DRL 20 LH 5 s with auditory feedback accompanying each response (solid lines), and from the following session on the same schedule but without auditory feedback (dashed lines). Data are from three different rats. (From Kelleher and others, 1959. Copyright by the Society for the Experimental Analysis of Behavior, Inc.)

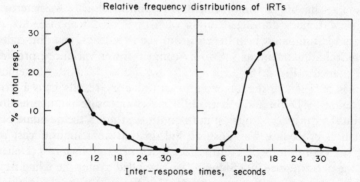

FIG. 8.15. Development of performance on DRL 20 s without auditory feedback. Left panel: session 1, right panel: session 30. Data reported are distributions of IRTs by 3 s classes for one rat. (From Kelleher and others, 1959. Copyright 1959 by the Society for the Experimental Analysis of Behavior, Inc.)

Differential response feedback stimuli provide the subject with some information about the position of its response in time, either directly in the form of *clock stimuli*, or indirectly through the use of different stimuli after reinforced versus non-reinforced responses. Clock stimuli contingent upon the response control behaviour as response-independent clocks discussed above. However, Segal (1962a) has described a level of control that is intermediary between classical FI situations and FI with added response-independent clocks (Fig. 8.1). Similar analyses have been proposed by Kendall (1972) and by Auge (1977). Clock stimuli associated with the early part of the interval acquire a negative value and punish responding, while stimuli associated with the terminal part acquire a positive value and increase responding. The second category of differential feedback is illustrated in an experiment by Uramoto (1971). The perform-

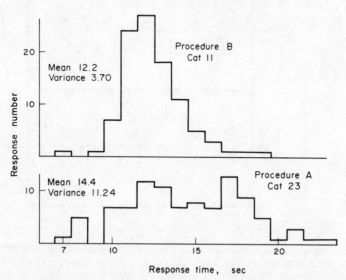

FIG. 8.16. Samples of response latencies distributions obtained in the final sessions on a reinforcement of 10 s latency schedule for two cats. Top: procedure B with a light stimulus contingent upon reinforced responses. Bottom: procedure A without the light stimulus. Mean response latencies and variances are given on the left. (From Uramoto, 1971. Copyright 1971 by Pergamon Press, Ltd.)

ance of cats in a schedule of reinforcement of response latencies was improved when a
light stimulus was made contingent upon each reinforced response. Acquisition was
not more rapid in animals which benefit from such feedback, but they reach a better
level of temporal adjustment, as shown by a much lower variance in the distribution
of response latencies (Fig. 8.16).

In schedules of negative reinforcement, contradictory results have been produced.
Thus, an attempt by Dunn and others (1971) to compensate the absence of temporal
"anchor points" by providing light stimuli contingent upon the avoidance response in
a shock-deletion procedure was unsuccessful, in that no temporal patterning developed under such conditions. In contrast, feedback of avoidance responses was shown
to improve the performance in a Sidman non-signalled avoidance schedule (Bolles and
Grossen, 1969).

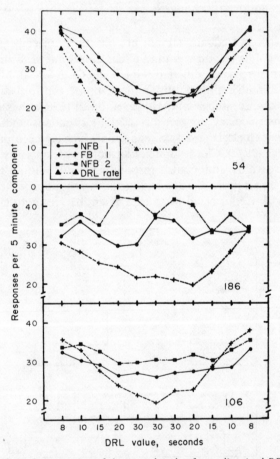

FIG. 8.17. Response rate in each component of the second cycle of a *cyclic mixed DRL schedule*. Different
conditions are first and second non-feedback (NFB-1 and 2) conditions and first feedback (FB-1) condition
for three pigeons. Abscissa: delay values of successive components ordered as they appear in the cyclic
schedule. Each point is the average from the last seven sessions under each condition. The curve labeled
"DRL rate" in the top panel indicates the number of responses that would be emitted in each component if
every response terminated an IRT just long enough to qualify for reinforcement. (From Staddon, 1969.
Copyright 1969 by the Society for the Experimental Analysis of Behavior, Inc.)

The positive effects of differential feedback could be due to their discriminative properties or to their acquired reinforcing value (secondary reinforcement). Staddon (1969a) has attempted to dissociate these two possible functions. In a DRL schedule with limited hold in pigeons, a red feedback stimulus was made contingent upon IRTs which were too long, and a green stimulus upon IRTs which were too short. In this manner, feedback stimuli occurred only after unreinforced responses, and were therefore never associated with the reinforcement. The possibility that they became conditioned reinforcers is therefore ruled out. The stimuli had no significant effects in simple DRL-LH schedules or in multiple schedules where phases with and phases without feedback stimuli alternated. In contrast, they showed an effect in a mixed-cyclic-DRL, a schedule described above (p. 67) in which conditions for obtaining food are more demanding (Fig. 8.17). Curiously enough, control by the stimuli was not exerted through some sort of negative feed-back mechanism that would adjust the next IRT as a consequence of the feedback obtained for the previous IRT, as one would expect in a step-by-step correcting system. In this case, the information "too long" should have induced a shortening, and conversely the information "too short" a lengthening, of the next IRT. Instead, it seems that birds relied on the general patterning of feedback stimuli through the successive cycles of the schedule and on the relative frequency of occurrence of each of the two possible stimuli. Staddon concluded that pigeons cannot be controlled by knowledge of results in the strict sense of the term, and that the limitations observed in the capacity of this species in adjusting to DRL delays beyond a certain value are not due to a lack of informative feedback since, when such a feedback is provided, the subjects are not able to use it.

IX. *Temporal Regulation and Inhibition*

Waiting is an active regulation of action.—Pierre Janet, 1928

IN THIS final chapter, we shall discuss what we think is at present the most acceptable hypothesis to account for many of the facts we have been reviewing: behavioural adjustment to time implies inhibitory processes. This is by no means a new hypothesis. It was already formulated by Pavlov, who saw in *inhibition of delay* a particularly convincing case for the active nature of inhibition. Resorting to the concept of active inhibition does not, of course, elucidate the nature or the locus of the timing device(s) underlying temporal regulations of behaviour. It does, however, specify some essential properties of the mechanisms still to be discovered. Wherever we shall look for it in the physiology or chemistry of the brain, we shall have to look for something more complex than pacemakers. We must look for a system that combines timer(s) with inhibitory regulators. Adjusting to or estimating duration for an animal organism is not simply putting landmarks in the flow of time; it means retaining and deferring action. We are dealing here with a relaxation phenomenon, not with a pendulum type oscillatory system (Wever, 1962). This means that temporal regulation involves some building up of tension that is suddenly discharged as Janet put it half a century ago, applying to temporal regulation a main theme of his psychological theory.

Though it was central in Pavlov's theoretical thinking and derived from very convincing experimental evidence, the concept of inhibition went out of favour with some American behaviourists after Skinner (1938) suggested that we could do without it, and for the sake of the principle of parsimony, explain by a lack or reduction of excitation everything that inhibition was supposed to explain. Inhibition has been rehabilitated since then on experimental and theoretical grounds (for a general picture of this rehabilitation, see Boakes and Halliday, 1972). Most of the facts and the discussions surrounding them, however, concern *conditioned inhibition*, that is inhibition associated with the negative stimulus in a discrimination experiment. Inhibition of delay has received comparatively little attention, though it would bring further support to the importance of inhibition in operant conditioning, as we have argued elsewhere (Richelle, 1972).

Inhibitory processes are evidenced by a variety of behavioural phenomena. Concern with conditioned inhibition has led most experimenters to favour techniques that detect and measure inhibition through manipulation of negative stimuli. One now classical demonstration consists in testing several stimuli for generalization along the physical continuum of the negative stimulus, and in describing a gradient of inhibition. This gradient is clearly distinct from the gradient of excitation provided that the negative stimulus has been chosen from another dimension than the positive stimulus (Brown and Jenkins, 1967). In another kind of test, the negative stimulus is combined with a positive stimulus and a reduction in responding is observed. This approach does not apply to temporal regulation, since time is not a stimulus in the usual sense of the word. Though it is a measurable dimension of events, it cannot be isolated and presented without the material support of some other dimension. However, the difficulty has been elegantly overcome in some experiments by associating visual or auditory stimuli with the passage of time. A study by Wilkie (1974) provides an illustration of this strategy. Pigeons were trained on a discrete trial Fixed-Interval schedule (3 or 6 min) in which each reinforcement was followed by a time-out of 1 min, and each trial was signalled by a visual stimulus: a vertical line that appeared on the response key and remained present throughout the interval. In a generalization test, lines of five different orientations including the original vertical line were presented randomly, each time for the whole interval. Separate measures of response rates were taken in the first, second, and last third of the interval. An inhibitory gradient was revealed, the minimum point of which was with the vertical line presented during the first third of the interval. In the case of the excitatory gradient, the maximum point was attained when the vertical line was presented during the last third of the interval (Fig. 9.1). The

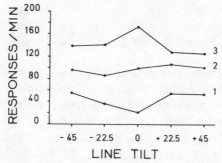

FIG. 9.1. Inhibitory and excitatory gradients for five stimulus-orientations (in angle degrees from the vertical 0° orientation) in a generalization test with pigeons trained on a discrete-trial FI 6 min schedule. Response rate per minute was computed separately for the first, second and last third of the interval, corresponding to curves numbered 1, 2 and 3 respectively. Stimulus 0 was present throughout the interval in training before generalization. (Drawn after numerical data from Wilkie, 1974.)

external stimulus exerts a differential control as a function of time, or more correctly phrased, time controls the effect—inhibitory or excitatory—of the stimulus. Similar conclusions can be drawn from a study by Auge (1977) where different stimuli were associated with successive phases of the interval in a Fixed-Interval schedule, and then tested at other phases (see Chapter 8). In DRL schedules with pigeons, Gray (1976) has similarly demonstrated differential control by a coloured stimulus associated with the contingencies. Generalization tests with other wavelength values show excitatory gradients for long IRTs and inhibitory gradients for short IRTs (leading to flat gradients when short and long IRTs are confused in a common treatment). The crucial controlling factor is time, not the wave-length of the stimulus.

Other examples of inhibition in temporal regulation pertain to a loose category that comes last in Hearst's (1968) classification of procedures available to the experimenter who wants to demonstrate inhibitory processes. This category is labelled "symptoms and by-products of inhibitory control" and the items it covers would appear to many students of operant behaviour less decisive than tests of conditioned inhibition. However, we see no reason why it should be so, except the particular history of research born from the renewed interest in inhibition. What Hearst presents as secondary "symptoms or by-products" are in fact no less convincing demonstrations of inhibition than generalization gradients. Disinhibition, induction, behavioural breakdowns (or generalized inhibition phenomena) were, in Pavlov's view, and still are essential facts on which the concept of active inhibition is founded.

9.1. DISINHIBITION

Because of its importance in Pavlov's arguments, the phenomenon of disinhibition was illustrated in the method section (2.2.5). A novel external stimulus presented during the latent phase of a trace or delayed conditioned reflex, or of a conditioned reflex to time, provokes a reappearance of conditioned responses, an effect opposite to the effect of the same stimulus presented during a phase of excitation. Though he developed some objections to Pavlov's interpretation of these and other data, Skinner admitted that disinhibition would support the view that inhibition is not reducible to a decrease or a lack of excitation; that "if true, it would invalidate" his own formulation. He failed to obtain disinhibition effects in operant situations, and as a result he did not change his view. Since then, disinhibition has been demonstrated in FI and DRL schedules.

Flanagan and Webb (1964) have obtained disinhibition effects in a FI 1 min schedule, by presenting auditory probe stimuli at various points in the interval. Using a procedure more similar to Pavlov's, Singh and Wickens (1968) presented a novel auditory stimulus (buzzer) either throughout the first half or the last half of a 3 min interval, after stabilized performance was reached. A clear-cut disinhibition effect was observed when the buzzer was presented during the first half of the interval (corresponding to the pause), while response rate in the activity phase was depressed when the novel stimulus was presented during the second half. Hinrichs (1968) trained pigeons under FI 1 min with the response key illuminated in white, red or green, depending upon the individual. In test trials, the colour of the key light was changed for the whole duration of the interval, six times per session. Subsequently, the subjects were trained with the new colour, and tests trials were run with a "novel" stimulus that was of the original colour experienced by the birds. The disinhibition effect was evident in all subjects and in both tests: a flat distribution of responses was observed in place of the scalloped pattern obtained prior to the test trials (Fig. 9.2). This is analogous to Pavlov's classical results in temporal conditioning (see p. 13). Malone (1971) used a more complicated procedure, involving less contrast between original and novel stimuli, and tested several orientations of a line on the response key (a technique borrowed by Wilkie in the experiment cited above). Though his results were not as clear-cut as those reported by Hinrichs, he insisted that they were in line with Pavlov's data, where disinhibition effects were not always perfect, but seem to have been influenced among other things by the kind of novel stimulus used and by the strength of the inhibition upon which it is checked. Working with rats, Contrucci and others (1971) presented a novel stimulus (4 s buzzer) four times in selected inter-reinforcement intervals of a DRL 20 s session, either after 6 s or after 12 s. The subject had reached an efficiency ratio of 0.50 or more before testing for disinhibition was started. Very clear disinhibition effects were obtained, that is the number of "error responses" (spaced by less than 20 s) was significantly larger for the inter-reinforcement period during which the novel stimulus had been presented. In both FI and

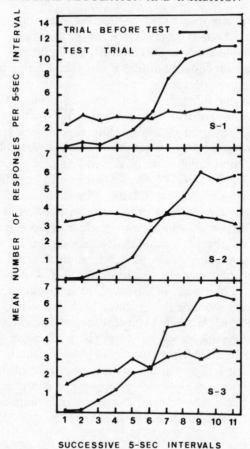

FIG. 9.2. Disinhibition in Fixed-Interval performance. Response pattern with and without novel stimuli (change in key colour). Each test and control curve is based on twenty-four spaced trials. (Redrawn, after Hinrichs, 1968. Copyright 1968 by the Psychonomic Society, Inc.)

DRL schedules, the disinhibition effect is transitory and it fades out with repetition, as already noted by Pavlov and as expected from the laws of habituation to neutral novel stimuli.

The failure to obtain disinhibition reported by Skinner, and more recently by Wolach and Ferraro (1969)—also to some extent by Malone (1971), despite the author's willingness to interpret his results along Pavlovian lines—might be due to a basic difference between the operant situation and the Pavlovian situation with regard to the possibility of detecting disinhibited responses. The salivary responses of Pavlov's dogs, being autonomic in nature, are recorded throughout the experiment wherever they occur, irrespective of the fact that the dog might be doing something else; the experimenter catches them in the very place where they occur in the organism. This is not the case for the operant response: it is recorded only if the organism makes the necessary physical contact with the recording apparatus. For the disinhibition effect to be detected, the subject must be in the place where the experimenter has chosen to record the response. As the subject is free to move in the experimental space, this condition is not likely to be fulfilled every time a disinhibition test is undertaken. This is especially important when probe stimuli are used rather than stimuli covering a whole interval. This difficulty might be avoided by testing animals under physical restriction designed in such a way that it would maintain a preparatory posture very close to the operant response.

Though it is somewhat different from the classical disinhibition test, the procedure known as *conditioned suppression* when applied to DRL schedules has led to interesting results related to our present concern. Superimposing a stimulus followed by an unescapable shock on some ongoing behaviour (such as Variable-Interval responding) results in partial or total suppression of responding (Estes and Skinner, 1941; for a recent review see Blackman, 1977). The importance of the effect is a function of a number of variables, such as shock intensity, baseline rates of responding, etc. The direction of the effect may be reversed (that is, responding is increased rather than suppressed) under certain conditions, for example when the baseline behaviour is continuous avoidance. Preshock stimuli superimposed on DRL have been shown to produce suppression in some cases and acceleration in others. When suppression is obtained, the longer the delay, the less pronounced it is (Randich and others, 1978). Facilitative effects are observed, at least with mild shocks, when the DRL schedule alternates with another schedule that maintains a higher rate of responding, such as a VI in a multiple DRL-VI. While the classical suppressing effect is observed in the VI component (using the same shock intensity), an acceleration is obtained during the stimulus in the DRL component (Blackman, 1968). This is best interpreted as a disinhibition effect. Interestingly, if the second component of the multiple schedule is not a VI but an Extinction schedule, a suppressive effect is observed in the DRL component (Randich and others, 1978). The degree of inhibition is not an absolute but a relative value, and it is dependent on the context.

9.2. GENERALIZED INHIBITION

Overloading inhibitory processes results in various types of disturbance of adapted behaviour. Pavlov observed that some dogs would suddenly become overexcited, their conditioned behaviour being of course severely disrupted, and that others would eventually fall asleep. The prevalence of one or the other type of consequence depended, according to Pavlov, upon the constitutional type of the animal. What was to Pavlov an important phenomenon, justifying exploration in depth and much discussion, was largely neglected in operant conditioning studies. In most experiments, extinction of conditioned behaviour, when not deliberately intended by the experimenter, is an unfortunate accident that spoils several weeks or months of hard work. It is also viewed as a failure to control behaviour by the schedule of reinforcement, a failure to which many experimenters might not be ready to readjust themselves. This is probably why systematic reports of generalized inhibition in operant conditioning studies of temporal regulation are very scarce though the phenomenon, according to our own experience, is far from unusual.

Lejeune (1971a) has described progressive extinction of operant responding in two cats trained under FI 5 min, after training under FI 2 min. The deterioration began after thirty or fifty sessions with an interval value of 5 min. These animals were females. No similar phenomenon was observed in other subjects in the same experiment, who adjusted to delays up to 15 min. It must be noted that the amount of reinforcement was proportional to the duration of the delay. Figure 9.3 shows one example of this deterioration. A very similar phenomenon was reported by Deliège (1975) in her study on the effects of removing the cues associated with reinforcement on a FI schedule in rats. As we have seen in Chapter 8, this modified the typical pattern of responding, and drastically reduced the response output. This reduction can be strong enough to completely suppress operant responding, as shown in Fig. 9.4.

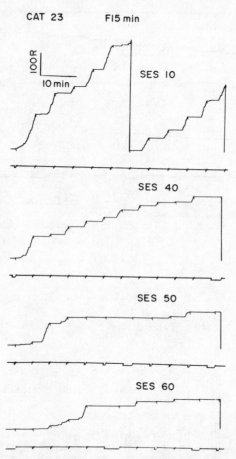

FIG. 9.3. Progressive deterioration of performance in a cat on FI 5 min, possibly due to generalization of inhibition. (After Lejeune, 1971a.)

If, as we suggest, the "inhibitory load" is much larger in DRL than in FI schedules, we would expect generalized inhibition and extinction of operant behaviour for much shorter delays in DRL than in FI. Macar (1969, 1971a) observed just this in cats trained on a two-lever DRL 40 or 60 s schedule. The performances of the three subjects had been stabilized fairly well for over twenty sessions. Then the behaviour deteriorated, abruptly in one animal and progressively in the other two, as shown in Fig. 9.5. Observation of the animals in the experimental situation and in their living quarters revealed a number of characteristics classically described in "experimental

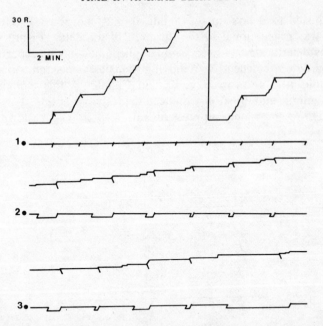

FIG. 9.4. Deterioration of performance of one rat in a Fixed-Interval 2 min schedule after removal of auditory and visual stimuli associated with the delivery of reinforcement. Cumulative records: 1, control session with associated cues; 2 and 3, at two stages of removal procedures. (After Deliège, 1975, unpublished data.)

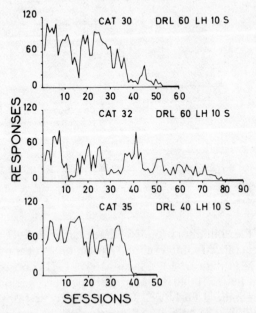

FIG. 9.5. Progressive or abrupt extinction of conditioned behaviour in three individual cats during exposure to a two-lever DRL 40 or 60 s. Ordinate: number of sequences Response on lever A-Response on lever B. (From Macar, 1969, unpublished data.)

neurosis". Two cats were observed to lay somnolently in the experimental chamber during the whole session; they sometimes would not consume the milk offered to them. Though they ate and drank normally in their home-cage, they ran away when humans approached them or attempted to touch or catch them, they shuddered at slight noises and their fur bristled at the sight of unfamiliar persons. These subjects had been quiet and gentle prior to the extinction. The third subject, of a most excited type at the outset of the study, exhibited increased agitation and aggressiveness from the time generalized inhibition of operant responding developed and became difficult to approach and handle. Several attempts to reinstate normal behaviour failed. Among other manipulations, Macar changed the amount of the reinforcer, the degree of deprivation, the length of the limited hold, introduced novel stimuli, and gave the subjects a rest. Finally, she succeeded in reconditioning two of her three cats on FR 1 to 60, and switched them back to DRL, but she did not follow up the long-term effect of this "cure".

Other phenomena of generalized inhibition have been observed in pigeons in a task of discriminating the duration of visual stimuli. When exposed to long (4 h) sessions of discrimination between a 10 s positive stimulus and a 5 s negative stimulus, pigeons developed numerous and long pauses, in which they failed to respond to the positive stimulus (Richelle, 1972).

9.3. AVERSIVENESS OF SCHEDULES INVOLVING TEMPORAL REGULATIONS

Another well-known by-product of inhibitory control is the aversiveness of the situation in which it takes place. Such aversiveness is suggested, if not always decisively evidenced, by several kinds of typical phenomena. A subject will normally produce responses that will allow it to escape or avoid an aversive event or situation. One should expect then, that given the possibility to do so, it will escape a schedule endowed with aversive properties. Schedules of positive reinforcement may, under certain conditions, have aversive properties, as a number of studies have shown in which the subject was given the possibility to respond to produce periods of Time-Out. For instance, Fixed-Ratio schedules become aversive when the required number of responses per reinforcement increases beyond a certain value (Azrin, 1961). Pigeons peck at a side key to produce Time-Out from a Fixed-Interval schedule (Brown and Flory, 1972). The duration and the frequency of Time-Out are a function of the length of the interval up to values between 240 and 480 s. Escape from DRL has not been clearly demonstrated up to now. Mantanus (1973) reports failure to obtain consistent responding on a side-key reinforced by TO from a DRL 20 s in pigeons. In a study by Zeiler (1972c), pigeons stopped pecking side-keys if resetting the timer of a DRL was the consequence of this behaviour. It has been argued that studies on TO from FR or FI schedules are not really conclusive because the Time-Out does not seriously affect the distribution of reinforcement; the contingencies are not interrupted by the Time-Out, and a simple explanation might be found in the reinforcing value of changing the environment in a monotonous situation. Responding reinforced by Time-Out would not be escape behaviour but another kind of interim activity.

Another sign of aversiveness is the attack behaviour induced by some schedules of reinforcement, including DRL and FI. Attacks directed toward a conspecific (real or dummy) are much more frequent under FI or DRL than under CRF (continuous reinforcement) or when no contingencies are in operation. This has been demonstrated by Richards and Rilling (1972) and by DeWeeze (1977) with Fixed-Interval schedules and by Knutson and Kleinknecht (1970) with DRL (see Fig. 9.6). These attacks are currently considered as one particular class of interim or adjunctive activities. If the opportunity to engage in aggressive behaviour is contingent upon the production of a specific response subjects will emit this response (Cherek and others, 1973).

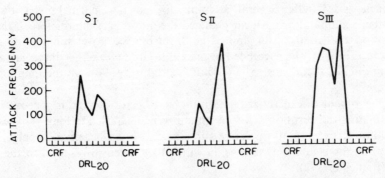

FIG. 9.6. Frequency of aggressive responses made against a target pigeon by pigeons 1, 2 and 3 during sessions of continuous reinforcement and DRL 20 s reinforcement. (From Knutson and Kleinknecht, 1970. Copyright 1970 by the Psychonomic Society, Inc.)

9.4. BEHAVIOURAL CONTRAST

Behavioural contrast (or induction in Pavlov's terminology) is another correlate of inhibition. In multiple schedules, two different contingencies alternate in association with different external stimuli. When a DRL schedule alternates with a VI schedule, it induces an increased rate of responding in the latter (though the rate of reinforcement may be equalized in both components of the multiple schedule). This effect is similar to what is observed when extinction periods alternate with positively reinforced responding (Terrace, 1968). Terrace has also proceeded to generalization tests with the coloured stimuli associated with the DRL and the VI components. He observed a shift of the generalization gradient in the direction opposite to the value of the DRL stimulus. This peak-shift is classically observed after discrimination training in the direction opposite to the negative stimulus. Weisman (1969) has obtained inhibitory gradients for a stimulus associated with the DRL components of a multiple schedule VI 1 min DRL 20 s.

Behavioural contrast is not a function of the frequency of reinforcement, but of the time between responses. This was elegantly demonstrated by Reynolds and Limpo (1968). They compared rates of responding in pigeons in the two components of a multiple DRL 35 DRL 35 s schedule. The two components were signalled by a red and a green light respectively. Training on this schedule resulted in comparable behaviour in both components as would be expected. An external visual clock—successive lighting of eight small lamps—was then added to one component (see Chapter 8, p. 200). This facilitated temporal adjustment and resulted in an improved spacing of responses and consequently, in a reduced number of responses together with an

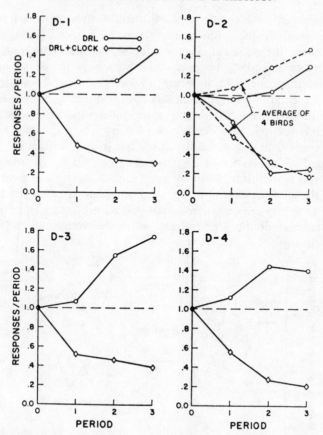

FIG. 9.7. Total number of responses per period of five sessions for each of four pigeons in each of two components of a multiple schedule. (From Reynolds and Limpo, 1968. Copyright 1968 by the Society for the Experimental Analysis of Behavior, Inc.)

increased number of reinforcements. At the same time, the number of responses increased in the other component, with the consequence that the number of reinforcements decreased (Fig. 9.7). This suggests that if response rates are lowered by added stimuli, there is a tendency for a compensatory (disinhibitory) increase to occur, and thus that inhibition in temporal schedules should be attributed to the spacing of responses in time rather than to the spacing of reinforcement.

9.5. OMISSION EFFECT

Further evidence of inhibition in Fixed-Interval schedules is provided by studies on the *omission effect*. If a short black-out period is occasionally presented in lieu of food reinforcement, the rate of responding following this omission of reinforcement is much higher than after food (Staddon and Innis, 1966). At first sight, this might be interpreted as a *frustration effect* (Amsel, 1958): the omission of the expected food would induce a state of increased arousal. A closer look at the data suggests a very different interpretation that has been documented by Staddon (1970b, 1974) and supported by a number of experiments. According to the frustration hypothesis, the more similarities there are between the usual reinforcing event and the signals presented in omission trials, the larger the omission effect should be, the higher the rate of responses. Experimental results invalidate this prediction. Signals presented on omission trials can be varied systematically as to the degree of similarity with the reinforcement. For instance, if the presentation of food is usually accompanied by a 3 s extinction of the key-light or of the house-light, by the noise and by the light of the grain dispenser, switching off the house-light and switching on the dispenser light will resemble more the reinforcing event than just switching off the house light. Presenting in lieu of any of these combined stimuli a stimulus that is not associated with reinforcement, say a click, would be still more distant from the usual reinforcement, and the distance would be maximal in the extreme case of unsignalled omission. It turns out that the more dissimilarity there is between omission signals and reinforcement, the larger the omission effect, the more responses are produced in the following interval (Fig. 9.8 Kello,

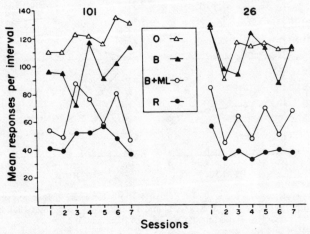

FIG. 9.8. Mean number of responses per second-component interval following reinforcement (R) and the various non-reinforcement events: B + ML (blackout + magazine light), B (blackout), 0 (unsignalled non-reinforcement). Daily means are plotted separately for each subject. (From Kello, 1972.)

1972). This would indicate that the reinforcement exerts a maximal inhibitory effect on responding in the subsequent interval, and that this effect decreases as an inverse function of the similarity between the stimuli presented and the original reinforcing event. Looked at in this way, the reinforcement omission effect might be viewed as a case of disinhibition.

9.6. THE INHIBITION—DISCHARGE BALANCE

Though the list might not be exhaustive, we have reviewed the main categories of data that suggest or strongly support the presence of inhibition in the situations where temporal regulations of behaviour are required or induced. Again, this confirms Janet's intuition (though he phrased it in different terms) and Pavlov's analysis. It is generally explicitly or implicitly accepted today by most experimenters in the field.

The work of Staddon has been especially important in reinstating the concept of inhibition in the study of temporal control (Staddon, 1969b, 1974, 1977), as has the work of Dews, though this author has not indulged to the same extent in theorizing. In one of his recent experimental contributions (Dews, 1978), in which he demonstrates the generality of the scalloped pattern in FI schedules, accumulating factual evidence of accelerated rate (rather than break-and-run pattern), Dews suggestively asks:

"Does acceleration start from the very beginning of the interval? In the absence of actual responding, no direct assessment of acceleration can be made, of course,...Starting from a level of suppressed responding at the beginning of the FI, acceleration may have to proceed by some time before actual responding starts. Perhaps acceleration is continuous and even fairly constant from the beginning of the FI, in spite of the fact that an initial pause in responding usually occurs".

What is postulated in such covert acceleration is, of course, inhibition below zero.

The view that active inhibitory processes are essential in all kinds of temporal regulation of behaviour help us in understanding a number of otherwise disparate and puzzling facts, and in accounting for several paradoxes in the frame of a unified hypothesis. Coming back to what has been suggested at the beginning of this chapter, we shall emphasize two fundamental aspects of the concept of inhibition as we see it in the present context. One aspect has been illustrated in the previous pages where we have argued, on experimental grounds, that behavioural inhibition is an active process *sui generis*, distinct from excitation, and not merely symmetrical to it. Whatever its physiological basis, behavioural inhibition would involve regulatory processes of a more elaborate type than excitation. The second aspect, admittedly less clearly demonstrated by experimental evidence (perhaps just because almost no experiment has been designed to test it), concerns the limits of inhibition, or of what we would call the *inhibitory load* that an organism can tolerate. Active inhibitory processes could not maintain their control beyond a certain point (for instance throughout long periods) unless the organism would have some opportunity for relief from the stress involved in inhibition. Inhibitory load should be counterbalanced by compensation activities. With both these aspects in mind, we may again look at some of the problems we have come across in our analysis of temporal regulations.

The contrast in the order of magnitude of the delay that can be mastered in situations inducing spontaneous regulations on the one hand, and in situations requiring time estimation of one's own behaviour or of some external events on the other, might be accounted for in terms of difference in the inhibitory load and in provision for compensation. Fixed-Interval contingencies allow for phases of high rates of operant responding alternating with pauses, and thus provide the necessary compensatory mechanisms. DRL schedules, or schedules of reinforcement of response duration, or discrimination of duration of external events, do not usually provide for such compensation. Hence, animals would adjust to much longer delays in the former as compared to the latter type of schedules. The fact that an organism is able to refrain from responding for a longer time when nothing requires it to do so than when pausing is the very condition for reinforcement would no longer appear as a paradox when we consider the mechanism of inhibition involved in each case.

The function of collateral behaviour, as already suggested in Chapter 7, is best understood in terms of compensation for inhibition. The long-prevailing reluctance of behaviourists to accept the concept of inhibition was responsible, as Staddon (1969b) correctly pointed out, for

> "...the strenuous attempts that are made to find positive behaviour (e.g. hom-ogenous or heterogenous chains) to 'mediate' spaced-responding and other time-related behaviours. The major impetus for this research comes from a more or less implicit acceptance of the exclusiveness of excitatory control".

The more general study of interim activities and adjunctive behaviour has led to such notions, as displacement, response competition, and inhibition, that are akin with our hypothesis of compensation for inhibition. What must be clearly emphasized is that temporal inhibition preceeds, or is a condition of collateral activities rather than a consequence of them, as sometimes suggested.

External events might also counterbalance inhibition. This might be part of the facilitating effect of various kinds of external clocks. More interestingly, it might be an important aspect of the control exerted by recurrent periodic stimuli—reinforcing or not—in Fixed-Interval or Fixed-Time schedules. We have compared such periodic stimuli with synchronizers of biological rhythms, which contribute very effectively to the cyclical patterns. Periodic external events would, so to speak, take over part of the organism's task in temporal regulation and alleviate the inhibitory load.

Inhibitory processes might also be the key to some, if not all, differences observed when different responses are used. Inhibition is of course always inhibition of some kind of behaviour. Not all responses are equally amenable to inhibition in a given context. Key-pecking in pigeons, because of its status in the natural repertoire, might not be amenable to the inhibition involved in DRL spacing, while more arbitrary responses (say lever-pressing) might be (see Chapter 5, Section 3.). Possibly, a similar reasoning might apply to species differences. Because of the level they occupy on the phyletic scale, or because of their ethological peculiarities, some species might be more prepared than others to inhibit certain responses in a given context, and this would account for the differences observed in adjustment to temporal schedules between pigeons and rats, or between mice and cats (see Chapter 4).

Whenever a compensation is needed to keep a mechanism operating, there is the danger that the equilibrium will be disrupted if the factors that must be compensated for are too strong, or if the compensating forces are too weak. Generalized inhibition would develop when compensating regulations fail. Depending upon the constitutional type of the organism, and possibly upon its individual history, generalized inhibition would be reached through excessive maladaptive inhibitory processes, or would mean more or less brutal switching to excitation and emotional by-products.

The preceding remarks must be taken as working hypotheses for future studies, not as statements of facts. Hopefully, they should lead to the designing of new experiments, aimed at providing new and decisive arguments for a unifying theory of temporal regulation based on the concept of active inhibition. The central theme of such a theory is that an organism that exhibits temporal regulation is not simply reading a clock dial (the working of which is unknown and in an unidentified location), it is organizing and deferring its own behaviour. Along these lines, it would seem advisable to concentrate on essential aspects of the problem and pay less attention to subtleties that might reveal unimportant, if not irrelevant. One example, now familiar to the reader, will suffice to make the point clear. It might not be crucial for a general theory of temporal regulation, to know which of the break-and-run or scallop patterns is observed more frequently in FI schedules, or which appears prior to the other; it would be more important, to show that both reflect—or do not reflect—the kind of compensation mechanism that we have hypothesized.

References

ADAMS, J. A. (1966) Some mechanisms of motor responding: An examination of attention. In E. A. Bilodeau (Ed.), *Acquisition of Skill*. Academic Press, New York, pp. 169–200.

ADAMS, J. A. (1968) Response feedback and learning. *Psychol. Bull.* **70**, 486–504.

ADAMS, J. A. (1976) Issues for a closed-loop theory of motor learning. In G. E. Stelmach (Ed.), *Motor Control: Issues and Trends*. Academic Press, New York, pp. 87–107.

ADAMS, J. A. (1977) Feedback theory of how joint receptors regulate the timing and positioning of a limb. *Psychol. Rev.* **84**, 504–523.

ADAMS, J. A. and L. R. CREAMER (1962a) Anticipatory timing of continuous and discrete responses. *J. exp. Psychol.* **63**, 84–90.

ADAMS, J. A. and L. R. CREAMER (1962b) Proprioception variables as determiners of anticipatory timing behavior. *Hum. Factors* **4**, 217–222.

ADAMS, J. A. and L. V. XHIGNESSE (1960) Some determinants of two-dimensional visual tracking behavior. *J. exp. Psychol.* **60**, 391–403.

AITKEN, W. C., J. T. BRAGGIO and P. ELLEN (1975). Effect of prefeeding on the DRL performance of rats with septal lesions. *J. comp. physiol. Psychol.* **89**, 546–555.

AJURIAGUERRA, J. DE (1968) Le passé de la chronophysiologie. In *Cycles Biologiques et Psychiatrie*. Georg et Cie, Genève; Masson et Cie, Paris, pp. 13–23.

ALLAN, L. G. and A. B. KRISTOFFERSON (1974) Psychophysical theories of duration discrimination. *Perception and Psychophysics* **16**, 26–34.

ALLEMAN, H. D. and J. R. PLATT (1973) Differential reinforcement of interresponse times with controlled probabilities of reinforcement per response. *Learn. Motiv.* **4**, 40–73.

ALLEMAN, H. D. and M. D. ZEILER (1974) Patterning with fixed-time schedules of response—independent reinforcement. *J. exp. Anal. Behav.* **22**, 135–141.

AMSEL, A. (1958) The role of frustrative nonreward in non-continuous reward situation. *Psychol. Bull.* **55**, 102–119.

AMSEL, A., R. LETZ and D. R. BURDETTE (1977) Appetitive learning and extinction in 11-day-old rat pups: effects of various reinforcement conditions. *J. comp. physiol. Psychol.* **91**, 1156–1167.

ANDERSON, A. C. (1932) Time discrimination in the white rat. *J. comp. Psychol.* **13**, 27–55.

ANDERSON, M. C. and S. J. SHETTLEWORTH (1977) Behavioral adaptation to fixed-interval and fixed-time food delivery in golden hamsters. *J. exp. Anal. Behav.* **27**, 33–49.

ANDERSON, O. D. (1976) *Time Series Analysis and Forecasting—The Box-Jenkins Approach*. Butterworths, London and Boston.

ANGER D. (1956) The dependence of interresponse times upon the relative reinforcement of different interresponse times. *J. exper. Psychol.* **52**, 145–161.

ANGER, D. (1963) The role of temporal discriminations in the reinforcement of sidman avoidance behavior. *J. exp. Anal. Behav.* **6**, 477–506.

APPEL, J. B. (1968) Fixed-interval punishment. *J. exp. Anal. Behav.* **11**, 803–808.

APPEL, J. B. and R. H. HISS (1962) The discrimination of contingent from non-contingent reinforcement. *J. comp. physiol. Psychol.* **55**, 37–39.

AUGE, R. J. (1977) Stimulus functions within a fixed-interval clock schedule: reinforcement, punishment, and discriminative stimulus control. *Anim. Learn. Behav.* **9**, 117–123.

AYRES, J. B., J. O. BENEDICT, R. GLACKENMEYER and W. MATTHEWS (1974) Some factors involved in the comparison of response systems: acquisition and transfer of headpoke and lever-press sidman avoidance. *J. exp. Anal. Behav.* **22**, 371–379.

AZRIN, N. H. (1956) Some effects of two intermittent schedules of immediate and non-immediate punishment. *Journal of Psychology* **42**, 3–21.

AZRIN, N. H. (1961) Time-out from positive reinforcement. *Science* **133**, 282–283.

BADDELEY, A. D. (1966) Time estimation at reduced body temperature. *Amer. J. Psychol.* **79**, 474–479.

BAGDONAS, A., V. B. POLYANSKY and E. N. SOKOLOV (1966) Participation of visual cortical units of rabbits in the conditioned reflex to time. *Zh. Vyssh. Nerv. Deiat.* **5**, 791–798.

BAHRICK, H. P. (1957) An analysis of stimulus variables influencing the proprioceptive control of movements. *Psychol. Rev.* **64**, 324–328.

BALSTER, R. L. and C. R. SCHUSTER (1973) Fixed-interval schedule of cocaine reinforcement: effect of dose and infusion duration. *J. exp. Anal. Behav.* **20**, 119–129.

BARD, L. (1922) Les bases physiologiques de la perception du temps. *J. Psychol. norm. pathol.* **2**, 119–146.

BARFIELD, R. J. and B. D. SACHS (1968) Sexual behavior: stimulation by painful electrical shock to skin in male rats. *Science* **161**, 392–395.

BAROFSKY, I. (1969) The effect of high ambient temperature on timing behavior in rats. *J. exp. Anal. Behav.* **12**, 59–72.

BARON, A. and M. GALIZIO (1976) Clock control of human performance on avoidance and fixed-interval schedules. *J. exp. Anal. Behav.* **26**, 165–180.

BARON, M. R. (1965) The stimulus, stimulus control, and stimulus generalization. In D. Mostofsky (Ed.), *Stimulus Generalization.* Stanford University Press, Stanford.

BAROWSKI, E. I. and D. E. MINTZ (1975) The effect of time-out locus during fixed-ratio reinforcement. *Bull. Psychon. Soc.* **5**, 137–140.

BAROWSKI, E. I. and D. E. MINTZ (1978) The effects of time-out duration during fixed-ratio reinforcement. *Bull. Psychon. Soc.* **11**, 215–218.

BARRETT, J. E. (1975) Conjunctive schedules of reinforcement II: response requirements and stimulus effects. *J. exp. Anal. Behav.* **24**, 23–31.

BARRETT, J. E. and J. R. GLOWA (1977) Reinforcement and punishment of behavior by the same consequent event. *Psychol. Rep.* **40**, 1015–1021.

BARTLEY, S. H. and G. H. BISHOP (1933) The cortical response to stimulation of the optic nerve in the rabbit. *Amer. J. Physiol.* **103**, 159–172.

BAUM, W. M. (1973) The correlation-based law of effect. *J. exp. Anal. Behav.* **20**, 137–153.

BEARD, R. R. and G. W. WERTHEIM (1966) Behavioral impairment associated with small doses of carbon monoxide. *Amer. J. Publ. Health* **57**, 2012–2022.

BEASLEY, J. L. and B. L. SELDEEN (1965) The effect of acceleration on food-reinforced DRL and FR. *J. exp. Anal. Behav.* **8**, 315–319.

BEATTY, W. W. (1973) Effects of gonadectomy on sex differences in DRL behavior. *Physiol. Behav.* **10**, 177–178.

BEATTY, W. W. (1977) Sex differences in DRL and active avoidance behaviors in the rat depend upon the day-night cycle. *Bull. Psychon. Soc.* **10**, 95–97.

BEATTY, W. W. (1978) DRL behavior in gerbils and hamsters of both sexes. *Bull. Psychon. Soc.* **11**, 41–42.

BEATTY, W. W., C. M. BIERLEY and J. M. GERTH (1975a). Effects of neonatal gonadectomy on DRL behavior. *Bull. Psychon. Soc.* **6**, 615.

BEATTY, W. W. and J. S. SCHWARTZBAUM (1968) Commonality and specificity of behavioral dysfunction following septal and hippocampal lesions in rats. *J. comp. physiol. Psychol.* **66**, 60–68.

BEATTY, W. W., D. R. STUDELSKA AND J. M. GERTH (1975b) Some aspects of the development of sex differences in DRL behavior. *Bull. Psychon. Soc.* **6**, 622–624.

BEECHER, M. D. (1971) Operant conditioning in the bat Phyllostomus Hastatus. *J. exp. Anal. Behav.* **16**, 219–223.

BEER B. and G. TRUMBLE (1965) Timing behavior as a function of amount of reinforcement. *Bull. Psychon. Soc.* **2**, 71–72.

BELING, I. (1929) *Z. vergl. Physiol.* **9**, 259–338.

BELL, C. R. (1965) Time estimation and increases in body temperature. *J. exp. Psychol.* **70**, 232–234.

BELL, C. R. (1975) Effects of lowered temperature on time estimation. *Q.J. exp. Psychol.* **27**, 531–538.

BELL, C. R. (1977) Time and temperature: a reply to Green and Simpson. *Q.J. exp. Psychol.* **29**, 341–344.

BELL, C. R. and K. A. PROVINS (1962) Effects of high temperature environmental conditions on human performance. *Journal of Occupational Medicine* **4**, 202–211.

BELL, C. R. and K. A. PROVINS (1963) Relation between physiological responses to environment heat and time judgement. *J. exp. Psychol.* **66**, 572–579.

BELLEVILLE, R. E., F. H. ROHLES, M. E. GRUNZKE and F. C. CLARK (1963) Development of a complex multiple schedule in the chimpanzee. *J. exp. Anal. Behav.* **6**, 549–556.

BERLYNE, D. E. (1957) Uncertainty and conflict: a point of contact between information-theory and behavior-theory concepts. *Psychol. Rev.* **64**, 329–339.

BERLYNE, D. E. (1960) *Conflict, Arousal and Curiosity.* McGraw Hill, New York.

BERRYMAN, R., W. WAGMAN and F. KELLER (1960) Chlorpromazine and the discrimination of response-produced cues. In L. Uhr and J. G. Miller (Eds.), *Drugs and Behavior,* John Wiley and Sons, New York, pp. 243–249.

BIRREN, J. E., S. L. CARDON and S. L. PHILIPS (1963) Reaction time as a function of the cardiac cycle in young adults. *Science* **140**, 195–196.

BITTERMAN, M. E. (1960) Toward a comparative psychology of learning. *Amer. Psychol.* **15**, 704–712.

BLACKMAN, D. E. (1968) Conditioned suppression or facilitation as a function of behavioral baseline. *J. exp. Anal. Behav.* **11**, 53–61.

BLACKMAN, D. E. (1970) Effects of a pre-shock stimulus on temporal control of behavior. *J. exp. Anal. Behav.* **14**, 313–319.

BLACKMAN, D. E. (1977) Conditioned suppression and the effects of classical conditioning on operant behavior. In W. K. Honig and J. E. R. Staddon (Eds.), *Handbook of Operant Conditioning*, Prentice-Hall, Inc., Englewood Cliffs, New Jersey, pp. 340–364.

BLACKMAN, D. E. and D. J. SANGER (1978) *Contemporary Research in Behavioral Pharmacology*. Plenum Press, New York and London, 1978.

BLANCHARD, R. (1977) Control of observing by antecedent stimuli. *Learn. Motiv.* **8**, 569–580.

BLANCHETEAU, M. (1965) Contribution à l'étude des estimations temporelles chez l'animal par la méthode du double évitement. *L'année Psychologique* **65**, 325–355.

BLANCHETEAU, M. (1967a) Les conduites collatérales dans l'estimation temporelle chez l'animal. *L'année Psychologique* **67**, 385–392.

BLANCHETEAU, M. (1967b) Effets séquentiels dans l'estimation du temps par double évitement chez le rat. *L'année Psychologique* **67**, 1–21.

BLONDIN, C. (1974) *Conditionnement operant de lémuriens malgaches*. University of Liège, Unpublished Master Thesis.

BLOUGH, D. S. (1958) New test for tranquillizers. *Science* **127**, 586–587.

BLOUGH, D. S. (1963) Interresponse times as a function of continuous variables: a new method and some data. *J. exp. Anal. Behav.* **6**, 237–246.

BLOUGH, D. S. (1966) The reinforcement of least-frequent interresponse times. *J. exp. Anal. Behav.* **9**, 581–591.

BOAKES, R. A. and M. S. HALLIDAY (1972) *Inhibition and Learning*. Academic Press, New York.

BOAKES, R. A., M. S. HALLIDAY and J. S. MOLE (1976) Successive discrimination training with equated reinforcement frequencies: failure to obtain behavioral contrast. *J. exp. Anal. Behav.* **26**, 65–78.

BOICE, R. and J. A. WITTER (1970) Motivating prairie dogs. *Psychon. Sci.* **20**, 287–289.

BOLLES, R. C. (1970) Species-specific defense reactions and avoidance learning. *Psychol. Rev.* **77**, 32–48.

BOLLES, R. C. (1975) *Theory and Motivation*, 2nd Ed. Harper and Row, Publishers, New York.

BOLLES, R. C. and DE LORGE, J. (1962) The rat's adjustment to a-diurnal feeding cycles. *J. comp. physiol. Psychol.* **55**, 760–762.

BOLLES, R. C. and N. E. GROSSEN (1969) Effects of an informational stimulus on the acquisition of avoidance behavior in rats. *J. comp. physiol. Psychol.* **68**, 90–99.

BOLLES, R. C. and R. J. POPP (1964) Parameters affecting the acquisition of Sidman avoidance. *J. exp. Anal. Behav.* **7**, 315–321.

BOLLES, R. C. and L. W. STOKES (1965) Rat's anticipation of diurnal and a-diurnal feeding. *J. comp. physiol. Psychol.* **60**, 290–294.

BONEAU, C. A. and J. L. COLE (1967) Decision theory, the pigeon and the psychophysical function. *Psychol. Rev.* **74**, 123–135.

BONVALLET, M., P. DELL and G. HIEBEL (1954) Tonus sympathique et activité électrique corticale. *Electroenceph. Clin. Neurophysiol.* **6**, 114–119.

BORDA, R. P. (1970) The effect of altered drive states on the contingent negative variation in rhesus monkeys. *Electroenceph. Clin. Neurophysiol.* **29**, 173–180.

BOREN, M. C. P. and L. R. GOLLUB (1972) Accuracy of the performance on a matching-to-sample procedure under interval schedules. *J. exp. Anal. Behav.* **18**, 65–77.

BOWER, G. H. (1961) Correlated delay of reinforcement. *J. comp. physiol. Psychol.* **54**, 196–203.

BRADY, J. V. and D. G. CONRAD (1960a) Some effects of brain stimulation on timing behavior. *J. exp. Anal. Behav.* **3**, 93–106.

BRADY, J. V. and D. G. CONRAD (1960b) Some effects of limbic system self-stimulation upon conditioned emotional behavior. *J. comp. physiol. Psychol.* **53**, 128–137.

BRADY, J. V. and W. J. H. NAUTA (1953) Subcortical mechanisms in emotional behavior: affective changes following septal forebrain lesions in the albino rat. *J. comp. physiol. Psychol.* **46**, 339–346.

BRADY, J. V. and W. J. H. NAUTA (1955) Subcortical mechanisms in emotional behavior: the duration of affective changes following septal and habenular lesions in the albino rat. *J. comp. physiol. Psychol.* **43**, 412.

BRAGGIO, J. T. and P. ELLEN (1974) Differential proprioceptive feedback and DRL performance in normal and septal rats. *J. comp. physiol. Psychol.* **87**, 80–89.

BRAGGIO, J. T. and P. ELLEN (1976) Cued DRL training: Effects on the permanence of lesion-induced responding. *J. comp. physiol. Psychol.* **90**, 694–703.

BRAKE, S. C. (1978) Discrimination training in infant, preweanling, and weanling rats: effects of prior learning experiences with the discriminanda. *Anim. Learn. Behav.* **6**, 435–443.

BRELAND K. and M. BRELAND (1961) The misbehavior of organisms. *Amer. Psychol.* **16**, 661–664.

BROWN, F. A. JR. (1976) Evidence for external timing of biological clocks. In J. D. Palmer, *An Introduction to Biological Rhythms*. Academic Press, New York, San Francisco, London, pp. 209–279.

BROWN, T. G. and R. K. FLORY (1972) Schedule induced escape from fixed-interval reinforcement. *J. exp. Anal. Behav.* **17**, 395–403.

BROWN, P. L. and H. M. JENKINS (1967) Conditioned inhibition and excitation in operant discrimination training. *J. exp. Psychol.* **75**, 225–266.

BROWN, P. L. and H. M. JENKINS (1968) Auto-shaping of the pigeon's keypeck. *J. exp. Anal. Behav.* **11**, 1–8.

BROWN, S. and J. A. TROWILL (1970) Lever-pressing performance for brain stimulation on FI and VI schedules in a single-lever situation. *Psychol. Rep.* **26**, 699–706.

BRUNER, A. and S. H. REVUSKY (1961) Collateral behavior in humans. *J. exp. Anal. Behav.* **4**, 849–850.

BUCHWALD, J. S., E. S. HALLAS and S. SCHRAMM (1966) Relationships of neuronal spike populations and EEG activity in chronic cats. *Electroenceph. Clin. Neurophysiol.* **21**, 227–238.

BUNNING, E. (1973) *The Physiological Clock*. Springer Verlag, Berlin.

BUNO, W. JR. and J. C. VELLUTI (1977) Relationships of hippocampal theta cycles with barpressing during self stimulation. *Physiol. Behav.* **19**, 615–621.

BURKETT, E. E. and B. N. BUNNEL (1966) Septal lesions and the retention of DRL performance in the rat. *J. comp. physiol. Psychol.* **62**, 468–471.

BUSH, R. R., E. GALANTER and R. D. LUCE (1963) Characterization and classification of choice experiments. In R. D. Luce, R. R. Bush and E. Galanter (Eds.), *Handbook of Mathematical Psychology*, vol. 1. John Wiley and Sons, Inc., New York, London, Sidney, pp. 77–102.

BUYTENDIJK, F. J. J., W. FISCHEL and P. B. TER LAAG (1935) Uber den Zeitsinn der Tiere. *Arch. Neerl. Physiol.* **20**, 123–154.

BYRD, L. D. (1969) Response in the cat maintained under response-independent electric shock and response-produced electric shock. *J. exp. Anal. Behav.* **12**, 1–10.

BYRD, L. D. (1972) Responding in the squirrel monkey under second order schedules of shock delivery. *J. exp. Anal. Behav.* **18**, 155–167.

BYRD, L. D. (1975) Contrasting effects of morphine on schedule-controlled behavior in the chimpanzee and baboon. *J. pharmac. exper. Therap.* **193**, 861–869.

BYRD, L. D. and M. J. MARR (1969) Relations between patterns of responding and the presentation of stimuli under second order schedules. *J. exp. Anal. Behav.* **12**, 713–722.

CAGGIULA, A. R. and R. EIBERGEN (1969) Copulation of virgin male rats evoked by painful peripheral stimulation. *J. comp. physiol. Psychol.* **69**, 414–419.

CALLAWAY, E. (1964) Response speed, the EEG alpha cycle and the autonomic cardiovascular cycle. In J. Birren and A. Wilford (Eds.), *Behavior, Aging, and the Nervous System*. C. Thomas, Springfield, Illinois.

CALLAWAY, E. (1965) *Some Effects of Respiratory and Cardiac Cycles*. Office of Naval Research Contract NONR 2931 (00). Progress Report, 46 pp.

CANTOR, M. B. (1971) Signalled reinforcing brain stimulation facilitates operant behavior under schedules of intermittent reinforcement. *Science* **174**, 610–613.

CAPLAN, H. J., J. KARPICKE and M. RILLING (1973) Effects of extended fixed-interval training on reversed scallops. *Anim. Learn. Behav.* **1**, 293–296.

CAPLAN, M. (1970) Effects of withheld reinforcement on timing behavior of rats with limbic lesions. *J. comp. physiol. Psychol.* **71**, 119–135.

CAPLAN, M. and J. STAMM (1967) DRL acquisition in rats with septal lesions. *Psychon. Sci.* **8**, 5–6.

CAREY, R. J. (1967a) Contrasting effects of increased thirst and septal ablation on DRL responding in rats. *Physiol. Behav.* **2**, 287–290.

CAREY, R. J. (1967b) Independence of effects of septal ablations on water intake and response inhibition. *Psychon. Sci.* **8**, 3–4.

CAREY, R. J. (1968) A further localization of inhibition deficits resulting from septal ablation. *Physiol. Behav.* **3**, 645–649.

CAREY, R. J. (1969) Contrasting effects of anterior and posterior septal injury on thirst motivated behaviour. *Physiol. Behav.* **4**, 759–764.

CARLSON, V. R. and I. FEINBERG (1976) Judgment of short time intervals following awakening from different EEG stages of sleep. *Physiol. Psychol.* **4**, 341–345.

CARLSON, V. R., I. FEINBERG and D. R. GOODENOUGH (1978) Perception of the duration of sleep intervals as a function of EEG sleep stage. *Physiol. Psychol.* **6**, 497–500.

CARLTON, P. L. (1961) The interacting effects of deprivation and reinforcement schedule. *J. exp. Anal. Behav.* **4**, 379–381.

CARRIGAN, P. F., J. D. BENEDICT and J. B. AYRES (1972) A comparison of lever-press and head-poke discriminated Sidman avoidance. *Behavior Research Methods and Instrumentation* **4**, 301–303.

CARTER, D. E. and J. J. BRUNO (1968) Extinction and reconditioning of behavior generated by a DRL contingency of reinforcement. *Psychon. Sci.* **11**, 19–20.

CARTER, D. E. and G. J. MACGRADY (1966) Acquisition of a temporal discrimination by human subjects. *Psychon. Sci.* **5**, 309–310.

CATANIA, A. C. (1970) Reinforcement schedules and psychophysical judgment: a study of some temporal properties of behavior. In W. N. Schoenfeld (Ed.), *The Theory of Reinforcement Schedules*, Prentice-Hall, Inc., Englewood Cliffs, New Jersey, pp. 1–42.

CATANIA, A. C. and D. CUTTS (1963) Experimental control of superstitious responding in humans. *J. exp. Anal. Behav.* **6**, 203–208.

CATANIA, A. C. and G. S. REYNOLDS (1968) A quantitative analysis of the responding maintained by interval schedules of reinforcement. *J. exp. Anal. Behav.* **11**, 327–383.

CHEREK, D. R., T. THOMPSON and G. T. HEISTAD (1973) Responding maintained by the opportunity to attack during an interval food reinforcement schedule. *J. exp. Anal. Behav.* **19**, 113–123.

CHIORINI, J. R. (1969) Slow potential changes from cat cortex and classical aversive conditioning. *Electroenceph. Clin. Neurophysiol.* **26**, 399–406.

CHRISTINA, R. W. (1970) Proprioception as a basis for the temporal anticipation of motor responses. *J. Mot. Behav.* **2**, 125–133.

CHRISTINA, R. W. (1971) Movement-produced feedback as a mechanism for the temporal anticipation of motor responses. *J. Mot. Behav.* **3**, 97–104.

CHUNG, S. H. and A. J. Neuringer (1967) Control of responding by a percentage reinforcement schedule. *Psychon. Sci.* **8**, 25–26.

CHURCH, R. M. and J. CARNATHAN (1963) Differential reinforcement of short latency responses in the white rat. *J. comp. physiol. Psychol.* **56**, 120–123.

CHURCH, R. M. and M. Z. DELUTY (1977) Bisection of temporal intervals. *J. exp. Psychol.: Anim. Behav. Proc.* **3**, 216–228.

CHURCH, R. M., D. J. GETTY and N. D. LERNER (1976) Duration discrimination by rats. *J. exp. Psychol.: Anim. Behav. Proc.* **2**, 303–312.

CLARK, F. C. (1962) Some observations on the adventitious reinforcement of drinking under food reinforcement. *J. exp. Anal. Behav.* **5**, 61–63.

CLARK, C. V. and R. L. ISSACSON (1965) Effect of bilateral hippocampal ablation on DRL performance. *J. comp. physiol. Psychol.* **59**, 137–140.

CLARK, F. C. and J. B. SMITH (1977) Schedules of food postponement: II. Maintenance of behavior by food postponement and effects of the schedule parameter. *J. exp. Anal. Behav.* **28**, 253–269.

CLOAR, T. and K. B. MELVIN (1968) Performance of two species of quail on basic reinforcement schedules. *J. exp. Anal. Behav.* **11**, 187–190.

COHEN, J. (1964) Psychological time. *Scientific American* **211**, 116–124.

COHEN, J. (1967) *Psychological Time in Health and Disease.* Charles Thomas, Publisher, Springfield, Illinois.

COHEN, P. S. (1970) DRL escape: effects of minimum duration and intensity of electric shock. *J. exp. Anal. Behav.* **13**, 41–50.

COHEN, S. L., J. E. HUGUES and D. A. STUBBS (1973) Second-order schedules: manipulation of brief-stimulus duration at component completion. *Anim. Learn. Behav.* **1**, 121–124.

COHEN, S. L. and S. A. STUBBS (1976) Discriminative properties of briefly presented stimuli. *J. exp. Anal. Behav.* **25**, 15–25.

COLERIDGE, H. M., J. C. G. COLERIDGE and F. ROSENTHAL (1976) Prolonged inactivation of cortical pyramidal tract neurones in cats by distension of the carotid sinus. *J. Physiol. (London)* **256**, 635–649.

COLLIER, G., E. HIRSCH and R. KANAREK (1977) The operant revisited. In W. K. Honig and J. E. R. Staddon (Eds.), *Handbook of Operant Behavior*, Prentice-Hall, Inc., Englewood Cliffs, New Jersey, pp. 28–53.

COLLIER, G. and L. MYERS (1961) The loci of reinforcement. *J. exp. Psychol.* **61**, 57–66.

COLLIER, G. and F. WILLIS (1961) Deprivation and reinforcement. *J. exp. Psychol.* **62**, 377–384.

CONRAD, D. G., M. SIDMAN and R. J. HERRNSTEIN (1958) The effect of deprivation upon temporally spaced responding. *J. exp. Anal. Behav.* **1**, 59–65.

CONTRUCCI, J. J., D. HOTHERSALL and D. D. WICKENS (1971) The effect of a novel stimulus introduced into a DRL schedule at two temporal placements. *Psychon. Sci.* **23**, 97–99.

COOK, L. and R. KELLEHER (1961) The interaction of drugs and behavior. In E. Rothlin (Ed.), *Neuropsychopharmacology*, vol. 2. Elsevier, Amsterdam, pp. 77–92.

COQUERY, J-M. and J. REQUIN (1967) Problèmes psychophysiologiques posés par la périodicité cardiaque. *Psychol. Belg.* **7**, 17–44.

CORFIELD-SUMNER, P. K. and D. E. BLACKMAN (1976) Fixed versus variable sequences of food and stimulus presentation in second-order schedules. *J. exp. Anal. Behav.* **26**, 405–413.

COUCH, J. V. and J. A. TROWILL (1971) Free operant DRL performance maintained by a single electrical brain stimulation reinforcement. *Psychon. Sci.* **25**, 303–304.

COURY, J. N. (1967) Neural correlates of food and water intake in the rat. *Science* **156**, 1763–1765.

COWLES, J. T. and J. L. FINAN (1941) An improved method for establishing temporal discrimination in white rats. *J. Psychol.* **11**, 335–342.

CREUTZFELDT, O. D., S. WATANABE and H. D. LUX (1966) Relations between EEG phenomena and potentials of single cortical cell. *Electroenceph. Clin. Neurophysiol.* **20**, 1–37.

CROSSMAN, E. K., R. S. HEAPS, D. L. NUNES and L. A. ALFERINK (1974) The effects of number of responses on pause length with temporal variables controlled. *J. exp. Anal. Behav.* **22**, 115–120.

CUMMING, W. W. and W. N. SCHOENFELD (1958) Behavior under extended exposure to a high value fixed-interval reinforcement schedule. *J. exp. Anal. Behav.* **1**, 245–263.

CUMMING, W. W. and W. N. SCHOENFELD (1960) Behavioral stability under extended exposure to a time-correlated reinforcement contingency. *J. exp. Anal. Behav.* **3**, 71–82.

DAVENPORT, J. W., C. F. FLAHERTY and J. P. DYRUD (1966) Temporal persistence of frustration effects in monkeys and rats. *Psychon. Sci.* **6**, 411–412.

DAVIS, H., J. MEMMOTT and H. M. B. HURWITZ (1975) Autocontingencies: a model for subtle behavioral control. *J. exp. Psychol.* **104**, 169–188.

DAVIS, H. and L. WHEELER (1967) The collateral pretraining of spaced responding. *Psychon. Sci.* **8**, 281–282.

DEADWYLER, S. A. and E. F. SEGAL (1965) Determinants of polydipsia: VII. Removing drinking solution midway through DRL session. *Psychon. Sci.* **3**, 195–196.

DELACOUR, J. (1971) Effects of medial thalamic lesions in the rat: a review and an interpretation. *Neuropsychologia* **9**, 157–174.

DELACOUR, J. and T. ALEXINSKY (1969) Rôle spécifique d'une structure thalamique médiane dans le comportement conditionné du rat blanc. *C.R. Acad. Sci., Paris* **268**, 569–572.

DELIÈGE, M. (1975) *Le comportement de régulation temporelle en programme à intervalle fixe.* Unpublished doctoral dissertation, University of Liège.

DE LORGE, J. (1967) Fixed-interval behavior maintained by conditioned reinforcement. *J. exp. Anal. Behav.* **10**, 271–276.

DE WEEZE, J. (1977) Schedule-induced biting under fixed-interval schedules of food or electric shock presentation. *J. exp. Anal. Behav.* **27**, 419–432.

DEWS, P. B. (1955) Studies on behavior. I. Differential sensitivity to pentobarbital of pecking performance in pigeons depending on the schedule of reward. *J. Pharmacol. exp. Therap.* **113**, 393–401.

DEWS, P. B. (1962) The effect of multiple S^Δ periods on responding on fixed-interval schedule. *J. exp. Anal. Behav.* **5**, 369–374.

DEWS, P. B. (1965a) The effect of multiple S^Δ periods on responding on a fixed-interval schedule: II. In a primate. *J. exp. Anal. Behav.* **8**, 53–54.

DEWS, P. B. (1965b) The effect of multiple S^Δ periods on responding on a fixed-interval schedule: III. Effects of changes in pattern of interruption, parameters and stimuli. *J. exp. Anal. Behav.* **8**, 427–435.

DEWS, P. B. (1966a) The effect of multiple S^Δ periods on responding on a fixed-interval schedule: IV. Effect of continuous S^Δ with only short S^D probes. *J. exp. Anal. Behav.* **9**, 147–151.

DEWS, P. B. (1966b) The effect of multiple S^Δ periods on responding on a fixed-interval schedule: V. Effect of periods of complete darkness and of occasional omission of food presentation. *J. exp. Anal. Behav.* **9**, 573–578.

DEWS, P. B. (1969) Studies on responding under fixed-interval schedules of reinforcement: the effect on the pattern of responding of changes in requirements at reinforcement. *J. exp. Anal. Behav.* **12**, 191–199.

DEWS, P. B. (1970) The theory of fixed-interval responding. In W. N. Schoenfeld (Ed.), *The Theory of Reinforcement Schedules.* Appleton Century Crofts, New York, pp. 43–62.

DEWS, P. B. (1978) Studies on responding under fixed-interval schedules of reinforcement: II. The scalloped pattern of the cumulative record. *J. exp. Anal. Behav.* **29**, 67–75.

DEWS, P. B. and G. R. WENGER (1977) Rate-dependency of the behavioral effects of amphetamine. In T. Thompson and P. B. Dews (Eds.), *Advances in Behavioral Pharmacology*, vol. 1. Academic Press, New York, pp. 167–277.

DINSMOOR, J. A., M. P. BROWNE and C. E. LAWRENCE (1972) A test of the negative discriminative stimulus as a reinforcer for observing. *J. exp. Anal. Behav.* **18**, 79–85.

DINSMOOR, J. A., G. A. FLINT, R. F. SMITH and N. F. VIEMEISTER (1969) Differential reinforcing effects of stimuli associated with the presence or absence of a schedule of punishment. In D. P. Hendry (Ed.), *Conditioned Reinforcement.* Dorsey, Homewood, Illinois, pp. 357–384.

DINSMOOR, J. A. and G. W. SEARS (1973) Control of avoidance by a response-produced stimulus. *Learn. Motiv.* **4**, 284–293.

DMITRIEV, A. S. and A. M. KOCHIGINA (1959) The importance of time as stimulus of conditioned reflex activity. *Psychol. Bull.* **56**, 106–132.

DONCHIN, E., D. OTTO, L. K. GERBRANDT and K. H. PRIBRAM (1971) While a monkey waits: electrocortical events recorded during the foreperiod of a reaction time study. *Electroenceph. Clin. Neurophysiol.* **31**, 115–127.

DONG, E. JR. and B. A. REITZ (1970) Effect of timing of vagal stimulation on heart rate in the dog. *Circulation Research* **27**, 635–646.

DONOVICK, P. J. (1968) Effects of localized septal lesions on hippocampal EEG activity and behavior in rats. *J. comp. physiol. Psychol.* **66**, 569–578.

DOTY, R. W. (1961) Conditioned reflexes formed and evoked by brain stimulation. In D. E. Sheer (Ed.). *Electrical Stimulation of the Brain*, Univ. Texas Press, Austin, pp. 397–412.

DOUGHERTY, J. and R. PICKENS (1973) Fixed-interval schedules of intravenous cocaine presentation in rats. *J. exp. Anal. Behav.* **20**, 111–118.

DUKICH, T. D. and A. E. LEE (1973) A comparison of measures of responding under fixed-interval schedules. *J. exp. Anal. Behav.* **20**, 281–290.

DUNHAM, P. (1977) The nature of reinforcing stimuli. In W. K. Honig, and J. E. R. Staddon (Eds.), *Handbook of Operant Behavior.* Prentice-Hall, Inc., Englewood Cliffs, New Jersey, pp. 98–124.

DUNN, M. E., W. S. FOSTER and H. M. B. HURWITZ (1971) Effects of cycle length on performance on a temporally defined avoidance schedule. *J. exp. Anal. Behav.* **16**, 263–268.

DURUP, G. and A. FESSARD (1935) L'électroencéphalogramme de l'homme. *L'Année Psychologique* **36**, 1–32.

EDWARDS, D. D., J. R. WEST and V. JACKSON (1968) The role of contingencies in the control of behavior. *Psychon. Sci.* **10**, 39–40.

EHRLICH, A. (1963) Effects of tegmental lesions on motivated behavior in rats. *J. comp. physiol. Psychol.* **56**, 390–396.

EISLER, H. E. (1975) Subjective duration and psychophysics. *Psychol. Rev.* **82**, 428–450.

ELLEN, P. and W. C. AITKEN (1971) Absence of temporal discrimination following septal lesions. *Psychon. Sci.* **22**, 129–131.

ELLEN, P. and J. BRAGGIO (1973) Reactions to DRL schedule change in rats with septal damage. *Physiol. Psychol.* **1**, 267–272.

ELLEN, P. and J. BUTTER (1969) External cue control of DRL performance in rats with septal lesions. *Physiol. Behav.* **4**, 1–6.

ELLEN, P., P. G. DORSETT and W. K. RICHARDSON (1977a) The effect of cue-fading on the DRL performance of septal and normal rats. *Physiol. Psychol.* **5**, 469–476.

ELLEN, P., G. GILLENWATER and W. K. RICHARDSON (1977b) Extinction responding by septal and normal rats following acquisition under four schedules of reinforcement. *Physiol. Behav.* **18**, 609–615.

ELLEN, P. and M. M. KELNHOFER (1971) Discrimination of response feedback following septal lesions. *Psychon. Sci.* **23**, 94–96.

ELLEN, P., L. MAKOHON and W. K. RICHARDSON (1978) Response suppression on DRL by rats with septal damage. *J. comp. physiol. Psychol.* **92**, 511–521.

ELLEN, P. and E. W. POWELL (1962a) Effects of septal lesions on behavior generated by positive reinforcement. *Exp. Neurol.* **6**, 1–11.

ELLEN, P. and E. W. POWELL (1962b) Temporal discrimination in rats with rhinencephalic lesions. *Exp. Neurol.* **6**, 538–547.

ELLEN, P. and E. W. POWELL (1963) Timing behavior after lesions of zona incerta and mamillary body. *Science* **141**, 828–830.

ELLEN, P. and E. W. POWELL (1966) Differential conditioning of septum and hippocampus. *Exp. Neurol.* **16**, 162–171.

ELLEN, P., A. S. WILSON and E. W. POWELL (1964) Septal inhibition and timing behavior in the rat. *Exp. Neurol.* **10**, 120–132.

ELLIS, M. J. (1969) Control dynamics and timing a discrete motor response. *J. Mot. Behav.* **1**, 119–134.

ELLIS, M. J., R. A. SCHMIDT and M. G. WADE (1968) Proprioception variables as determinants of lapsed-time estimation. *Ergonomics* **11**, 577–586.

ELSMORE, T. F. (1971a) Control of responding by stimulus duration. *J. exp. Anal. Behav.* **16**, 81–87.

ELSMORE, T. F. (1971b) Independence of post-reinforcement pause length and running rate on FI pacing reinforcement schedules. *Psychon. Sci.* **23**, 371–372.

ELSMORE, T. F. (1972) Duration discrimination: effects of probability of stimulus presentation. *J. exp. Anal. Behav.* **18**, 465–469.

EMERY, F. E. (1929) Effects of the heart rate on the tonus of skeletal muscle. *Amer. J. Physiol.* **88**, 529–533.

ESKIN, R. M. and M. E. BITTERMAN (1960) Fixed-interval and fixed ratio performance in the fish as function of prefeeding. *Amer. J. Psychol.* **73**, 417–423.

ESTES, W. K. and B. F. SKINNER (1941) Some quantitative properties of anxiety. *J. exp. Psychol.* **29**, 390–400.

EVANS, H. L. (1971) Rat's activity: influence of light-dark cycle, food presentation and deprivation. *Physiol. Behav.* **7**, 455–459.

FALK, J. (1961) Production of polydipsia in normal rats by an intermittent food schedule. *Science* **133**, 195–196.

FALK, J. L. (1966) Schedule-induced polydipsia as a function of fixed interval length. *J. exp. Anal. Behav.* **9**, 37–39.

FALK, J. L. (1969) Conditions producing polydipsia in animals. *Annals of the New York Academy of Sciences* **157**, 569–593.

FALK, J. L. (1977) The origin and functions of adjunctive behavior. *Anim. Learn. Behav.* **5**, 325–335.

FANTINO, E. (1977) Conditioned reinforcement: choice and information. In W. K. Honig and J. E. R. Staddon (Eds.), *Handbook of Operant Behavior*, Prentice-Hall, Inc., Englewood Cliffs, New Jersey, pp. 313–340.

FARMER, J. and W. N. SCHOENFELD (1964a) Inter-reinforcement times for the barpressing response of white rats on two DRL schedules. *J. exp. Anal. Behav.* **7**, 119–122.

FARMER, J. and W. N. SCHOENFELD (1964b) Effects of a DRL contingency added to a fixed-interval reinforcement. *J. exp. Anal. Behav.* **7**, 391–399.

FARMER, J. and W. N. SCHOENFELD (1966) Varying temporal placement of an added stimulus in a fixed-interval schedule. *J. exp. Anal. Behav.* **9**, 369–375.

FARTHING, G. W. and E. HEARST (1972) Stimulus generalization along the click-frequency (flutter) continuum in pigeons. *Perceptions and Psychophysics* **12**, 176–182.

FECHNER, G. T. (1860) *Elemente der Psychophysik.* Breitkopf und Hartel, Leipzig.

FELTON, M. and D. O. LYON (1966) The post-reinforcement pause. *J. exp. Anal. Behav.* **9**, 131–134.

FEOKRITOVA, I. P. (1912) *Le temps, excitateur conditionnel de la glande salivaire.* Thèse, Saint-Petersbourg.

FERRARO D. P. and D. M. GRILLY (1970) Response differentiation: a psychophysical method for response-produced stimuli. *Perception and Psychophysics* **7**, 206–208.

FERRARO, D. P., W. N. SCHOENFELD and G. A. SNAPPER (1965) Sequential response effects in the white rat during conditioning and extinction on a DRL schedule. *J. exp. Anal. Behav.* **8**, 255–260.

FERSTER, C. B. and B. F. SKINNER (1957) *Schedules of reinforcement.* Appleton Century Crofts, New York.

FERSTER, C. B. and J. ZIMMERMAN (1963) Fixed-interval performances with added stimuli in monkeys. *J. exp. Anal. Behav.* **6**, 317–322.

FINDLEY, J. D. (1963) *Establishment and maintenance of long term S^{Δ} avoidance in a chimpanzee.* Paper presented at the Eastern Psychological Association, New York, April 13.

FLANAGAN, B. and W. B. WEBB (1964) Disinhibition and external inhibition in fixed interval conditioning. *Psychon. Sci.* **1**, 123–124.

FLORY, R. K. (1969) Attack behavior as a function of minimal inter-food interval. *J. exp. Anal. Behav.* **12**, 825–828.

FLYNN, J. P., P. D. MacLEAN and C. KIM (1961) Effects of hippocampal after-discharges on conditioned responses. In D. E. Sheer (Ed.), *Electrical Stimulation of the Brain*, Univ. of Texas Press, Austin, pp. 380—386.

FONTAINE, O., A. BEAUJOT, M. DIDELEZ and D. LECLERCQ (1966) Etude d'une nouvelle technique d'évitement sans signal avertisseur chez le rat. *Psychol. Belg.* **6**, 11–17.

FONTAINE, O., A. HAUGLUSTAINE, PH. LIBON and M. RICHELLE (1975) Action du sulpiride sur le comportement operant chez l'animal. *Therapie* **30**, 573–584.

FONTAINE, O., K. KLAUSER, S. KLOCKER, C. LACREMANS and M. LICOT (1977) Comparaison de trois réponses chez le rat. Unpublished data, Laboratory of Experimental Psychology, University of Liège.

FONTAINE, O. and M. RICHELLE (1967) Antagonisme des effets centraux et périphériques de la trémorine par deux parasympathicolytiques, l'atropine et la scopolamine. *Psychopharmacologia* **11**, 154–164.

FOREE, D. D. and V. M. LOLORDO (1970) Signalled and unsignalled free-operant avoidance in the pigeon. *J. exp. Anal. Behav.* **13**, 283–290.

FORSYTH, R. P. (1966) Influence of blood pressure on patterns of voluntary behavior. *Psychophysiology* **2**, 98–102.

FOWLER, S. C., C. MORGENSTEIN and J. M. NOTTERMAN (1972) Spectral analysis of variations in force during barpressing time discrimination. *Science* **176**, 1126–1127.

FOX, R. H., P. A. BRADBURY and I. F. G. HAMPTON (1967) Time judgment and body temperature. *J. exp. Psychol.* **75**, 88–96.

FRAISSE, P. (1948) Etude comparée de la perception et de l'estimation de la durée chez les enfants et les adultes. *Enfance* **1**, 199–211.

FRAISSE, P. (1967) Perception et estimation du temps. In P. Fraisse and J. Piaget (Eds.), *Traité de psychologie expérimentale*, n° 6, P.U.F., Paris, pp. 63–99.

FRAISSE, P. and F. ORSINI (1958) Etude expérimentale des conduites temporelles: étude génétique de l'estimation de la durée. *L'Année Psychologique* **58**, 1–6.

FRAISSE, P., M. SIFFRE, G. OLERON and N. ZUILI (1968) Le rythme veille-sommeil et l'estimation du temps. In *Cycles biologiques et psychiatrie.* Georg et Cie, Genève; Masson, Paris, pp. 257–265.

FRANÇOIS, M. (1927) Contribution à l'étude du sens du temps. La température comme facteur de variation de l'appréciation subjective des durées. *L'Année Psychologique* **28**, 186–204.

FRANÇOIS, M. (1928) Influence de la température sur notre appréciation du temps. *C.R. Soc. Biol., Paris* **98**, 201–203.

FRANK, J. and J. E. R. STADDON (1974) Effects of restraint on temporal discrimination behavior. *Psychol. Rec.* **24**, 123–130.

FREY, P. W. and J. A. COLLIVER (1973) Sensitivity and responsivity measures for discrimination learning. *Learn. Motiv.* **4**, 327–342.

FRIED, P. A. (1972) Septum and behavior: a review. *Psychol. Bull.* **78**, 292–310.

FRY, W., R. T. KELLEHER and L. COOK (1960) A mathematical index of performance on fixed interval schedule of reinforcement. *J. exp. Anal. Behav.* **3**, 193–199.

GAHÉRY, Y. and D. VIGIER (1974) Inhibition effects in the cuneate nucleus produced by vago-aortic afferent fibers. *Brain Research* **75**, 241–246.

GAMZU, E. and D. R. WILLIAMS (1971) Classical conditioning of a complex skeletal act. *Science* **171**, 923–925.

GENTRY, W. D. (1968) Fixed-ratio schedule-induced aggression. *J. exp. Anal. Behav.* **11**, 813–817.

GIBBON, J. (1977) Scalar expectancy theory and Weber's law in animal timing. *Psychol. Rev.* **84**, 279–325.

GINSBURG, N. (1960) Conditioned vocalization in the budgerigar. *J. comp. physiol. Psychol.* **53**, 183–186.

GINSBURG, N. and V. NILSON (1971) Measuring flicker threshold in the budgerigar. *J. exp. Anal. Behav.* **15**, 189–192.

GIURGEA, C. (1953) *Elaborarea reflexuleu conditionat prin excitare directa a scoartei cerebrale.* Bucharest Ed. Acad. rep. pop. Romane.

GIURGEA, C. (1955) Die dynamik der Ausarbeitung einer zeitlichen Bezichung durch direkte Reizung der Hirnrinde. *Ber. Ges. Physiol.* **175**, 80.

GLAZER, H. and D. SINGH (1971) Role of collateral behavior in temporal discrimination performance and learning. *J. exp. Psychol.* **91**, 78–84.

GLENCROSS, D. J. (1977) Control of skilled movements. *Psychol. Bull.* **84**, 14–29.

GLICK, S. D. and R. D. COX (1976) Differential effects of unilateral and bilateral lesions on side preference and timing behavior in rats. *J. comp. physiol. Psychol.* **90**, 528–535.

GLICKSTEIN, M., W. A. QUIGLEY and W. C. STEBBINS (1965) Effects of frontal lesions and parietal lesions on timing behavior in monkeys. *Psychon. Sci.* **1**, 265–266.

GODEFROID, J. (1968) *Essai de mesure d'un comportement instinctif à l'aide du conditionnement operant.* Unpublished Master Thesis, University of Liège.

GODEFROID, J., E. CHARLIER and J. P. MOREAU (1969) *Conditionnement de hamsters dorés sur un programme à composante temporelle (FI) à partir de la motivation d'amassement.* Unpublished data, Laboratory of experimental Psychology, University of Liège.

GOL, A. P., P. KELLAWAY, M. SHAPIRO and C. M. HURST (1963) Studies of hippocampectomy in the monkey, baboon and cat. *Neurology* **3**, 1031–41.

GOLDBERG, S. R. (1973) Control of behavior by stimuli associated with drug injections. In L. Goldberg and F. Hoffmeister (Eds.), *Psychic Dependence.* Springer Verlag, Berlin, pp. 106–109.

GOLDBERG, S. R. (1975) Stimuli associated with drug injection as events that control behavior. *Pharmacological Reviews* **27**, 325–340.

GOLDBERG, S. R. (1976) Second-order schedules of morphine or cocaine injection. *Pharmacologist* **18**, 197.

GOLDBERG, S. R. and R. T. KELLEHER (1976) Behavior controlled by scheduled injections of cocaine in squirrel and rhesus monkeys. *J. exp. Anal. Behav.* **25**, 93–104.

GOLDBERG, S. R., R. T. KELLEHER and W. H. MORSE (1975) Second-order schedules of drug injection. *Fedn. Proc.* **34**, 1771–1776.

GOLDBERG, S. R., W. H. MORSE and D. M. GOLDBERG (1976) Behavior maintained under a second-order schedule by intramuscular injection of morphine or cocaine in rhesus monkeys. *J. Pharm. exp. Therap.* **199**, 278–286.

GOLDBERG, S. R., and A. H. TANG (1977) Behavior maintained under second-order schedules of intravenous morphine injection in squirrel and rhesus monkeys. *Psychopharmacology* **51**, 235–242.

GOLDFARB, J. L. and S. GOLDSTONE (1963) Proprioceptive involvement, psychophysical method and temporal judgment. *Percept. Mot. Skills* **17**, 286.

GOLDSTONE, S., W. K. BOARDMAN and W. T. LHAMON (1958) Kinesthetic cues in the development of time concepts. *Journal of Genetic Psychology* **93**, 185–190.

GOLLUB, L. R. (1964) The relation among measures of performance on fixed-interval schedules. *J. exp. Anal. Behav.* **7**, 337–343.

GOLLUB, L. (1977) Conditioned reinforcement: schedule effects. In W. K. Honig and J. E. R. Staddon (Eds.), *Handbook of Operant Behavior*, Prentice-Hall, Inc., Englewood Cliffs, New Jersey, pp. 288–313.

GONZALEZ, F. A. and R. J. NEWLIN (1976) Effects of a delay reinforcement procedure on performance under IRT > t schedules. *J. exp. Anal. Behav.* **26**, 221–235.

GONZALEZ, R. C., M. E. ROCHELLE and M. E. BITTERMAN (1962) Extinction in the fish after partial and consistent reinforcement with number of reinforcements equated. *J. comp. physiol. Psychol.* **55**, 381–386.

GRABOWSKI, J. and T. THOMPSON (1972) Response patterning on an avoidance schedule as a function of time-correlated stimuli. *J. exp. Anal. Behav.* **18**, 525–534.

GRANJON, M. and J. REQUIN (1970) Un rôle de la périodicité cardiaque dans la régulation temporelle de l'activité sensori-motrice? *Psychol. Belg.* **2**, 141–167.

GRAY, V. A. (1976) Stimulus control of differential-reinforcement-of-low-rate responding. *J. exp. Anal. Behav.* **25**, 199–207.

GREEN, T. R. G. and A. J. SIMPSON (1977) Time and temperature: a note on Bell. *Q.J. exp. Psychol.* **29**, 337–340.

GREENWOOD, P. (1977) *Contribution à l'étude des régulations temporelles acquises chez le chat (felis domesticas).* Unpublished Master Thesis, University of Liège.

GREENWOOD, P. (1978) *Régulation temporelle chez le chat.* Paper read at the Annual Meeting of the Belgian Psychological Society, University of Leuven.

GREENWOOD, P. (1978b). Retroregulation proprioceptive et regulation temporelle acquise chez le chat. *C.R. Soc. Biol.* **172**, 1013–1016.

GREER, K. and N. HARVEY (1978) Timing and positioning of limb movements: comments on Adam's theory. *Psychol. Rev.* **85**, 482–484.

GRICE, G. E. (1948) The relation of secondary reinforcement to delayed reward in visual discrimination learning. *J. exp. Psychol.* **38**, 1–16.

GRIER, J. B. (1971) Non parametric indexes for sensitivity and bias: computing formulas. *Psychol. Bull.* **75**, 424–429.

GROSSMAN, K. E. (1973) Continuous, fixed-ratio and fixed interval reinforcement in honey bees. *J. exp. Anal. Behav.* **20**, 105–109.

GUILFORD, J. P. (1954) *Psychometric Methods.* McGraw-Hill, New York.

GUTTMAN, N. (1953) Operant conditioning, extinction and periodic reinforcement in relation to concentrations of sucrose used as reinforcing agent. *J. exp. Psychol.* **46**, 213–224.

GUTTMAN, N. and H. I. KALISH (1956) Discriminability and stimulus generalization. *J. exp. Psychol.* **51**, 79–88.

HABLITZ, J. J. (1973) Operant conditioning and slow potential change from monkey cortex. *Electroenceph. Clin. Neurophysiol.* **34**, 399–408.

HAINSWORTH, F. R. (1967) Saliva spreading, activity and body temperature regulation in the rat. *Amer. J. Physiol.* **212**, 1288–1292.

HAKE, D. F. and R. L. CAMPBELL (1972) Characteristics of response-displacement effects of shock-generated responding during negatively reinforced procedures: pre-shock responding and post-shock aggressive responding. *J. exp. Anal. Behav.* **17**, 303–323.

HAMILTON, C. L. (1963) Interaction of food intake and temperature regulation in the rat. *J. comp. physiol. Psychol.* **56**, 476–488.

HAMILTON, C. L. and J. R. BROBECK (1964) Food intake and temperature regulation in the rats with rostral hypothalamic lesions. *Amer. J. Physiol.* **207**, 291.

HANEY, R. R. (1972) Response force distributions within a fixed-interval schedule. *Psychol. Rec.* **22**, 515–521.

HANEY, R. R., J. A. BEDFORD and R. BERRYMAN (1971) Schedule control in the whitenecked raven (corvus Cryptoleucus). *Psychon. Sci.* **23**, 104–105.

HARVEY, G. A. and H. F. HUNT (1965) Effect of septal lesions on thirst in the rat as indicated by water consumption and operant responding for water reward. *J. comp. physiol. Psychol.* **59**, 49–56.

HARZEM, P. (1969) Temporal discrimination. In R. M. Gilbert and N. S. Sutherland (Eds.), *Animal Discrimination Learning.* Academic Press, London, pp. 299–319.

HARZEM, P., C. F. LOWE and G. C. L. DAVEY (1975) After-effects of reinforcement magnitude: dependence upon context. *Q.J. exp Psychol.* **27**, 579–584.

HARZEM, P., C. F. LOWE and P. T. SPENCER (1978) Temporal control of behavior: schedule interaction. *J. exp. Anal. Behav.* **30**, 255–270.

HAUGLUSTAINE, A. (1972) *Etude de quelques programmes de renforcement positif chez le cobaye (Cavia Cobaya).* Unpublished Master Thesis, University of Liège.

HAWKES, L. and C. P. SHIMP (1975) Reinforcement of behavior patterns: shaping a scallop. *J. exp. Anal. Behav.* **23**, 3–16.

HAWKES, S. R., R. J. JOY and W. O. EVANS (1962) Autonomic effects on estimates of time: evidence for a physiological correlate of temporal experience. *J. Psychol.* **53**, 183–191.

HAWKINS, T. D. and S. S. PLISKOFF (1964) Brain stimulation intensity, rate of self-stimulation and reinforcing strength: an analysis through chaining. *J. exp. Anal. Behav.* **7**, 285–288.

HEARST, E. (1965) Approach, avoidance and stimulus generalization. In D. Mostofsky (Ed.), *Stimulus Generalization.* Stanford University Press, Stanford, pp. 331–355.

HEARST, E. (1968) Discrimination learning as the summation of excitation and inhibition. *Science* **162**, 1303–1306.

HEARST, E. and H. M. JENKINS (1974) Sign-tracking: The stimulus-reinforcer relation and directed action. *Psychonomic Society Monographs.*

HEARST, E. M., B. KORESKO and R. POPPEN (1964) Stimulus generalization and the response-reinforcement contingency. *J. exp. Anal. Behav.* **7**, 369–380.

HEMMES, N. S. (1973) Behavioral contrast in pigeons depends upon the operant. *J. comp. physiol. Psychol.* **85**, 171–178.

HEMMES, N. S. (1975) Pigeon's performance under differential reinforcement of low rate schedule depends upon the operant. *Learn. Motiv.* **6**, 344–357.

HENDRICKS, J. (1966) Flicker threshold as determined by a modified conditioned suppression procedure. *J. exp. Anal. Behav.* **9**, 501–506.

HENDRY, D. P. (1969) Reinforcing value of information: fixed ratio schedules. In D. P. Hendry (Ed.), *Conditioned Reinforcement.* Dorsey Press, Homewood, Illinois.

HENDRY, D. P. and J. N. COULBOURN (1967) Reinforcing effects of an informative stimulus that is not a positive discriminative stimulus. *Psychon. Sci.* **7**, 241–242.

HENDRY, D. P. and P. V. DILLOW (1966) Observing behavior during interval schedules. *J. exp. Anal. Behav.* **9**, 337–349.

HENDRY, D. P., R. SWITALSKY and M. YARCZOWER (1969) Generalization of conditioned suppression after differential training. *J. exp. Anal. Behav.* **12**, 799–806.

HERNANDEZ-PEON, R. (1960) Neurophysiological correlates of habituation and other manifestations of plastic inhibition (internal inhibition). In H. H. Jasper and G. D. Smirnov (Eds.), *The Moscow Colloquium on Electroencephalography of Higher Nervous Activity, Electroenceph. Clin. Neurophysiol.* suppl. **13**, 101–114.

HERON, W. T. (1949) Time discrimination in the white rat. *J. comp. physiol. Psychol.* **42**, 27–31.

HERRNSTEIN, R. J. (1966) Supersition: A corollary of the principles of operant conditioning. In W. K. Honig (Ed.), *Operant Behavior: Areas of Research and Application.* Appleton Century Crofts, New York, pp. 33–51.

HERRNSTEIN, R. J. (1970) On the law of effect. *J. exp. Anal. Behav.* **13**, 242–266.

HERRNSTEIN, R. J. and D. H. LOVELAND (1972) Food avoidance in hungry pigeons and other perplexities. *J. exp. Anal. Behav.* **18**, 369–383.

HERRNSTEIN, R. J. and W. H. MORSE (1957) Effects of pentobarbital on intermittently reinforced behavior. *Science* **125**, 929–931.

HERRNSTEIN, R. J. and W. H. MORSE (1958) A conjunctive schedule of reinforcement. *J. exp. Anal. Behav.* **1**, 15–24.

HEUCHENNE, C., H. MANTANUS and D. DEFAYS (1979) *La mesure de l'adaptation temporelle dans le programme DRL.* Unpublished report, Department of Mathematics applied to Psychology, University of Liège.

HEYMANS, C. and E. NEIL (1958) *Reflexogenic Areas of the Cardiovascular System.* Little-Brown, Boston.

HIENZ, R. D. and D. A. ECKERMAN (1974) Latency and frequency of responding under discrete-trial fixed-interval schedules of reinforcement. *J. exp. Anal. Behav.* **21**, 341–355.

HINDE, R. A. (1969) Control of movement patterns in animals. *Q.J. exp. Psychol.* **21**, 105–126.

HINELINE, P. (1977) Negative reinforcement and avoidance. In W. K. Honig and J. E. R. Staddon (Eds.), *Handbook of operant Behavior,* Prentice-Hall, Inc., Englewood Cliffs, New Jersey, pp. 364–415.

HINRICHS, J. V. (1968) Disinhibition of delay in fixed-interval instrumental conditioning. *Psychon. Sci.* **12**, 313–314.

HOAGLAND, H. (1933) The physiologic control of judgments of duration: evidence for a chemical clock. *J. gen. Psychol.* **9**, 267–287.

HOAGLAND, H. (1935) *Pacemakers in Relation to Aspects of Behavior.* Macmillan, New York.

HOAGLAND, H. (1943) The chemistry of time. *Science Monographs* **56**, 56–61.

HOAGLAND, H. (1951) Consciousness and the chemistry of time. In H. A. Abramson (Ed.), *Problems of Consciousness,* Macy, New York.

HODOS, W., G. S. ROSS and J. V. BRADY (1962) Complex response patterns during temporally spaced responding. *J. exp. Anal. Behav.* **5**, 473–479.

HODOS, W., SMITH, L. and J. C. BONBRIGHT (1976) Detection of the velocity of movement of visual stimuli by pigeons. *J. exp. Anal. Behav.* **25**, 143–156.

HOLLOWAY, F. A. (1978) State dependent retrieval based on time of day. In B. T. Ho, D. W. Richards III and D. L. Chute (Eds.), *Drug Discrimination and State Dependent Learning,* Academic Press, New York, pp. 319–343.

HOLLOWAY, F. A. and F. D. JACKSON (1976) Differential operant behavior on time of day. *Bull. Psychon. Soc.* **8**, 94–96.

HOLZ, W. C. and N. H. AZRIN (1963) A comparison of several procedures for eliminating behavior. *J. exp. Anal. Behav.* **6**, 399–406.

HOLZ, W. C., N. H. AZRIN and R. E. ULRICH (1963) Punishment of temporally spaced responding. *J. exp. Anal. Behav.* **6**, 115–122.

HONIG, W. K. and J. E. R. STADDON (Eds.) (1977) *Handbook of Operant Behavior.* Prentice-Hall, Inc., Englewood Cliffs, New Jersey.

HOTHERSALL, D., D. A. JOHNSON and A. COLLEN (1970) Fixed-ratio responding following septal lesions in the rat. *J. comp. physiol. Psychol.* **73**, 470–476.

HSIAO, S. and M. A. LLOYD (1969) Do rats drink water in excess of apparent need when they are given food? *Psychon. Sci.* **15**, 155–156.

HULL, C. L. (1943) *Principles of Behavior.* Prentice-Hall, Englewood Cliffs, N.J.

HURWITZ, H. B. M. (1963) Facilitation of counting-like behavior. *Animal Behavior* **11**, 439–454.

HURWITZ, H. B. M. and J. R. MILLENSON (1961) Maintenance of avoidance behavior under temporally defined contingencies. *Science* **133**, 284–285.

HUSTON, J. P. and K. ORNSTEIN (1975, April) *Effects of various diencephalic lesions on hypothalamic self-stimulation.* Paper presented at the 1st international conference on self-stimulation reward, Beerse (Belgium).

HUTCHINSON, R. R. (1977) By-products of aversive control. In W. K. Honig and J. E. R. Staddon (Eds.), *Handbook of Operant Behavior,* Prentice-Hall, Inc., Englewood Cliffs, New Jersey, pp. 415–432.

HUTCHINSON, R. R. and G. S. EMLEY (1977) Electric shock produced drinking in the squirrel monkey. *J. exp. Anal. Behav.* **28**, 1–12.

HUTT, P. J. (1954) Rate of bar pressing as a function of quality and quantity of food reward. *J. comp. physiol. Psychol.* **47**, 235–239.

INNIS, N. K. and J. E. R. STADDON (1971) Temporal tracking on cyclic-interval reinforcement schedules. *J. exp. Anal. Behav.* **16**, 411–423.

IRWIN, D. A. and C. S. REBERT (1970) Slow potential changes in cat brain during classical appetitive conditioning of jaw movements using two levels of reward. *Electroenceph. Clin. Neurophysiol.* **28**, 119–126.

IVERSEN, L. L. and S. D. IVERSEN (Eds.) (1978) *Handbook of Psychopharmacology*, vols. 7 and 8. Plenum Press, New York.

JACKSON, F. B. and J. A. GERGEN (1970) Acquisition of operant schedules by squirrel monkeys lesioned in the hippocampal area. *Physiol. Behav.* **5**, 543–547.

JANET, P. (1928) *L'evolution de la mémoire et de la notion de temps*. A Chahine, Paris.

JARRARD, L. E. (1965) Hippocampal ablation and operant behavior in the rat. *Psychon. Sci.* **2**, 115–116.

JASPER, H. H. and C. SHAGASS (1941) Conscious time judgments related to conditioned time intervals and voluntary control of alpha rhythms. *J. exp. Psychol.* **28**, 503–508.

JENKINS, H. M. (1970) Sequential organization in schedules of reinforcement. In W. N. Schoenfeld (Ed.), *The Theory of Reinforcement Schedules*. Appleton Century Crofts, New York, pp. 63–109.

JENKINS, H. M. (1973) Effects of the stimulus-reinforcer relation on selected and unselected responses. In R. A. Hinde and J. Stevenson-Hinde (Eds.), *Constraints on Learning*. Academic Press, New York, pp. 189–206.

JENKINS, H. M. and R. A. BOAKES (1973) Observing stimulus sources that signal food or no food. *J. exp. Anal. Behav.* **20**, 197–207.

JENKINS, H. M. and R. H. HARRISON (1960) Effect of discrimination training on auditory generalization. *J. exp. Psychol.* **59**, 246–253.

JENKINS, H. M. and B. R. MOORE (1973) The form of the auto-shaped response with food or water reinforcers. *J. exp. Anal. Behav.* **20**, 163–181.

JENSEN, C. and D. FALLON (1973) Behavioral after-effects of reinforcement and its omission as a function of reinforcement magnitude. *J. exp. Anal. Behav.* **19**, 459–468.

JOHANSON, C. E. (1978) Drugs as reinforcers. In D. E. Blackman and D. J. Sanger (Eds.), *Contemporary Research in Behavioral Pharmacology*. Plenum Press, New York and London, pp. 325–390.

JOHN, E. R. and K. F. KILLAM (1960) Electrophysiological correlates of differential approach-avoidance conditioning in cats. *J. Nerv. Ment. Dis.* **136**, 183–201.

JOHNSON, D. A., L. A. BIELIAUSKAS and J. LANCASTER (1973) DRL training and performance following anterior, posterior, or complete septal lesions in infant and adult rats. *Physiol. Behav.* **11**, 661–669.

JOHNSON, C. T., D. S. OLTON, F. D. GAGE and P. G. JENKO (1977) Damage to hippocampus and hippocampal connections: effects on DRL and spontaneous alternation. *J. comp. physiol. Psychol.* **91**, 508–522.

JONES, B. (1973) Is there any proprioceptive feedback? Comments on Schmidt (1971). *Psychol. Bull.* **79**, 386–388.

JONES, B. (1974) Is proprioception important for skilled performance? *J. Mot. Behav.* **6**, 33–45.

JONES, E. and R. NARVER (1966) Effects of voluntary and involuntary mediation on estimates of short-time intervals. *Psychon. Sci.* **6**, 505–506.

JWAIDEH, A. R. and D. E. MULVANEY (1976) Punishment of observing by a stimulus associated with the lower of two reinforcement frequencies. *Learn. Motiv.* **7**, 211–222.

KADDEN, R. M. (1973) Facilitation and suppression of responding under temporally defined schedules of negative reinforcement. *J. exp. Anal. Behav.* **19**, 469–480.

KADDEN, R. M., W. N. SCHOENFELD and A. G. SNAPPER (1974) Aversive schedules with independent probabilities of reinforcement for responding and not responding by rhesus monkeys: II. Without signal. *J. comp. physiol. Psychol.* **87**, 1189–1197.

KAMP, A., A. J. VAN RIJN and J. M. ZWART (1969) Slow potential changes recorded from cerebral cortex of dogs (abstract). *Electroenceph. Clin. Neurophysiol.* **27**, 678.

KAPLAN, J. (1965) Temporal discrimination in rats during continuous brain stimulation. *Psychon. Sci.* **2**, 255–256.

KAPOSTINS, E. E. (1963) The effects of DRL schedules on some characteristics of word utterance. *J. exp. Anal. Behav.* **6**, 281–290.

KASPER, P. M. (1964) Attenuation of passive avoidance by continuous septal stimulation. *Psychon. Sci.* **1**, 219–220.

KATZ, H. N. (1976) A test of the reinforcing properties of stimuli correlated with nonreinforcement. *J. exp. Anal. Behav.* **26**, 45–56.

KEEHN, J. D. and R. RIUSECH (1979) Schedule-induced drinking facilitates schedule-controlled feeding. *Anim. Learn. Behav.* **7**, 41–44.

KEELE, S. W. (1968) Movement control in skilled motor performance. *Psychol. Bull.* **70**, 387–403.

KEESEY, R. E. (1962) The relationship between pulse frequency, intensity and duration and the rate of responding for intracranial stimulation. *J. comp. physiol. Psychol.* **55**, 671–678.

KEESEY, R. E. and J. W. KLING (1961) Amount of reinforcement and free-operant responding. *J. exp. Anal. Behav.* **4**, 125–132.

KELLEHER, R. T. (1966) Conditioned reinforcement in second-order schedules. *J. exp. Anal. Behav.* **9**, 475–485.

KELLEHER, R. T. (1975) Characteristics of behavior controlled by scheduled injections of drugs. *Pharmacol. Rev.* **27**, 307–323.

KELLEHER, R. T. and W. T. FRY (1962) Stimulus functions in chained fixed-interval schedules. *J. exp. Anal. Behav.* **5**, 167–173.

KELLEHER, R. T., W. FRY and L. COOK (1959) Inter-response time distribution as a function of differential reinforcement of temporally spaced responses. *J. exp. Anal. Behav.* **2**, 91–106.

KELLEHER, R. T. and S. R. GOLDBERG (1977) Fixed interval responding under second-order schedules of food presentation or cocaine injection. *J. exp. Anal. Behav.* **28**, 221–232.

KELLEHER, R. T. and W. H. MORSE (1968a) Schedules using noxious stimuli: III. Responding maintained with response-produced electric shocks. *J. exp. Anal. Behav.* **11**, 819–838.

KELLEHER, R. T. and W. H. MORSE (1968b) Determinants of the specificity of behavioral effects of drugs. *Ergebnisse der Physiologie* **60**, 1–56.

KELLEHER, R. T. and W. H. MORSE (1969) Determinants of the behavioral effects of drugs. In D. H. Tedeschi and R. E. Tedeschi (Eds.), *Importance of Fundamental Principles in Drug Evaluation*. Raven Press, New York, pp. 383–405.

KELLEHER, R. T., W. C. RIDDLE and L. COOK (1963) Persistent behavior maintained by unavoidable shock. *J. exp. Anal. Behav.* **6**, 507–517.

KELLO, J. E. (1972) The reinforcement—omission effect on fixed-interval schedules: frustration or inhibition? *Learn. Motiv.* **3**, 138–147.

KELSO, J. A. S. (1978) Joint receptors do not provide a satisfactory basis for motor timing and positioning. *Psychol. Rev.* **85**, 474–481.

KENDALL, S. B. (1972) Some effects of response dependent clock stimuli in a fixed-interval schedule. *J. exp. Anal. Behav.* **17**, 161–168.

KILLEEN, P. (1969) Reinforcement frequency and contingency as factors in fixed-ratio behavior. *J. exp. Anal. Behav.* **12**, 391–395.

KILLEEN, P. (1975) On the temporal control of behavior. *Psychol. Rev.* **82**, 89–115.

KILLEEN, P., S. J. HANSON and S. R. OSBORNE (1978) Arousal: its genesis and manifestation as response rate. *Psychol. Rev.* **85**, 571–581.

KINCHLA, J. (1970) Discrimination of two auditory durations by pigeons. *Perception and Psychophysics* **8**, 299–307.

KING, G. D. (1974) Wheel running in the rat induced by a fixed-time presentation of water. *Anim. Learn. Behav.* **2**, 325–328.

KINTSCH, W. and R. S. WITTE (1962) Concurrent conditioning of a bar press and salivation response. *J. comp. physiol. Psychol.* **55**, 963–968.

KISCH, G. B. (1966) Theories of sensory reinforcement. In W. K. Honig (Ed.), *Operant Behavior: Areas of Research and Application*. Appleton Century Crofts, New York, pp. 109–159.

KNAPP, H. D., E. TAUB and A. J. BERMAN (1958) Effects of deafferentation on a conditioned avoidance response. *Science* **128**, 842–843.

KNAPP, H. D., E. TAUB and A. J. BERMAN (1963) Movements in monkeys with deafferented forelimbs. *Exp. Neurol.* **7**, 305–315.

KNUTSON, J. F. and R. A. KLEINKNECHT (1970) Attack during differential reinforcement of low rate responding. *Psychon. Sci.* **19**, 289–290.

KOCH, E. (1932) Die Irradiation der pressoreceptorischen Kreislaufreflexe. *Klin. Wschr.* **11**, 225–227.

KOPYTOVA, F. V. and L. K. KULIKOVA (1970) Conditioned reactions to time in the neurons of the dorsal hippocampus. *Zh. Vyssh. Nerv. Deiat.* **20**, 1221–1230.

KOPYTOVA, F. V. and T. U. S. MEDNIKOVA (1972) Conditioned reactions to time in neurons of the amygdaloid complex. *Zh. Vyssh. Nerv. Deiat.* **22**, 930–939.

KRAMER, T. J. (1968) *Effects of timeout on spaced responding in pigeons*. Unpublished Master Thesis, Michigan State University.

KRAMER, T. J. and M. RILLING (1969) Effects of time-out on spaced responding in pigeons. *J. exp. Anal. Behav.* **12**, 283–288.

KRAMER, T. J. and G. M. RILLING (1970) Differential reinforcement of low rates: A selective critique. *Psychol. Bull.* **74**, 225–254.

KRAMER, T. J. and M. RODRIGUEZ (1971) The effect of different operants on spaced responding. *Psychon. Sci.* **25**, 177–178.

KUCH, D. O. (1974) Differentiation of press duration with upper and lower limits of reinforced values. *J. exp. Anal. Behav.* **22**, 275–283.

KUCH, D. O. and J. R. PLATT (1976) Reinforcement rate and interresponse time differentiation. *J. exp. Anal. Behav.* **26**, 471–486.

LA BARBERA, J. D. and R. M. CHURCH (1974) Magnitude of fear as a function of the expected time to an aversive event. *Anim. Learn. Behav.* **2**, 199–202.

LACEY, J. I. and B. C. LACEY (1970) Some autonomic-central nervous system interrelationships. In P. Black (Ed.), *Physiological Correlates of Emotion*, Academic Press, New York.

LACEY, B. C. and J. I. LACEY (1977) Change in heart period: a function of sensorimotor event timing within the cardiac cycle. *Physiol. Psychol.* **5**, 383–393.

LACEY, B. C. and J. I. LACEY (1978) Two-way communication between the heart and the brain. *Amer. Psychol.* 99–113.

LAMING, D. (1973) *Mathematical Psychology.* Academic Press, London.

LANE, H. (1961) Operant control of vocalizing in the chicken. *J. exp. Anal. Behav.* **4**, 171–177.

LASHLEY, K. S. (1917) The accuracy of movement in the absence of excitation from the moving organ. *Amer. J. Phys.* **43**, 169–194.

LASHLEY, K. S. (1951) The problem of serial order in behavior. In L. A. Jeffress (Ed.), *Cerebral Mechanisms in Behavior*, Wiley, New York.

LATIES, V. G. and B. WEISS (1962) Effects of alcohol on timing behavior. *J. comp. physiol. Psychol.* **55**, 85–91.

LATIES, V. G. and B. WEISS (1966) Influence of drugs on behavior controlled by internal and external stimuli. *J. Pharm. exp. Therap.* **152**, 388–396.

LATIES, V. G., B. WEISS, R. L. CLARK and M. D. REYNOLDS (1965) Overt "mediating" behavior during temporally spaced responding. *J. exp. Anal. Behav.* **8**, 107–116.

LATIES, V. G., B. WEISS and A. B. WEISS (1969) Further observation of overt "mediating" behavior and the discrimination of time. *J. exp. Anal. Behav.* **12**, 43–57.

LATTAL, K. A. (1972) Response-reinforcer independance and conventional extinction after fixed-interval and variable-interval schedules. *J. exp. Anal. Behav.* **18**, 133–140.

LATTAL, K. and A. J. BRYAN (1976) Effects of concurrent response-independent reinforcement on fixed interval schedule performance. *J. exp. Anal. Behav.* **26**, 495–504.

LE CAO, K. and R. LINCÉ (1974) *L'immobilité en DRL.* Unpublished data. Laboratory of Experimental Psychology, University of Liège.

LEJEUNE, H. (1971a) Note sur les régulations temporelles acquises en programme à intervalle fixe chez le chat. *Revue du Comportement Animal* **5**, 123–129.

LEJEUNE, H. (1971b) Lésions thalamiques médianes et régulation temporelle acquise en programme FI chez le rat albinos. *Physiol. Behav.* **7**, 575–582.

LEJEUNE, H. (1972) Action de masse au niveau du thalamus médian et programme à intervalle fixe chez le rat albinos. *Physiol. Behav.* **9**, 863–865.

LEJEUNE, H. (1974) *Structures thalamiques médianes et apprentissages operants.* Unpublished doctoral dissertation, University of Liège.

LEJEUNE, H. (1976) Une expérience de conditionnement operant chez un prosimien: Perodicticus Potto Edwarsi. *Psychologica Belgica* **16**, 199–208.

LEJEUNE, H. (1977a) Lésions thalamiques médianes et effets différentiels dans les apprentissages operants. *Physiol. Behav.* **18**, 349–356.

LEJEUNE, H. (1977b) Type de réponse operante et lésions thalamiques médianes. *Physiol. Behav.* **18**, 357–359.

LEJEUNE, H. (1978) Sur un paradoxe dans l'estimation du temps chez l'animal. *L'Année Psychologique* **78**, 163–181.

LEJEUNE, H. and D. DEFAYS (1979) *Différences inter-individuelles dans les programmes FI et DRL chez le rat albinos.* Unpublished data, Laboratory of Experimental Psychology, University of Liège.

LEJEUNE, H. and J. DELACOUR (1970) Etude de l'effet de lésions thalamiques médianes sur la régulation temporelle acquise par un conditionnement "operant" chez le rat albinos. *C.R. Acad. Sci., Paris* **270**, 166–169.

LEJEUNE, H. and H. MANTANUS (1977) *Some variables affecting DRL acquisition in the pigeon.* Paper read at the Annual Meeting of the Belgian Psychology Society, University of Leuven.

LÉVITSKY, D. and G. COLLIER (1968) Schedule-induced wheel running. *Physiol. Behav.* **3**, 571–573.

LEVY, M. N., P. J. MARTIN, T. IANO and H. ZIESKE (1970) Effects of single vagal stimuli on heart rate and atrioventricular conduction. *Amer. J. Physiol.* **218**, 1256–1262.

LEVY, M. N. and H. ZIESKE (1972) Synchronization of the cardiac pacemaker with repetitive stimulation of the carotid sinus nerve in the dog. *Circulation Res.* **30**, 634–664.

LIBERSON, W. T. and P. ELLEN (1960) Conditioning of the driven brain wave rhythm in the cortex and the hippocampus in the rat. In J. Wortis (Ed.), *Recent Advances in Biological Psychiatry*, Grune and Stratton, New York.

LINCÉ, R. (1976) *Activité collatérale et régulation temporelle en DRL.* Unpublished Master Thesis, University of Liège.

LINDLEY, D. B. (1952) Psychological phenomena and the electroencephalogram. *Electroenceph. Clin. Neurophysiol.* **4**, 443–456.

LINWICK, D. C. and L. MILLER (1978) Acquisition of leverpressing without experimenter assistance by rats on differential reinforcement of low-rate-schedules. *Bull. Psychon. Soc.* **12**, 193–195.

LOCKART, J. M. (1967) Ambient temperature and time estimation. *J. exp. Psychol.* **73**, 286–291.

LOCKART, J. M. (1971) Ambient temperature and the flickerfusion threshold. *J. exp. Psychol.* **87**, 314–319.

LOGAN, F. A. (1960) *Incentive: How the Conditions of Reinforcement Affect the Performances of Rats.* Yale University Press, New Haven.

LOGAN, F. A. (1961) Discrete-trials DRL. *J. exp. Anal. Behav.* **4**, 277–279.

LOGAN, F. A. (1967) Variable DRL. *Psychon. Sci.* **9**, 393–394.

LOGAN, F. A., E. M. BEIER and R. A. ELLIS (1955) The effect of varied reinforcement on speed of location. *J. exp. Psychol.* **49**, 260–266.

LORENS, S. A. and C. Y. KONDO (1969) Effects of septal lesions on food and water intake and operant responding for food. *Physiol. Behav.* **4**, 729–732.

LOW, M. D., R. P. BORDA and P. KELLAWAY (1966) Contingent negative variation in rhesus monkeys: an EEG sign of a specific mental process. *Percept. Mot. Skills* **22**, 443–446.

LOW, M. D., J. D. FROST, R. P. BORDA and P. KELLAWAY (1966) Surface negative slow potential shift associated with conditioning in man. *Neurology* **16**, 771–782.

LOWE, C. F., G. C. L. DAVEY and P. HARZEM (1974) Effects of reinforcement magnitude on interval and ratio schedules. *J. exp. Anal. Behav.* **22**, 553–560.

LOWE, C. F. and P. HARZEM (1977) Species differences in temporal control of behavior. *J. exp. Anal. Behav.* **28**, 189–201.

R. D. LUCE and E. GALANTER (1963) Discrimination. In R. D. Luce, R. R. Bush and E. Galanter (Eds.), *Handbook of Mathematical Psychology*, vol. 1. John Wiley and Sons, Inc., New York, London and Sidney, pp. 191–243.

LUND, C. A. (1976) Effects of variations in the temporal distribution of reinforcements on interval schedule performance. *J. exp. Anal. Behav.* **26**, 155–164.

LYDERSEN, T. and D. PERKINS (1972) Effects of duration of DRL correlated stimuli on timing and collateral responses. *Psychon. Sci.* **29**, 149–150.

MACAR, F. (1969) *Etude de quelques problèmes dans le cadre des régulations temporelles acquises.* Unpublished Master Thesis, University of Liège.

MACAR, F. (1970) *Périodicité cardiaque et régulation temporelle: étude préliminaire dans un programme de conditionnement DRL chez le chat.* Unpublished dissertation.

MACAR, F. (1971a) Névrose expérimentale dans un programme de renforcement des débits de réponses lents chez le chat. *Journal de Psychologie normale et pathologique* **2**, 191–205.

MACAR, F. (1971b) Addition d'une horloge externe dans un programme de conditionnement au temps chez le chat. *Journal de Psychologie normale et pathologique* **1**, 89–100.

MACAR, F. (1977) Signification des variations contingentes négatives dans la dimension temporelle du comportement. *L'Année Psychologique* **77**, 439–474.

MACAR, F. and N. VITTON (1978) Contingent negative variation and accuracy of time estimation: a study on cats. *Electroenceph. Clin. Neurophysiol.* In press.

MACAR, F., N. VITTON and J. REQUIN (1973) Effects of time uncertainty on the time-course of preparatory processes in cats performing a RT task. *Acta Psychologica* **37**, 113–124.

MACAR, F., N. VITTON and J. REQUIN (1976) Slow cortical potential shifts recorded in cats performing a time estimation task: preliminary results. *Neuropsychologia* **14**, 353–361.

MACKINTOSH, N. J. (1977) Stimulus control: attentional factors. In Honig, W. K. and J. E. R. Staddon (Eds.), *Handbook of Operant Behavior*, Prentice-Hall, Inc., Englewood Cliffs, New Jersey, pp. 481–513.

MADIGAN, R. J. (1978) Reinforcement context effects on fixed-interval responding. *Anim. Learn. Behav.* **6**, 193–197.

MALAGODI, E. F., J. DE WEESE and J. M. JOHNSTON (1973a) Second order schedules: a comparison of chained brief stimulus, and tandem procedures. *J. exp. Anal. Behav.* **20**, 447–460.

MALAGODI, E. F., J. DE WEESE, F. M. WEBBE and G. PALERMO (1973b) Responding maintained by schedules of electric shock presentation. *Bull. Psychon. Soc.* **2**, 331.

MALAGODI, E. F., M. L. GARDNER and G. PALERMO (1978) Responding maintained under fixed-interval and fixed-time schedules of electric shock presentation. *J. exp. Anal. Behav.* **30**, 271–280.

MALONE, J. C. (1971) Properties of the fixed-interval S^D. *Psychon. Sci.* **23**, 57–59.

MALOTT, R. W. and W. W. CUMMING (1964) Schedules of interresponse time reinforcement. *Psychol. Rec.* **14**, 211–252.

MALOTT, R. W. and W. W. CUMMING (1966) Concurrent schedules of interresponse time reinforcement: probability of reinforcement and the lower bound of the reinforced interresponse time interval. *J. exp. Anal. Behav.* **9**, 317–326.

MANNING, F. J. (1973) Performance under temporal schedules by monkeys with partial ablations of prefrontal cortex. *Physiol. Behav.* **11**, 563–569.

MANTANUS, H. (1973) *Echappement au DRL.* Unpublished Master Thesis, University of Liège.

MANTANUS, H. (1979) *Influences environnementales sur la discrimination temporelle chez le pigeon.* Unpublished data, University of Liège.

MANTANUS, H., C. FAYT and B. FERETTE (1977a) Nature de l'operant et renforcement de débits de réponses lents chez le pigeon. *Psychol. Belg.* **17**, 127–134.

MANTANUS, H., I. JULIEN and I. PITZ (1977b) Comparaison de deux réponses operantes chez le pigeon. *Psychologie Française* **22**, 61–68.

MARCUCELLA, H. (1974) Signalled reinforcement in differential-reinforcement-of-low-rate schedules. *J. exp. Anal. Behav.* **22**, 381–390.

MARCUCELLA, H., J. S. McDONALL, I. MUNRO and V. MOSELEY (1977) Presignaled behavior as a predictor of signaled DRL performance. *Bull. Psychon. Soc.* **10**, 335–338.

MARLEY, E. and W. H. MORSE (1966) Operant conditioning in the newly hatched chicken. *J. exp. Anal. Behav.* **9**, 95–103.

MARR, M. J. and M. D. ZEILER (1974) Schedules of response-independent conditioned reinforcement. *J. exp. Anal. Behav.* **21**, 433–444.

MAURISSEN, J. (1970) *Régulation temporelle acquise en programme FI et DRL chez la souris.* Unpublished Master Thesis, University of Liège.

McADAM, D. W. (1966) Slow potential changes recorded from human brain during learning of a temporal interval. *Psychon. Sci.* **6**, 435–436.

McCALLUM, W. C., D. PAPAKOSTOPOULOS, R. GOMBI, A. L. WINTER, R. COOPER and H. B. GRIFFITH (1973) Event related slow potential changes in human brain stem. *Nature* **242**, 5398, 465–467.

McCLEARY, R. A. (1966) Response-modulating functions of the limbic system: Initiation and suppression. In E. Stellar and J. M. Spague (Eds.), *Progress in Physiological Psychology*, vol. 1. Academic Press, New York, pp. 209–272.

McFARLAND, D. J. (1974) Time sharing as a behavioural phenomenon. *Advances in the study of behavior* **5**, 201–224.

McKEARNEY, J. W. (1968) Maintenance of responding under a fixed-interval schedule of electric shock presentation. *Science* **160**, 1249–1251.

McKEARNEY, J. W. (1969) Fixed-interval schedules of electric shock presentation: extinction and recovery of performances under different shock intensities and fixed-interval durations. *J. exp. Anal. Behav.* **12**, 301–313.

McKEARNEY, J. W. (1970a) Responding under fixed-ratio and multiple fixed-interval fixed-ratio schedules of electric shock presentation. *J. exp. Anal. Behav.* **14**, 1–16.

McKEARNEY, J. W. (1970b) Rate-dependent effects of drugs: modification by discriminative stimuli of the effects of amorbarbitol on schedule-controlled behavior. *J. exp. Anal. Behav.* **14**, 167–175.

McKEARNEY, J. W. (1972) Maintenance and suppression of responding under schedules of electric shock presentation. *J. exp. Anal. Behav.* **17**, 425–432.

McKEARNEY, J. W. (1974a) Differences in responding under fixed-time and fixed-interval schedules of electric shock presentation. *Psychol. Rep.* **34**, 907–914.

McKEARNEY, J. W. (1974b) Responding under constant probability schedule of electric shock presentation. *Psychol. Rep.* **35**, 907–914.

McKEARNEY, J. W. AND J. E. BARRETT (1978) Schedule-controlled behavior and the effect of drugs. In D. E. Blackman and D. J. Sanger (Eds.), *Contemporary Research in Behavioral Pharmacology*. Plenum Press, New York and London, pp. 1–68.

McMILLAN, D. E. (1969) Reinforcement contingencies maintaining collateral responding under a DRL schedule. *J. exp. Anal. Behav.* **12**, 413–422.

McMILLAN, D. E. and R. A. PATTON (1965) Differentiation of a precise timing-response. *J. exp. Anal. Behav.* **8**, 219–226.

McMILLAN, J. C. (1974) Average uncertainty as a determinant of observing behavior. *J. exp. Anal. Behav.* **22**, 401–408.

McSWEENEY, F. K. (1975) Matching and contrast on several concurrent treadle-press schedules. *J. exp. Anal. Behav.* **23**, 193–198.

MECHNER, F. and L. GUEVREKIAN (1962) Effects of deprivation upon counting and timing in the rat. *J. exp. Anal. Behav.* **5**, 463–466.

MECHNER, F., L. GUEVREKIAN and V. MECHNER (1963) A fixed-interval schedule in which the interval is initiated by a response. *J. exp. Anal. Behav.* **6**, 323–330.

MECHNER, F. and M. LATRANYI (1963) Behavioral effects of caffeine, metamphetamine, and methylphenidate in the rat. *J. exp. Anal. Behav.* **6**, 331–342.

MEDNIKOVA, T. U. S. (1975) Conditioned reactions to time of hypothalamic neurons. The perifornical nucleus. *Zh. Vyssh. Nerv. Deiat.* **25**, 1022–1030.

MELTZER, D. and J. A. BRAHLEK (1968) Quantity of reinforcement and fixed-interval performance. *Psychon. Sci.* **12**, 207–208.

MELTZER, D. and J. A. BRAHLECK (1970) Quantity of reinforcement and fixed-interval performance within subject effects. *Psychon. Sci.* **20**, 30–31.

MELTZER, D. and D. L. Howerton (1973) Sequential effects of reinforcement magnitude on fixed-interval performance in rats. *J. comp. physiol. Psychol.* **85**, 361–366.

MELTZER, D. and D. L. HOWERTON (1975) Sequential effects of signalled and unsignalled variation in reinforcement magnitude on fixed-interval performance. *Bull. Psychon. Soc.* **6**, 461–464.

MELTZER, D. and D. L. HOWERTON (1976) Interval duration and the sequential effects of reinforcement magnitude on FI performance. *Bull. Psychon. Soc.* **8**, 303–305.

MIGLER, B. and J. V. BRADY (1964) Timing behavior and conditioned fear. *J. exp. Anal. Behav.* **7**, 247–251.

MILLENSON, J. R. (1966) Probability of response and probability of reinforcement in a response-defined analogue of an interval schedule. *J. exp. Anal. Behav.* **9**, 87–94.

MILLER, L. and D. C. LINWICK (1979) Acquisition of lever-pressing without assistance by rats maintained on interval and ratio schedules. *Bull. Psychon. Soc.* **13**, 103–104.

MILSTEIN, V. (1965) Contingent alpha blocking: conditioning or sensitization? *Electroenceph. Clin. Neurophysiol.* **18**, 272–277.

MOFFITT, M. and C. P. SHIMP (1971) Two-key concurrent paced variable-interval schedules of reinforcement. *J. exp. Anal. Behav.* **16**, 39–49.

MOGENSON, G. and J. CIOE (1977) Central reinforcement: a bridge between brain and behavior. In W. K. Honig and J. E. R. Staddon (Eds.), *Handbook of Operant Behavior*, Prentice-Hall, Inc., Englewood Cliffs, New Jersey, pp. 570–596.

MOLLIVER, M. E. (1963) Operant control of vocal behavior in the cat. *J. exp. Anal. Behav.* **6**, 197–202.

MOORE, B. R. (1973) The role of directed pavlovian reactions in simple instrumental learning in the pigeon. In R. A. Hinde and J. Stevenson-Hinde (Eds.), *Constraints on Learning*, Academic Press, New York, pp. 159–186.

MORELL, F. and H. H. JASPER (1956) Electrographic studies of the formation of temporary connections in the brain. *Electroenceph. Clin. Neurophysiol.* **8**, 201–215.

MORELL, F., L. ROBERTS and H. H. JASPER (1956) Effect of focal epileptogenic lesions and their ablation upon conditioned electrical responses of the brain in the monkey. *Electroenceph. Clin. Neurophysiol.* **8**, 217–235.

MORGAN, M. J. (1970) Fixed-interval schedules and delay of reinforcement. *Q.J. exp. Psychol.* **22**, 663–673.

MORSE, W. H. (1966) Intermittent reinforcement. In W. K. Honig (Ed.), *Operant Behavior: Areas of Research and Application*. Appleton Century Crofts, New York, pp. 52–108.

MORSE, W. H. and R. T. KELLEHER (1966) Schedules using noxious stimuli: I. multiple fixed-ratio and fixed-interval termination of schedule complexes. *J. exp. Anal. Behav.* **9**, 267–290.

MORSE, W. H. and R. T. KELLEHER (1970) Schedules as fundamental determinants of behavior. In W. N. Schoenfeld (Ed.), *The Theory of Reinforcement Schedules*, Appleton Century Crofts, New York, pp. 139–185.

MORSE, W. H. and R. T. KELLEHER (1977) Determinants of reinforcement and punishment. In W. K. Honig and J. E. R. Staddon (Eds.), *Handbook of Operant Behavior*. Prentice-Hall, Inc., Englewood Cliffs, New Jersey, 174–200.

MORSE, W. H., R. N. MEAD and R. T. KELLEHER (1967) Modulation of elicited behavior by a fixed-interval schedule of electric shock presentation. *Science* **157**, 215–217.

MOSTOFSKY, D. I., SHURTLEFF, D. A. and G. MARGOLIUS (1964) Comparative sensitivity of rats and humans to changes in auditory click rate. *J. comp. physiol. Psychol.* **58**, 436–440.

MOTT, F. W. and C. S. SHERRINGTON (1895) Experiments upon the influence of sensory nerves upon movement and nutrition of the limbs. *Proc. Roy. Soc.* **57**, 481–488.

MOWRER, O. H. (1960) *Learning Theory and the Symbolic Processes*. Wiley, New York.

MULVANEY, D. E., J. A. DINSMOOR, A. R. JWAIDEH and L. H. HUGUES (1974) Punishment of observing by the negative discriminative stimulus. *J. exp. Anal. Behav.* **21**, 37–44.

MÜNSTERBERG, H. (1889) *Beiträge zur Experimentellen Psychologie*, Heft 2, Siebeck, Freiburg.

MYERS, R. D. and D. C. MESKER (1960) Operant responding in a horse under several schedules of reinforcement. *J. exp. Anal. Behav.* **3**, 161–164.

NELSON, T. D. (1974) *Interresponse time as a stimulus: discrimination and emission of interresponse times by pigeons.* Unpublished doctoral dissertation, University of Maine, Orono.

NEURINGER, A. J. (1970) Superstitious key-pecking after three peck-produced reinforcements. *J. exp. Anal. Behav.* **13**, 127–134.

NEURINGER, A. J. and B. A. SCHNEIDER (1968) Separating the effects of interreinforcement time and number of interreinforcement responses. *J. exp. Anal. Behav.* **11**, 661–667.

NEVIN, J. A. and R. BERRYMAN (1963) A note on chaining and temporal discrimination. *J. exp. Anal. Behav.* **6**, 109–113.

NONNEMAN, A. J., J. VOIGT and B. E. KOLB (1974) Comparisons of behavioral effects of hippocampal and prefrontal cortex lesions in the rat. *J. comp. physiol. Psychol.* **87**, 249–260.

NOTTERMAN, J. M. and D. E. MINTZ (1965) *Dynamics of Response*, Wiley, New York.

NUMAN, R. and J. F. LUBAR (1974) Role of the proreal gyrus and septal area in response modulation in the cat. *Neuropsychologia* **12**, 219–234.

NUMAN, R., R. A. SEIFERT and J. F. LUBAR (1975) Effects of mediocortical frontal lesions on DRL performance in the rat. *Physiol. Psychol.* **3**, 390–394.

O'HANLON, J. F. JR., J. E. DANISCH and J. J. McGRATH (1968) Body temperature and rate of subjective time. *United States Army Medical Research and Developmental Command Contract.* D4-49-193-MD-2743. Human Factors Research Incorporated, technical report, 719–8.

O'HANLON, J. F., J. J. McGRATH and M. E. McCAULEY (1974) Body temperature and temporal acuity. *J. exp. Psychol.* **102**, 788–794.

OLDS, J. and P. MILNER (1954) Positive reinforcement produced by electrical stimulation of septal area and other regions of the rat brain. *J. comp. physiol. Psychol.* **47**, 419–427.

OWENS, J. A. and D. L. BROWN (1968) ICS reinforcement of DRL behavior in the rat. *Psychon. Sci.* **10**, 309–310.

PALMER, J. D. (1976) *An Introduction to Biological Rhythms.* Academic Press, New York and San Francisco.

PAVLOV, I. P. (1927) *Conditioned Reflexes.* Oxford University Press, 2nd Ed. 1960. Dover Publications, Inc., New York.

PELLEGRINO, L. (1965) The effects of amygdaloid stimulation on passive avoidance. *Psychon. Sci.* **2**, 189–190.

PEPLER, R. D. (1971) Ambient temperature. In E. Furchtgott (Ed.), *Pharmacological and Biophysical Agents and Behavior.* Academic Press, New York and London, pp. 143–179.

PERIKEL, J. J., M. RICHELLE and J. MAURISSEN (1974) Control of key-pecking by stimulus duration. *J. exp. Anal. Behav.* **22**, 131–134.

PERSINGER, M. A., N. J. CARREY, G. F. LAFRENIÈRE and A. MAZZUCHIN (1978) Step-like DRL schedule change effects on blood chemistry, leukocytes and tissue in rats. *Physiol. Behav.* **21**, 899–904.

PETERSON, W. W., T. G. BIRDSALL and W. C. FOX (1954) Theory of signal detectability. *TRE Trans. Professional Group on Information Theory* **4**, 171–212.

PFAFF, D. (1968) Effects of temperature and time of day on time judgments. *J. exp. Psychol.* **76**, 419–422.

PIAGET, J. (1946) *Le développement de la notion de temps chez l'enfant.* P.U.F., Paris.

PIAGET, J. (1966) Time perception in children. In J. T. Fraser (Ed.), *The Voices of Time.* George Brasiller, New York, pp. 202–216.

PICKENS, R. and T. THOMPSON (1972) Simple schedules of drug self-administration in animals. In J. M. Singh, L. H. Miller and H. Lal (Eds.), *Drug Addiction: vol. 1, Experimental Pharmacology.* Futura Publishing Co., Mount Kisco, New York, pp. 107–120.

PIERON, H. (1923) Les problèmes psychophysiologiques de la perception du temps. *L'Année Psychologique* **24**, 1–25.

PINOTTI, O. and L. GRANATA (1954) Azione inhibitrice dei pressocettori carotidei sul riflesso linguo-mandibolare. *Boll. della Soc. Ital. Biol. Sperim.* **30**, 486–488.

PITTENDRIGH, C. S., P. C. CALDAROLA and E. S. COSBEY (1973) A differential effect of heavy water on temperature-dependent and temperature-independent aspects of the circadian system of *Drosophila pseudoobscura. Proc. Nat. Acad. Sci.* **70**, 2037–2041.

PIZZI, W. J. and S. A. LORENS (1967) Effect of lesions in the amygdalo-hippocampo-septal system on food and water intake in the rat. *Psychon. Sci.* **7**, 187–188.

PLATT, J. R. (1973) Percentile reinforcement: Paradigm for experimental analysis of response shaping. In G. H. Bower (Ed.), *The Psychology of Learning and Motivation: Advances in Theory and Research*, vol. 7. Academic Press, New York, pp. 271–296.

PLATT, J. R., D. O. KUCH and S. C. BITGOOD (1973) Rat's lever-press duration as psychophysical judgment of time. *J. exp. Anal. Behav.* **19**, 239–250.

PLISKOFF, S. S. and T. D. HAWKINS (1967) A method for increasing the reinforcement magnitude of intracranial stimulation. *J. exp. Anal. Behav.* **10**, 281–289.

PLISKOFF, S. S. and T. J. TIERNEY, JR. (1979) On Reynolds' *Discrimination and emission of temporal intervals by pigeons. Bull. Psychon. Soc.* **13**, 173–174.

PLISKOFF, S. S., J. E. WRIGHT and D. T. HAWKINS (1965) Brain stimulation as a reinforcer: intermittent schedules. *J. exp. Anal. Behav.* **8**, 75–88.

POOLE, E. W. (1961) Nervous activity in relation to the respiratory cycle. *Nature* **189**, 579–581.

POPOV, N. A. (1944) Zur Frage der Bedeutung der Zeitfactors für die Auslegung der höchsten Nerventätigkeit. Princip der Zyclochronic. *Physiographica Slovaca*, t. 1, Academia Scientarum et artium slovaca.

POPOV, N. A. (1948) Le facteur temps dans la théorie des réflexes conditionnés. *C.R. Soc. Biol., Paris* **142**, 156–158.

POPOV, N. A. (1950) Action prolongée surle cortex cérébral après stimulation rythmique. *J. Physiol, Paris* **42**, 51–72.

POULTON, E. C. (1952) Perceptual anticipation in tracking with two-pointer and one-pointer displays. *Brit. J. Psychol.* **43**, 222–229.

POULTON, E. C. (1957) On prediction in skilled movements. *Psychol. Bull.* **54**, 467–478.

POUTHAS, V. (1974) Régulation temporelle et histoire du conditionnement. *L'Année Psychologique* **74**, 109–124.

POUTHAS, V. (1979) Analyse des conduites observées au cours de conditionnements au temps chez l'animal. In *Du Temps Biologique au Temps Psychologique.* P.U.F. Paris, pp. 149–160.

POUTHAS, V. (1980) Histoire de conditionnement et longs délais d'attente. *L'Année Psychologique.* In press.

POUTHAS, V. and C. CAVE (1972) Evolution de deux conduites collatérales en cours de conditionnement temporel chez le chat. *L'Année Psychologique* **72**, 17–24.

POWELL, R. W. (1968) The effect of small sequential changes in fixed-ratio size upon the post-reinforcement pause. *J. exp. Anal. Behav.* **11**, 589–593.

POWELL, R. W. (1971) *Responding under DRL schedules in the crow*. Paper presented at the 12th annual meeting of the Psychonomic Society, November 1971, Chase-Park Plaza, St. Louis, Missouri.

POWELL, R. W. (1972a) Responding under basic schedules of reinforcement in the crow. *J. comp. physiol. Psychol.* **79**, 156–164.

POWELL, R. W. (1972b) The effect of deprivation upon fixed-interval responding: A two-state analysis. *Psychon. Sci.* **26**, 31–34.

POWELL, R. W. (1974) Comparison of differential reinforcement of low rates (DRL) performance in pigeons (*Columba livia*) and crows (*Corvus brachyrhynchos*). *J. comp. physiol. Psychol.* **86**, 736–746.

PUBOLS, L. M. (1966) Changes in food-motivated behavior of rats as a function of septal and amygdaloid lesions. *Exper. Neurol.* **15**, 240–254.

QUESADA, D. C. and R. A. SCHMIDT (1970) A test of the Adams-Creamer decay hypothesis for motor response timing. *J. Mot. Behav.* **2**, 273–283.

RACHLIN, H. and B. BURKHARD (1978) The temporal triangle: response substitution in instrumental conditioning. *Psychol. Rev.* **85**, 22–47.

RANDICH, A., W. J. JACOBS, V. M. LOLORDO and J. R. SUTTERER (1978) Conditioned suppression of DRL responding: effects of UCS intensity, schedule parameter and schedule context. *Q.J. exp. Psychol.* **30**, 141–150.

RANDOLPH, J. J. (1965) A further examination of collateral behavior in humans. *Psychon. Sci.* **3**, 227–228.

RAY, R. C. and W. J. McGILL (1964) Effects of class interval size upon certain frequency distributions of interresponse times. *J. exp. Anal. Behav.* **7**, 125–127.

RAZRAN, G. (1971) *Mind in Evolution: An East-West Synthesis of Learned Behavior and Cognition*. Houghton Mifflin Company, Boston.

REBERG, D., N. K. INNIS, B. MANN and C. EIZENGA (1978) "Superstitious" behaviors resulting from periodic response-independent presentation of food and water. *Anim. Behav.* **26**, 507–519.

REBERG, D., B. MANN and N. K. INNIS (1977) Superstitious behavior for food and water in the rat. *Physiol. Behav.* **19**, 803–806.

REBERT, C. S. (1972) Cortical and subcortical slow potentials in the monkey's brain during a preparatory interval. *Electroenceph. Clin. Neurophysiol.* **33**, 389–402.

REESE, E. P. and T. W. REESE (1962) The quail *Coturnix coturnix* as a laboratory animal. *J. exp. Anal. Behav.* **5**, 265–270.

REINBERG, A. (1977) *Des Rythmes Biologiques à la Chronobiologie*. Gauthier Villars, Bordas, Paris.

REQUIN, J. (1965) Rôle de la périodicité cardiaque dans la latence d'une réponse motrice simple. *Psychol. Fr.* **10**, 155–163.

REQUIN, J. and M. BONNET (1968) Quelques données expérimentales contradictoires sur la distribution de l'activité motrice spontanée dans le cycle cardiaque. *Cah. Psychol.* **11**, 23–34.

REQUIN, J. and M. GRANJON (1968) Données expérimentales préliminaires sur le rôle de la périodicité cardiaque dans l'appréciation de la durée d'un stimulus auditif. *Psychol. Fr.* **13**, 71–86.

RESCORLA, R. A. (1968) Pavlovian conditioned fear in Sidman avoidance. *J. comp. physiol. Psychol.* **65**, 55–60.

REYNOLDS, G. S. (1964a) Temporally spaced responding by pigeons: development and effects of deprivation and extinction. *J. exp. Anal. Behav.* **7**, 415–421.

REYNOLDS, G. S. (1964b) Accurate and rapid reconditioning of spaced responding. *J. exp. Anal. Behav.* **7**, 273–275.

REYNOLDS, G. S. (1966) Discrimination and emission of temporal intervals by pigeons. *J. exp. Anal. Behav.* **9**, 65–68.

REYNOLDS, G. S. and A. C. CATANIA (1962) Temporal discrimination in pigeons. *Science* **135**, 314–315.

REYNOLDS, G. S. and A. J. LIMPO (1968) On some causes of behavioral contrast. *J. exp. Anal. Behav.* **11**, 543–547.

REYNOLDS, G. S. and A. McLEOD (1970) On the theory of inter-response time reinforcement. In G. H. Bower (Ed.), *The Psychology of Learning and Motivation*. Academic Press, London, pp. 85–107.

RICCI, J. A. (1973) Key-pecking under response-independent food presentation after long, simple and compound stimuli. *J. exp. Anal. Behav.* **19**, 509–516.

RICHARDS, R. W. and M. RILLING (1972) Aversive aspects of a fixed-interval schedule of food reinforcement. *J. exp. Anal. Behav.* **17**, 405–411.

RICHARDSON, W. K. (1973) A test of the affectiveness of the differential reinforcement of low-rate schedule. *J. exp. Anal. Behav.* **20**, 385–391.

RICHARDSON, W. K. (1976) The sensitivity of the pigeon's key-peck to the differential reinforcement of long interresponse times. *Anim. Learn. Behav.* **4**, 231–240.

RICHARDSON, W. K. and D. B. CLARK (1976) A comparison of the key-peck and treadle-press operants in the pigeon: differential-reinforcement-of-low-rate schedule of reinforcement. *J. exp. Anal. Behav.* **26**, 237–256.

RICHARDSON, W. K. and T. E. LOUGHEAD (1974a) The effect of physical restraint on behavior under differential-reinforcement-of-low-rates schedule. *J. exp. Anal. Behav.* **21**, 455–461.

RICHARDSON, W. K. and T. E. LOUGHEAD (1974b) Behavior under large values of the differential-reinforce-ment-of-low-rates schedule. *J. exp. Anal. Behav.* **22**, 121–129.

RICHELLE, M. (1962) Action du chlordiazepoxide sur les régulations temporelles dans un comportement conditionné chez le chat. *Arch. Int. Pharmacodyn.* **140**, 434–449.

RICHELLE, M. (1968) Notions modernes de rythme biologiques et régulations temporelles acquises. In J. de Ajuriaguerra (Ed.), *Cycles Biologiques et Psychiatrie.* Georg et Cie., Genève; Masson, Paris, pp. 233–255.

RICHELLE, M. (1969) Combined action of diazepam and d-amphetamine on fixed-interval performance in cats. *J. exp. Anal. Behav.* **12**, 989–998.

RICHELLE, M. (1972) Temporal regulation of behaviour and inhibition. In R. A. Boakes and M. S. Halliday (Eds.), *Inhibition and Learning.* Academic Press, London, pp. 229–251.

RICHELLE, M. (1978) Action des benzodiazepines sur le comportement acquis chez l'animal. *Psychologie Medicale* **10**, A (hors série).

RICHELLE, M. (1979) Same or different? An exploration of the behavioral effects of benzamides. In press.

RICHELLE, M., C. CARPENTIER, F. CORNIL, J. P. BRONKART and C. LALIÈRE (1967) L'amassement comme motivation dans le conditionnement du hamster. *Psychol. Belg.* **7**, 67–74.

RICHELLE, M. and B. DJAHANGUIRI (1964) Effet d'un traitement prolongé au chlordiazepoxide sur un conditionnement temporel chez le rat. *Psychopharmacologia* **5**, 106–114.

RICHELLE, M. and H. LEJEUNE (1979) L'animal et le temps. In *Du Temps Biologique au Temps Psychologique.* P.U.F., Paris, pp. 73–128.

RICHELLE, M., B. XHENSEVAL, O. FONTAINE and L. THONE (1962) Action of chlordiazepoxide on two types of temporal conditioning in rats. *Int. J. Neuropharmacol.* **1**, 381–391.

RICHTER, C. P. (1965) *Biological Clocks in Medicine and Psychiatry.* Charles C. Thomas Publisher, Spring-field, Illinois, U.S.A.

ROBINSON, S., W. G. BLATT and C. TEPLITZ (1968) Heat tolerance of the resting and exercising rat. *Canadian Journal of Biochemistry and Physiology* **46**, 189–194.

ROPER, T. J. (1978) Diversity and substitutability of adjunctive activities under fixed-interval schedules of food reinforcement. *J. exp. Anal. Behav.* **30**, 83–96.

ROSENKILDE, C. E. and I. DIVAC (1976a) Time discrimination performance in cats with lesions in prefrontal cortex and caudate nucleus. *J. comp. physiol. Psychol.* **90**, 343–352.

ROSENKILDE, C. E. and I. DIVAC (1976b) Discrimination of time intervals in cats. *Acta Neurobiol. Exp.* **36**, 311–317.

ROSENKILDE, C. E. and W. LAWICKA (1977) Effects of medial and dorsal prefrontal ablation on a go left-go right time discrimination task in dogs. *Acta Neurobiol. Exp.* **37**, 209–221.

ROSS, G. S., W. HODOS and J. V. BRADY (1962) Electroencephalographic correlates of temporally spaced responding and avoidance behavior. *J. exp. Anal. Behav.* **5**, 467–472.

ROSS, J. F. and S. P. GROSSMAN (1975) Septal influences on operant responding in the rat. *J. comp. physiol. Psychol.* **89**, 523–536.

ROWLAND, V. and H. GLUCK (1960) Electroencephalographic arousal and its inhibition as studied by auditory conditioning. In. J. Wortis (Ed.), *Recent Advances in Biological Psychiatry.* Grune and Stratton, New York.

ROWLAND, V. and M. GOLDSTONE (1963) Appetitively conditioned and drive-related bioelectric baseline shift in cat cortex. *Electroenceph. Clin. Neurophysiol.* **15**, 474–485.

ROZIN, P. (1965) Temperature independence of an arbitrary temporal discrimination in the goldfish. *Science* **149**, 561–564.

RUBIN, H. B. and H. J. BROWN (1969) The rabbit as a subject in behavioral research. *J. exp. Anal. Behav.* **12**, 663–667.

RUCH, F. L. (1931) L'appréciation du temps chez le rat blanc. *L'Année Psychologique* **32**, 118–130.

RUCHKIN, D. S., M. G. MCCALLEY and E. H. GLASER (1977) Event related potentials and time estimation. *Psychophysiology* 14, **5**, 451–455.

RUZAK, B. and I. ZUCKER (1975) Biological rhythms and animal behavior. *Annual Rev. Psychol.* **26**, 137–171.

SAMS, C. F. and E. C. TOLMAN (1925) Time discrimination in white rats. *J. comp. physiol. Psychol.* **5**, 255–263.

SASLOW, C. A. (1968) Operant control of response latency in monkeys: Evidence for a central explanation. *J. exp. Anal. Behav.* **11**, 89–98.

SCHAUB, R. E. (1969) Response-cue contingency and cue effectiveness. In D. P. Hendry (Ed.), *Conditioned Reinforcement.* Dorsey Press, Homewood, Illinois, pp. 342–356.

SCHMALTZ, L. W. and R. L. ISAACSON (1966a) The effect of preliminary training condition upon DRL performance in the hippocampectomized rat. *Physiol. Behav.* **1**, 175–182.

SCHMALTZ, L. W. and R. L. ISAACSON (1966b) Retention of a DRL schedule by hippocampectomized and partially neodecorticate rate. *J. comp. physiol. Psychol.* **62**, 128–132.

SCHMALTZ, L. W. and R. L. ISAACSON (1968a) The effect of blindness on DRL 20 performances exhibited by animals with hippocampal destruction. *Psychon. Sci.* **11**, 241–242.

SCHMALTZ, L. W. and R. L. ISAACSON (1968b) Effects of caudate and frontal lesions on retention and relearning of a DRL schedule. *J. comp. physiol. Psychol.* **65**, 343–347.

SCHMIDT, R. A. (1971) Proprioception and the timing of motor responses. *Psychol. Bull.* **76**, 383–392.

SCHMIDT, R. A. (1973) Proprioception versus motor outflow in timing: A reply to Jones. *Psychol. Bull.* **79**, 389–390.

SCHMIDT, R. A. and R. W. CHRISTINA (1969) Proprioception as a mediator in the timing of motor responses. *J. exp. Psychol.* **81**, 303–307.

SCHNEIDER, B. A. (1969) A two-state analysis of fixed-interval responding in the pigeon. *J. exp. Anal. Behav.* **12**, 677–687.

SCHNEIDER, B. A. and A. J. NEURINGER (1972) Responding under discrete-trial fixed-interval schedules of reinforcement. *J. exp. Anal. Behav.* **18**, 187–199.

SCHNELLE, J. F., S. F. WALKER and H. M. B. HURWITZ (1971) Concurrent performance in septally operated rats: one and two response extinction. *Physiol. Behav.* **6**, 649–654.

SCHOENFELD, W. N., W. W. CUMMING and E. HEARST (1956) On the classification of reinforcement schedules. *Proceedings of the National Academy of Sciences* **42**, 563–570.

SCHWARTZ, B. (1972) The role of positive conditioned reinforcement in the maintenance of key pecking which prevents delivery of primary reinforcement. *Psychon. Sci.* **28**, 277–278.

SCHWARTZ, B. (1977) Two types of pigeon key-pecking: suppression of long—but not short—duration key pecks by duration-dependent shock. *J. exp. Anal. Behav.* **27**, 393–398.

SCHWARTZ, B. and E. GAMZU (1977) Pavlovian control of operant behavior. An analysis of autoshaping and its implications for operant conditioning. In W. K. Honig and J. E. R. Staddon (Eds.), *Handbook of Operant Behavior*. Prentice-Hall, Inc., Englewood Cliffs, New Jersey, pp. 53–98.

SCHWARTZ, B. and D. R. WILLIAMS (1971) Discrete-trials spaced responding in the pigeon: the dependence of efficient performance on the availability of a stimulus for collateral pecking. *J. exp. Anal. Behav.* **16**, 155–160.

SCHWARTZ, B. and D. R. WILLIAMS (1972a) The role of the response-reinforcer contingency in negative auto-maintenance. *J. exp. Anal. Behav.* **71**, 351–357.

SCHWARTZ, B. and D. R. WILLIAMS (1972b) Two different kinds of key peck in the pigeon: some properties of responses maintained by negative and positive response-reinforcer contingencies. *J. exp. Anal. Behav.* **18**, 201–216.

SCHWARTZBAUM, J. S. and P. E. GAY (1966) Interacting behavioral effects of septal and amygdaloid lesions in the rat. *J. comp. physiol. Psychol.* **61**, 59–65.

SCHWARTZBAUM, J. S., M. H. KELLICUT, T. M. SPIETH and J. B. THOMPSON (1964) Effects of septal lesions in rats on response inhibition associated with food-reinforced behavior. *J. comp. physiol. Psychol.* **58**, 217–224.

SCOBIE, S. R. and D. C. GOLD (1975) Differential reinforcement of low rates in goldfish. *Anim. Learn. Behav.* **3**, 143–146.

SEGAL, E. F. (1962a) Exteroceptive control of fixed-interval responding. *J. exp. Anal. Behav.* **5**, 49–57.

SEGAL, E. F. (1962b) Effects of dl-amphetamine under concurrent VI-DRL performance. *J. exp. Anal. Behav.* **5**, 105–112.

SEGAL, E. F. (1969) Transformation of polydipsic drinking into operant drinking: A paradigm? *Psychon. Sci.* **16**, 133–135.

SEGAL, E. F. (1972) Induction and the provenance of operants. In R. M. Gilbert and J. R. Millenson (Eds.), *Reinforcement. Behavioral Analyses*. Academic Press, New York, pp. 1–34.

SEGAL, E. F. and S. A. DEADWYLER (1964) Amphetamine differentially affects temporally spaced bar pressing and collateral water drinking. *Psychon. Sci.* **1**, 349–350.

SEGAL, E. F. and S. A. DEADWYLER (1965a) Determinants of polydipsia: II. DRL extinction. *Psychon. Sci.* **2**, 203–204.

SEGAL, E. F. and S. A. DEADWYLER (1965b) Determinants of polydipsia: VI. Taste of the drinking solution. *Psychon. Sci.* **3**, 101–102.

SEGAL, E. F. and S. M. HOLLOWAY (1963) Timing behavior in rats with water drinking as a mediator. *Science* **140**, 888–889.

SEGAL, E. F., D. L. ODEN and S. A. DEADWYLER (1965) Determinants of polydipsia: V. Effect of amphetamine and phenobarbital. *Psychon. Sci.* **2**, 33–34.

SEGAL-RECHTSCHAFFEN, E. (1963) Reinforcement of mediating behavior on a spaced-responding schedule. *J. exp. Anal. Behav.* **6**, 39–46.

SEITZ, C. P. and F. S. KELLER (1940) Oxygen deprivation and conditioning in the whole rat: I. The effect of deprivation on the lever pressing response. *J. Aviat. Med.* **11**, 210–213.

SHANAB, M. E. and J. L. PETERSON (1969) Polydipsia in the pigeon. *Psychon. Sci.* **15**, 51–52.

SHEPARD, R. N. (1965) Approximation to uniform gradients of generalization by monotone transformations of scale. In D. Mostofsky (Ed.), *Stimulus Generalization*. Stanford University Press, Stanford, pp. 94–110.

SHERMAN, J. G. (1959) *The temporal distribution of responses on fixed-interval schedules*. Unpublished doctoral dissertation, Columbia University.

SHERRINGTON, C. S. (1906) *The Integrative Action of the Nervous System.* Scribner's, New York.

SHIMP, C. P. (1967) The reinforcement of short interresponse times. *J. exp. Anal. Behav.* **10**, 425–434.

SHIMP, C. P. (1968) Magnitude and frequency of interresponse times. *J. exp. Anal. Behav.* **11**, 525–535.

SHIMP, C. P. (1971) The reinforcement of four interresponse times in a two-alternative situation. *J. exp. Anal. Behav.* **16**, 385–399.

SHIMP, C. P. (1973) Sequential dependencies in free-responding. *J. exp. Anal. Behav.* **19**, 491–497.

SHIMP, C. P. (1976) Short-term memory in the pigeon: relative recency. *J. exp. Anal. Behav.* **25**, 55–61.

SHULL, R. L. (1970a) A response-initiated FI schedule of reinforcement. *J. exp. Anal. Behav.* **13**, 13–15.

SHULL, R. L. (1970b) The response-reinforcement dependency in fixed-interval schedules of reinforcement. *J. exp. Anal. Behav.* **14**, 55–60.

SHULL, R. L. (1971a) Sequential patterns in post-reinforcement pauses on fixed-interval schedules of food. *J. exp. Anal. Behav.* **15**, 221–232.

SHULL, R. L. (1971b) Postreinforcement pause duration on fixed-interval and fixed-time schedules of food reinforcement. *Psychon. Sci.* **23**, 77–78.

SHULL, R. L. and A. J. BROWNSTEIN (1975) The relative proximity principle and the postreinforcement pause. *Bull. Psychon. Soc.* **5**, 129–131.

SHULL, R. L. and M. GUILKEY (1976) Food deliveries during the pause on fixed-interval schedules. *J. exp. Anal. Behav.* **26**, 415–423.

SHULL, R. L., M. GUILKEY and P. T. BROWN (1978) Latencies on response-initiated fixed-interval schedules: effects of signalling food availability. *Bull. Psychon. Soc.* **12**, 207–210.

SHULL, R. L., M. GUILKEY and W. WITTY (1972) Changing the response unit from a single peck to a fixed number of pecks in fixed-interval schedules. *J. exp. Anal. Behav.* **17**, 193–200.

SIDMAN, M. (1953) Two temporal parameters in the maintenance of avoidance behavior by the white rat. *J. comp. physiol. Psychol.* **46**, 253–261.

SIDMAN, M. (1955) Technique for assessing the effects of drugs on timing behavior. *Science* **122**, 925.

SIDMAN, M. (1956) Time discrimination and behavioral interaction in free operant situation. *J. comp. physiol. Psychol.* **49**, 469–473.

SIDMAN, M. (1960) *Tactics of Scientific Research.* Basic Books, New York.

SIDMAN, M. (1961) Stimulus generalization in an avoidance situation. *J. exp. Anal. Behav.* **4**, 157–169.

SIDMAN, M. (1962) Classical avoidance without a warning stimulus. *J. exp. Anal. Behav.* **5**, 97–104.

SIDMAN, M. (1966) Avoidance behavior. In W. K. Honig (Ed.), *Operant Behavior: Areas of Research and Application.* Appleton Century Crofts, New York, pp. 448–498.

SIDMAN, M., J. V. BRADY, D. G. CONRAD and A. SCHULMAN (1955) Reward schedules and behavior maintained by intracranial self-stimulation. *Science* **122**, 830–831.

SIEGEL, R. K. (1970) Apparent movement detection in the pigeon. *J. exp. Anal. Behav.* **14**, 93–97.

SILBY, R. and D. J. MCFARLAND (1974) A state-space approach to motivation. In D. J. McFarland (Ed.), *Motivational Control Systems.* Academic Press, London.

SILVERMAN, P. J. (1971) Chained and tandem fixed-interval schedules of punishment. *J. exp. Anal. Behav.* **16**, 1–13.

SINGER, G., M. J. WAYNER, J. STEIN, K. CIMINO and K. KING (1974) Adjunctive behavior induced by wheel running. *Physiol. Behav.* **12**, 493–495.

SINGH, D. and D. D. WICKENS (1968) Desinhibition in instrumental conditioning. *J. comp. physiol. Psychol.* **66**, 557–559.

SKINNER, B. F. (1938) *The Behavior of Organisms.* Appleton Century Crofts, New York.

SKINNER, B. F. (1948) Superstition in the pigeon. *J. exp. Psychol.* **38**, 168–172.

SKINNER, B. F. (1953) *Science and Human Behavior.* Macmillan, New York.

SKINNER, B. F. (1956) A case history in scientific method. *Amer. Psychol.* **11**, 221–233.

SKINNER, B. F. and W. H. MORSE (1957) Concurrent activity under fixed-interval reinforcement. *J. comp. physiol. Psychol.* **50**, 279–281.

SKINNER, B. F. and W. H. MORSE (1958) Sustained performance during very long experimental sessions. *J. exp. Anal. Behav.* **1**, 235–244.

SKUBAN, W. E. and W. K. RICHARDSON (1975) The effect of the size of the test environment on behavior under two temporally defined schedules. *J. exp. Anal. Behav.* **23**, 271–275.

SLONAKER, R. L. and D. HOTHERSALL (1972) Collateral behaviors and the DRL deficit of rats with septal lesions. *J. comp. physiol. Psychol.* **80**, 91–96.

SMITH, J. B. and F. C. CLARK (1972) Two temporal parameters of food postponement. *J. exp. Anal. Behav.* **18**, 1–12.

SMITH, J. B. and F. C. CLARK (1974) Intercurrent and reinforced behavior under multiple spaced responding. *J. exp. Anal. Behav.* **21**, 445–454.

SMITH, J. B. and F. C. CLARK (1975) Effects of d-amphetamine, chlorpromazine and chlordiazepoxide and intercurrent behavior during spaced responding schedules. *J. exp. Anal. Behav.* **24**, 241–248.

SMITH, R. F. and F. KELLER (1970) Free operant avoidance in pigeons using a treadle response. *J. exp. Anal. Behav.* **13**, 211–214.

SMITH, R. F. and L. W. SCHMALTZ (1979) Acquisition of appetitively and aversively motivated tasks in rats following lesions in the mamillary bodies. *Physiol. Psychol.* **7**, 43–48.

SNAPPER, A. G., D. A. RAMSAY and W. N. SCHOENFELD (1969) Disruption of a temporal discrimination under response-independent shock. *J. exp. Anal. Behav.* **12**, 423–430.

SNAPPER, A. G., E. H. SHIMOFF and W. N. SCHOENFELD (1971) Response effects of response-dependant and clock-dependent fixed-interval schedules of reinforcement. *Psychon. Sci.* **23**, 65–67.

SODETZ, F. J. (1970) Septal ablation and free-operant avoidance behavior in the rat. *Physiol. Behav.* **5**, 773–777.

SOKOLOV, E. V. (1966) Neuronal mechanisms of the orientating reflex. *XVIIIth Internat. Congr. Psychol. Symp.* **5**, 31–36.

SPENCE, K. W. (1947) The role of secondary reinforcement in delayed reward learning. *Psychol. Rev.* **54**, 1–8.

SPENCE, K. W. (1956) *Behavior Theory and conditioning.* Yale University Press, New Haven.

SQUIRES, N., J. NORBORG and E. FANTINO (1975) Second-order schedules: discrimination of components. *J. exp. Anal. Behav.* **24**, 157–171.

STADDON, J. E. R. (1965) Some properties of spaced responding in pigeons. *J. exp. Anal. Behav.* **8**, 19–27.

STADDON, J. E. R. (1969a) The effect of informative feedback on temporal tracking in the pigeon. *J. exp. Anal. Behav.* **12**, 27–38.

STADDON, J. E. R. (1969b) Inhibition and the operant. *J. exp. Anal. Behav.* **12**, 481–487.

STADDON, J. E. R. (1970a) Effect of reinforcement duration on fixed-interval responding. *J. exp. Anal. Behav.* **13**, 9–11.

STADDON, J. E. R. (1970b) Temporal effects of reinforcement: a "negative frustration" effect. *Learn. Motiv.* **1**, 227–247.

STADDON, J. E. R. (1972a) Reinforcement omission on temporal Go–No-Go schedules. *J. exp. Anal. Behav.* **18**, 223–229.

STADDON, J. E. R. (1972b) Temporal control and the theory of reinforcement schedules. In R. M. Gilbert and J. R. Millenson (Eds.), *Reinforcement: Behavioral Analyses.* Academic Press, New York and London, pp. 209–262.

STADDON, J. E. R. (1974) Temporal control, attention and memory. *Psychol. Rev.* **81**, 375–391.

STADDON, J. E. R. (1975) Autocontingencies: special contingencies or special stimuli? A review of Davis, Memmott and Hurwitz. *J. exp. Psychol.* **104**, 189–191.

STADDON, J. E. R. (1976) Learning as adaptation. In W. K. Estes (Ed.), *Handbook of Learning and Cognitive Processes. Vol. II. Conditioning and Behavior Theory.* Erlbaum Associates, New York, pp. 37–98.

STADDON, J. E. R. (1977) Schedule-induced behavior. In W. K. Honig and J. E. R. Staddon (Eds.), *Handbook of Operant Behavior.* Prentice-Hall, Inc. Englewood Cliffs, New Jersey, pp. 125–153.

STADDON, J. E. R. and S. L. AYRES (1975) Sequential and temporal properties of behavior induced by a schedule of periodic food delivery. *Behaviour* **54**, 26–49.

STADDON, J. E. R. and J. A. FRANK (1975) The role of the peckfood contingency on fixed-interval schedules. *J. exp. Anal. Behav.* **23**, 17–23.

STADDON, J. E. R. and N. K. INNIS (1966) An effect analogous to "frustration" on interval reinforcement schedules. *Psychon. Sci.* **4**, 287–288.

STADDON, J. E. R. and N. K. INNIS (1969) Reinforcement omission on fixed-interval schedules. *J. exp. Anal. Behav.* **12**, 689–700.

STADDON, J. E. R. and V. L. SIMMELHAG (1971) The "superstition" experiment: a reexamination of its implications for the principles of adaptative behavior. *Psychol. Rev.* **78**, 3–43.

STAMM, J. S. (1963) Function of prefrontal cortex in timing behavior of monkeys. *Exper. Neurol.* **7**, 87–97.

STAMM, J. S. (1964) Function of cingulate and prefrontal cortex in frustrative behavior. *Acta Biol. Exper.* (Warszava) **24**, 27–36.

STEBBINS, W. C., P. B. MEAD and J. M. MARTIN (1959) The relation of amount of reinforcement to performance under a fixed-interval schedule. *J. exp. Anal. Behav.* **2**, 351–356.

STEIN, N. (1977) Effects of reinforcement history on the mediation of human DRL performance. *Bull. Psychon. Soc.* **9**, 93–96.

STEIN, N. and S. FLANAGAN (1974) Human DRL performance, collateral behavior, and verbalization of the reinforcement contingency. *Bull. Psychon. Soc.* **3**, 27–28.

STEIN, N. and R. LANDIS (1973) Mediating role of human collateral behavior during spaced-responding schedule of reinforcement. *J. exp. Psychol.* **97**, 28–33.

STEINMAN, W. M. and C. M. BUTTER (1968) Response rate, suppression, and extinction in septal rats. *Proc. 76th Ann. Conv. Am. Psychol. Ass.* **3**, 277–278.

STEVENSON, J. A. F. and R. H. RIXON (1957) Environmental temperature and deprivation of food and water in the spontaneous motor activity of rats. *Yale Journal of Biology and Medicine* **29**, 575–584.

STEVENSON, J. G. and F. L. CLAYTON (1970) A response duration schedule: effects of training, extinction and deprivation. *J. exp. Anal. Behav.* **13**, 359–367.

STOTT, L. H. and F. L. RUCH (1939) Establishing time-discriminatory behavior in the white rat by use of an automatically controlled training and recording apparatus. *J. comp. Psychol.* **27**, 491–503.

STRETCH, R., E. R. ORLOFF and S. D. DALRYMPLE (1968) Maintenance of responding by a fixed-interval schedule of electric-shock presentation in squirrel monkeys. *Science* **12**, 583–586.

STUBBS, A. (1969) Contiguity of briefly presented stimuli with food reinforcement. *J. exp. Anal. Behav.* **12**, 271–278.

STUBBS, D. A. (1968) The discrimination of stimulus duration by pigeons. *J. exp. Anal. Behav.* **11**, 223–238.

STUBBS, D. A. (1971) Second order schedules and the problem of conditioned reinforcement. *J. exp. Anal. Behav.* **16**, 289–313.

STUBBS, D. A. (1976) Response bias and the discrimination of stimulus duration. *J. exp. Anal. Behav.* **25**, 243–250.

STUBBS, D. A. and S. L. COHEN (1972) Second order schedules: comparison of different procedures for scheduling paired and non-paired brief stimuli. *J. exp. Anal. Behav.* **18**, 403–413.

STUBBS, D. A. and P. J. SILVERMAN (1972) Second-order schedules: a brief shock at the completion of each component. *J. exp. Anal. Behav.* **17**, 201–212.

STUBBS, D. A. and J. R. THOMAS (1974) Discrimination of stimulus duration and d-amphetamine in pigeons: a psychophysical analysis. *Psychopharmacologia* **36**, 313–322.

STUBBS, D. A., S. J. VAUTIN, H. M. REID and D. L. DELEHANTY (1978) Discriminative functions of schedule stimuli and memory: a combination of schedule and choice procedures. *J. exp. Anal. Behav.* **29**, 167–180.

SWETS, J. A. (1973) The relative operating characteristic in psychology. *Science* **182**, 990–1000.

TAUB, E. (1977) Movement in nonhuman primates deprived of somatosensory feedback. In J. KEOGH AND R. S. HUTTON (Eds.), *Exercise and Sports Sciences Reviews*, Vol. 4. Journal Publishing Affiliates, Santa Barbara, California, U.S.A., pp. 335–374.

TAUB, E., R. C. BACON and A. J. BERMAN (1965) Acquisition of a trace-conditioned avoidance response after deafferentation of the responding limb. *J. comp. physiol. Psychol.* **59**, 275–279.

TAUB, E. and A. J. BERMAN (1963) Avoidance conditioning in the absence of relevant proprioceptive and exteroceptive feedback. *J. comp. physiol. Psychol.* **56**, 1012–1016.

TAUB, E. and A. J. BERMAN (1964) *The effect of massive somatic deafferentation on behavior and wakefulness in monkeys.* Paper read at Psychonomic Society, Niagara Falls, Ontario.

TAUB, E., S. J. ELLMAN and A. J. BERMAN (1966) Deafferentation in monkeys: effects on conditioned grasp response. *Science* **151**, 593–594.

TAUB, E., P. PERRELLA and G. BARRO (1972) Behavioral development in monkeys following bilateral fore-limb deafferentation on the first day of life. *Transactions of the American Neurological Association* **97**, 101–104.

TECCE, J. J. (1972) Contingent negative variation (CNV) and psychological processes in man. *Psychol. Bull* **77**, 73–108.

TERMAN, J. S. (1974) The control of interresponse time probabilities by the magnitudes of reinforcing brain stimulation. *Physiol. Behav.* **12**, 219–229.

TERMAN, J. S., M. TERMAN and J. KLING (1970) Some temporal properties of intracranial self-stimulation. *Physiol. Behav.* **5**, 183–191.

TERRACE, H. S. (1968) Discrimination learning, the peak shift, and behavioral contrast. *J. exp. Anal. Behav.* **11**, 727–741.

THOMAS, J. R. (1965) Discriminated time-out avoidance in pigeons. *J. exp. Anal. Behav.* **8**, 329–338.

THOR, D. H. (1963) Diurnal variability in time estimation. *Psychol. Abstracts* **37**, 7431.

THOR, D. H. and R. O. BALDWIN (1965) Time of day estimates at six times of day under normal conditions. *Percept. Mot. Skills* **21**, 904–906.

THORNE, B. M., K. RAGER and J. S. TOPPING (1976) DRL performances in rats following damages to the septal area, olfactory bulbs or olfactory tubercle. *Physiol. Psychol.* **4**, 493–497.

TINBERGEN, N. (1952) "Derived" activities: Their causation, biological significance, origin, and emancipation during evolution. *Quarterly Review of Biology* **27**, 1032.

TODD, G. E. and D. C. COGAN (1978) Selected schedules of reinforcement in the black-tailed prairie dog. *Anim. Learn. Behav.* **6**, 429–434.

TODOROV, J. C., E. A. M. FERRARI and D. G. DE SOUZA (1974) Key-pecking as a function of response-shock and shock-shock interval in unsignalled avoidance. *J. exp. Anal. Behav.* **22**, 215–218.

TOPPING, J. S., J. W. PICKERING and J. A. JACKSON (1971) Efficiency of DRL responding as a function of response effort. *Psychon. Sci.* **24**, 149–150.

TRAPOLD, M. A., J. G. CARLSON and W. A. MYERS (1965) The effect of non-contingent fixed- and variable-interval reinforcement upon subsequent acquisition of the fixed-interval scallop. *Psychon. Sci.* **2**, 261–262.

TREISMAN, M. (1963) Temporal discrimination and the indifference interval: implications for a model of the internal clock. *Psychological Monographs* **77**, 1–31 (Whole n° 576).

TREISMAN, M. (1965) The Psychology of Time. *Discovery*, October, 41–45.

URAMOTO, I. (1971) Effects upon conditioned response time of contingent photic stimulation in cats. *Physiol. Behav.* **6**, 203–204.

URAMOTO, I. (1973) Timing process controlled by low frequency flashes in cats. *Physiol. Behav.* **10**, 171–173.

URBAIN, C., A. POLING and T. THOMPSON (1979) Differing effects of intermittent food delivery on interim behavior in guinea pigs. *Physiol. Behav.* **22**, 621–625.

URBAN, F. M. (1907) On the method of just perceptible difference. *Psychol. Rev.* **14**, 244–253.

VALLIANT, P. M. (1978) *The effects of FR schedule shifts on physiochemical parameters of the albino rat.* Unpublished dissertation, University of Windsor, Windsor, Ontario, Canada.

VAN HOESEN, G. W., J. M. MACDOUGALL, J. R. WILSON and J. C. MITCHELL (1971) Septal lesions and the acquisition and maintenance of a discrete-trial DRL task. *Physiol. Behav.* **7**, 471–475.

VILLAREAL, J. (1967) *Schedule-induced pica.* Paper read at Eastern Psychological Association Meeting, Boston.

VORONIN L. L. (1971) Microelectrode study of cellular analog to a conditioned reflex to time. *Zh. Vyssh. Nerv. Deiat.* **21**, 1238–1246.

WAGMAN, A. M. I. (1968) The effect of frontal lobe lesions upon behavior requiring use of response-produced cues. *J. comp. physiol. Psychol.* **66**, 69–76.

WALL, A. M. (1965) Discrete-trial analysis of fixed-interval discrimination. *J. comp. physiol. Psychol.* **60**, 70–75.

WALLACE, M. and A. RABIN (1960) Temporal experience. *Psychol. Bull.* **57**, 213–236.

WALLER, M. B. (1961) Effects of chronically administered chlorpromazine on multiple schedule performance. *J. exp. Anal. Behav.* **4**, 351–359.

WALTER, W. G. (1966) Electric signs of expectancy and decision in the human brain. In H. L. Oestreicher and D. R. Moore (Eds.), *Cybernetic Problems in Bionics.* Gordon and Breach, New York, pp. 361–396.

WALTER, W. G., R. COOPER, V. J. ALDRIDGE, W. C. MACCALLUM and A. L. WINTER (1964) Contingent negative variation: an electric sign of sensorimotor association and expectancy in the human brain. *Nature* **203**, 380–384.

WASSERMAN, E. A. (1977) Conditioning of withintrial patterns of key pecking in pigeons. *J. exp. Anal. Behav.* **28**, 213–220.

WATANABE, M. (1976) *Neuronal correlates of timing behavior in the monkey.* Paper read at the XXIst International Congress of Psychology, Paris, July.

WEINBERG, H. (1968) Temporal discrimination of cortical stimulation. *Physiol. Behav.* **3**, 297–300.

WEINBERG, H., W. G. WALTER, R. COOPER and V. J. ALDRIDGE (1974) Emitted cerebral events. *Electroenceph. Clin. Neurophysiol.* **36**, 449–456.

WEISMAN, R. G. (1969) Some determinants of inhibitory stimulus control. *J. exp. Anal. Behav.* **12**, 443–450.

WEISS, B. (1970) The fine structure of operant behavior during transition states. In W. N. Schoenfeld (Ed.), *The Theory of Reinforcement Schedules.* Appleton Century Crofts, New-York, pp. 277–311.

WEISS, B. and V. G. LATIES (1964) Drug effects on the temporal patterning of behavior. *Fedn. Proc.* **23**, 801–807.

WEISS, B. and V. G. LATIES (1965) Reinforcement schedules generated by on-line digital computer. *Science* **148**, 658–661.

WEISS, B., V. G. LATIES, L. SIEGEL and D. GOLDSTEIN (1966) A computer analysis of serial interactions in spaced responding. *J. exp. Anal. Behav.* **9**, 619–626.

WERTHEIM, G. A. (1965) Some sequential aspects of IRT's emitted during Sidman avoidance behavior in the white rat. *J. exp. Anal. Behav.* **8**, 9–15.

WESTBROOK, R. F. (1973) Failure to obtain positive contrast when pigeons press a bar. *J. exp. Anal. Behav.* **20**, 499–510.

WETHERINGTON, C. L. and A. J. BROWNSTEIN (1979) Schedule control of eating by fixed-time schedules of water presentation. *Anim. Learn. Behav.* **7**, 38–40.

WEVER, R. (1962) Zum Mechanismus der biologischen 24-Stunden-Periodik. *Kybernetik* **1/4**, 139.

WILKIE, D. M. (1974) Stimulus control of responding during a fixed-interval reinforcement schedule. *J. exp. Anal. Behav.* **21**, 425–432.

WILKIE, D. M. and J. J. PEAR (1972) Intermittent reinforcement of an interresponse time. *J. exp. Anal. Behav.* **17**, 67–74.

WILLIAMS, B. A. (1973) The failure of stimulus control after presence-absence discrimination of click-rate. *J. exp. Anal. Behav.* **20**, 23–27.

WILLIAMS, D. R. and H. WILLIAMS (1969) Automaintenance in the pigeon: Sustained pecking despite contingent non-reinforcement. *J. exp. Anal. Behav.* **12**, 511–520.

WILSON, M. P. and F. S. KELLER (1953) On the selective reinforcement of spaced responding. *J. comp. physiol. Psychol.* **46**, 190–193.

WINOGRAD, E. (1965) Maintained generalization testing of conditioned suppression. *J. exp. Anal. Behav.* **8**, 47–51.

WITOSLAWSKI, J. J., R. B. ANDERSON and H. M. HANSON (1963) Behavioral studies with a block vulture, *Coragyps atratus. J. exp. Anal. Behav.* **6**, 605–606.

WOLACH, A. H. and D. P. FERRARO (1969) A failure to obtain disinhibition in fixed-interval conditioning. *Psychon. Sci.* **15**, 47–48.

WOLF, M. and D. M. BAER (1963) *Amer. Psychol.* **18**, 444.

WOODROW, H. (1928) Temporal discrimination in the monkey. *J. comp. physiol. Psychol.* **8**, 395–427.

WRIGHT, A. A. (1974) Psychometric and psychophysical theory within a framework of response bias. *Psychol. Rev.* **81**, 322–347.

WYCKOFF, L. B. JR. (1952) The role of observing responses in discrimination learning. *Psychol. Rev.* **59**, 431–442.

YAGI, B. (1962) The effect of motivating conditions on "the time estimation" in rats. *Jap. J. Psychol.* **33**, 8–24.

ZEIER, H. (1969) DRL performance and timing behavior of pigeons with archistriatal lesions. *Physiol. Behav.* **4**, 189–193.

ZEILER, M. D. (1968) Fixed and variable schedules of response-independent reinforcement. *J. exp. Anal. Behav.* **11**, 405–414.

ZEILER, M. D. (1972a) Fixed-interval behavior: effects of percentage reinforcement. *J. exp. Anal. Behav.* **17**, 177–189.

ZEILER, M. D. (1972b) Time limits for completing fixed-ratios: II. Stimulus specificity. *J. exp. Anal. Behav.* **18**, 243–252.

ZEILER, M. D. (1972c) Reinforcement of spaced-responding in a simultaneous discrimination. *J. exp. Anal. Behav.* **18**, 443–451.

ZEILER, M. D. (1976) Positive reinforcement and the elimination of reinforced response. *J. exp. Anal. Behav.* **26**, 37–44.

ZEILER, M. D. (1977) Schedules of reinforcement: the controlling variables. In W. K. Honig and J. E. R. Staddon (Eds.), *Handbook of Operant Behavior*. Prentice-Hall, Inc., Englewood Cliffs, New Jersey, pp. 201–233.

ZEILER, M. D. and E. R. DAVIS (1978) Clustering in the output of behavior. *J. exp. Anal. Behav.* **29**, 363–374.

ZIMMERMAN, J. (1961) Spaced responding in rats as a function of some temporal variables. *J. exp. Anal. Behav.* **4**, 219–224.

ZIMMERMAN, J. and C. R. SCHUSTER (1962) Spaced responding in multiple DRL schedules. *J. exp. Anal. Behav.* **5**, 497–504.

ZIRIAX, J. M. and A. SILBERBERG (1978) Discrimination and emission of different key-peck durations in the pigeon. *J. exp. Psychol.*: Animal Behavior Processes, **4**, 1–21.

ZUILI, N. and P. FRAISSE (1969) Le rôle des indices perçus dans la comparaison des durées chez l'enfant. *L'Année Psychologique* **69**, 17–36.

ZURIFF, G. E. (1969) Collateral responding during differential reinforcement of low rates. *J. exp. Anal. Behav.* **12**, 971–976.

ADDITIONAL REFERENCES

HATTEN, J. L. (1974) *The effect of feeder duration on Fixed-Interval performance: context dependencies.* Master's Thesis, University of North Carolina at Greensboro.

KELSEY, J. E. (1976). Behavioral effects of intraseptal injections of adrenergic drugs. *Physiol. Psychol.,* **4**, 433–438.

KING, G. D., R. W. SCHAEFFER and S. C. PIERSON (1974). Reinforcement schedule preference in a raccoon (*Procyon Lotor*). *Bull. Psychon. Soc.,* **4**, 97–99.

ROBERTS, S. and R. M. CHURCH (1978). Control of an internal clock. *J. exp. Psychol.: Anim. Behav. Proc.,* **4**, 318–337.

SCULL, J. and R. F. WESTBROOK (1973). Interactions in multiple schedules with different responses in each component. *J. exp. Anal. Behav.* **20**, 511–519.

Index of Names

Adams, J. A.   169, 170, 173, 176
Aitken, W. C.   119, 148, 150, 166
Ajuriaguerra, J. de   6
Alexinsky, T.   152
Allan, L. G.   42, 73
Alleman, H. D.   61, 126
Amsel, A.   124, 231
Anderson, A. C.   9
Anderson, M. C.   93, 135, 193, 195
Anderson, O. D.   27, 86
Anger, D.   24, 30, 32, 53, 192, 218
Appel, J. B.   22, 128, 129, 210, 211
Auge, R. J.   200, 201, 202, 206, 219, 223
Ayres, J. B.   121
Azrin, N. H.   64, 129, 210, 229

Baddeley, A. D.   183
Baer, D. M.   87, 99
Bagdonas, A.   160
Bahrick, H. P.   170
Baldwin, R. O.   135
Balster, R. L.   111, 116
Bard, L.   181
Barfield, R. J.   188
Barofsky, I.   183
Baron A.   207
Baron, M. R.   72
Barowski, E. I.   212
Barrett, J. E.   108, 109, 111, 211
Bartley, S. H.   156
Baum, W. M.   62
Beard, R. R.   187
Beasley, J. L.   168, 186
Beatty, W. W.   86, 135, 145, 187
Beecher, M. D.   86
Beer, B.   118
Beling, I.   138
Bell, C. R.   181, 182, 183
Belleville, R. E.   192
Berlyne, D. E.   207
Berman, A. J.   176
Berryman, R.   28, 175, 189
Birren, J. E.   161
Bishop, G. H.   156
Bitterman, M. E.   87, 98, 100
Blackman, D. E.   28, 60, 177, 213, 226
Blanchard, R.   207
Blancheteau, M.   10, 188, 189
Blondin, C.   86, 94
Blough, D. S.   23, 29, 60, 64
Boakes, R. A.   59, 207, 222
Boice, R.   94

Bolotina, O. P.   106
Bolles, R. C.   121, 138, 139, 217, 218, 220
Boneau, C. A.   42
Bonnet, M.   162
Bonvallet, M.   164
Borda, R. P.   159
Boren, M. C. P.   201, 204
Bower, G. H.   9
Brady, J. V.   62, 111, 145, 153, 154
Braggio, J. T.   148, 150, 166, 167
Brahlek, J.   114, 115
Brake, S. C.   124
Breland, K.   121
Breland, M.   121
Brobeck, J. R.   183
Brown, D. L.   111
Brown, F. A. Jr   3, 200
Brown, H. J.   86, 93
Brown T. G.   229
Brown, P. L.   120, 195, 222
Brown, S.   64, 111, 112
Brownstein, A. J.   128, 188, 211
Bruner, A.   189, 191
Bruno, J. J.   60
Bryan, A. J.   211
Buchwald, J. S.   160
Bunnel, B. N.   145, 166
Bunning, E.   3
Buño, W. Jr   157
Burkett, E. E.   145, 166
Burkhard, B.   93
Bush, R. R.   38
Butter, J.   148, 151, 152, 166
Buytendijk, F. J. J.   10
Byrd, L. D.   86, 97, 108, 109, 129, 130, 212, 214

Caggiula, A. R.   188
Callaway, E.   161, 162
Campbell, R. L.   130
Cantor, M. B.   111
Caplan, H. J.   20, 200, 201, 202, 204
Caplan, M.   145, 146, 147, 148, 150, 152, 166
Carey, R. J.   145, 151
Carlson, V. R.   157
Carlton, P. L.   117, 119
Carnathan, J.   30
Carrigan, P. F.   121
Carter, D. E.   60, 189
Catania, A. C.   30, 33, 36, 37, 38, 60, 65, 66, 71, 73, 74, 83, 114, 189, 192, 210
Cave, C.   188
Cherek, D. R.   189, 190, 230

Chiorini, J. R.   159
Christina, R. W.   170, 171, 175
Chung, S. H.   21, 211
Church, R. M.   30, 36, 37, 42, 50, 73, 74, 75, 78, 83, 84, 210
Cioe, J.   112
Clark, C. V.   145, 151
Clark, D. B.   123
Clark, F. C.   110, 191
Clayton, F. L.   29
Cloar, T.   87, 97, 98
Cogan, D. C.   86, 94
Cohen, J.   182
Cohen, P. S.   60
Cohen, S. L.   212
Cole, J. L.   42
Coleridge, H. M.   164
Collier, G.   107, 114, 188, 190
Colliver, J. A.   42
Conrad, D. G.   60, 111, 119, 153, 154
Contrucci, J. J.   218, 224
Cook, L.   18, 50, 168, 218
Coquery, J. M.   161
Corfield-Sumner, P. K.   213
Couch, J. V.   111
Coulbourn, J. N.   207
Coury, J. N.   145
Cowles, J. T.   9
Cox, R. D.   122, 152
Creamer, L. R.   170, 173
Creutzfeldt, O. D.   160
Crossman, E. K.   212
Cumming, W. W.   16, 19, 57, 64, 66
Cutts, D.   189

Davenport, J. W.   120
Davis, H.   57, 191, 210
Deadwyler, S. A.   188, 190, 193
Defays, D.   132
Delacour, J.   152
Deliège, M.   20, 215, 216, 218, 227
de Lorge, J.   138, 212
Deluty, M. Z.   37, 75, 76, 83
De Weeze, J.   20, 109, 130, 230
Dews, P. B.   15, 16, 18, 19, 21, 49, 50, 53, 54, 55, 56, 57, 82, 177, 198, 206, 208, 233
Dillow, P. V.   201, 207
Dinsmoor, J. A.   30, 207
Divac, I.   9, 152, 189
Djahanguiri, B.   179
Dmitriev, A. S.   8, 106
Donchin, E.   159
Dong, E. Jr   164
Donovick, P. J.   157
Doty, R. W.   155
Dougherty, J.   111, 117
Dukich, T. D.   18
Dunham, P.   100
Dunn, M. E.   33, 210, 220
Durup, G.   157
Dyrud, J. P.   120

Eckerman, D. A.   21, 53, 198
Edwards, D. D.   128
Ehrlich, A.   145
Eibergen, R.   188
Eisler, H. E.   43
Ellen, P.   23, 150, 59, 60, 61, 145, 147, 148, 150, 151, 152, 157, 166, 167, 168
Ellis, M. J.   170, 173, 175
Elsmore, T. F.   36, 37, 40, 41, 72, 74, 128, 210
Emery, F. E.   161
Emley, G. S.   130
Eskin, R. M.   87, 97
Estes, W. K.   226
Evans, H. L.   136

Falk, J. L.   93, 194, 198
Fallon, D.   114, 116
Fantino, E.   207
Farmer, J.   58, 71, 128, 200, 202, 206, 210
Farthing, G. W.   44, 72
Fechner, G. T.   42, 78
Feinberg, I.   157
Felton, M.   212
Feokritova, I. P.   12
Ferraro, D. P.   23, 29, 64, 71, 225
Ferster, C. B.   22, 57, 58, 200, 202
Fessard, A.   157
Finan, J. L.   9
Findley, J. D.   200, 201
Flaherty, C. F.   120
Flanagan, B.   224
Flanagan, S.   189
Flory, R. K.   189, 229
Flynn, J. P.   153
Fontaine, O.   121, 123, 180
Foree, D. D.   121
Forsyth, R. P.   162
Fowler, S. C.   171, 174
Fox, R. H.   183
Fraisse, P.   79, 124, 140
François, M.   182, 183
Frank, J.   128, 191, 192, 210, 211
Frey, P. W.   42
Fried, P. A.   145, 146
Fry, W.   18, 50, 200, 201, 218

Gahéry, Y.   164
Galanter, E.   38, 39, 41
Galizio, M.   207
Gamzu, E.   100, 195
Gay, P. E.   166
Gentry, W. D.   244
Gergen, I. A.   146, 151
Gibbon, J.   42, 83
Ginsburg, N.   44, 86, 97
Giurgea, C.   155
Glazer, H.   191, 192, 197, 198
Glencross, D. J.   176
Glick, S. D.   122, 152
Glickstein, M.   152
Glowa, J. R.   108, 111

Gluck, H. 157
Godefroid, J. 86, 93, 135
Gol, A. P. 151
Gold, D. C. 87, 99
Goldberg, S. R. 111, 214
Goldfarb, J. L. 175
Goldstone, M. 159
Goldstone, S. 175
Gollub, L. R. 16, 19, 20, 57, 201, 204, 212
Gonzalez, F. A. 59
Gonzalez, R. C. 87
Grabowski, J. 200, 201, 202
Granata, L. 164
Granjon, M. 161, 163, 165
Gray, V. A. 223
Green, T. R. G. 183
Greenwood, P. 28, 29, 60, 80, 108, 122, 132, 171, 172, 174, 175, 189, 191
Greer, K. 170
Grice, G. E. 166
Grier, J. B. 42
Grilly, D. M. 29, 71
Grossen, N. E. 217, 220
Grossman, K. E. 99, 107, 138
Guevrekian, L. 28, 120
Guilford, J. P. 78
Guilkey, M. 21, 211
Guttman, N. 72, 114

Hablitz, J. J. 159
Hainsworth, F. R. 184
Hake, D. F. 130
Halliday, M. S. 222
Hamilton, C. L. 183
Haney, R. R. 87, 98, 168
Harrison, R. H. 73
Harvey, G. A. 145, 148
Harvey, N. 170
Harzem, P. 23, 51, 57, 60, 68, 100, 106, 114, 115, 116, 128, 210, 211
Hatten, J. L. 210
Hauglustaine, A. 86, 88, 90
Hawkes, L. 69
Hawkes, S. R. 161
Hawkins, T. D. 112
Hearst, E. 44, 71, 72, 175, 189, 223
Hemmes, N. S. 121
Hendricks, J. 44
Hendry, D. P. 44, 200, 207
Hernandez-Peon, R. 156
Heron, W. T. 9
Herrnstein, R. J. 17, 57, 62, 128, 193
Heuchenne, C. 24
Heymans, C. 164
Hienz, R. D. 21, 53, 198
Hinde, R. A. 176
Hineline, P. 31, 32, 33
Hinrichs, J. V. 224
Hiss, R. H. 22, 128, 211
Hoagland, H. 182, 183
Hodos, W. 44, 45, 64, 189, 190, 192
Holloway, F. A. 137

Holloway, S. M. 188
Holz, W. C. 60, 64, 189, 193
Honig, W. K. 107
Hothersall, D. 147, 189, 190
Howerton, D. L. 114
Hsiao, S. 188
Hull, C. L. 9
Hunt, H. F. 145, 148
Hurwitz, H. B. M. 33, 175
Huston, J. P. 121
Hutchinson, R. R. 129, 130
Hutt, P. J. 114

Innis, N. K. 68, 212, 231
Irwin, D. A. 159
Isaacson, R. L. 125, 146, 148, 150, 151, 152
Iversen, L. L. 177
Iversen, S. D. 177

Jackson, F. B. 146, 151
Jackson, F. D. 137
Janet, P. 7
Jarrard, L. E. 145
Jasper, H. J. 157
Jenkins, H. M. 54, 55, 73, 120, 121, 195, 207, 222
Jensen, C. 114, 116
Johanson, C. E. 111, 117, 118
John, E. R. 157
Johnson, D. A. 152
Johnson, C. T. 151
Jones, B. 176
Jones, E. 161
Jwaideh, A. R. 207

Kadden, R. M. 33, 108, 109, 210
Kalish, H. I. 72
Kamp, A. 159
Kaplan, J. 153
Kapostins, E. E. 189
Kasper, P. M. 153
Katz, H. N. 207
Keehn, J. D. 188
Keele, S. W. 176
Keesey, R. E. 114, 154
Kelleher, R. T. 18, 22, 23, 50, 60, 71, 108, 109, 110, 111, 126, 129, 168, 192, 200, 201, 212, 213, 214, 218
Keller, F. S. 23, 64, 71, 121, 187, 188, 190, 193
Kello, J. E. 212, 232
Kelnhofer, M. M. 150
Kelsey, J. E. 9
Kelso, J. A. S. 170
Kendall, S. B. 200, 201, 206, 207, 219
Killam, K. F. 157
Killeen, P. 19, 50, 82, 93, 100, 212
Kinchla, J. 36, 37, 74
King, G. D. 86, 188
Kintsch, W. 166
Kisch, G. B. 217
Kleinknecht, R. A. 189, 230

Kling, J. W.   114
Knapp, H. D.   176
Knutson, J. F.   189, 230
Koch, E.   164
Kochigina, A. M.   8, 106
Kopytova, F. V.   160
Kramer, T. J.   23, 60, 64, 121, 189, 191
Kristofferson, A. B.   42, 73
Kuch, D. O.   29, 61, 62, 64, 65, 66, 217
Kulikova L. K.   160

La Barbera, J. D.   50, 210
Lacey, B. C.   162, 163, 165
Lacey, J. I.   162, 163, 165
Laming, D.   43
Landis, R.   189, 190
Lane, H.   86
Lashley, K. S.   53, 176
Laties, V. G.   26, 188, 189, 190, 191, 192, 200, 201
Latranyi, M.   28, 188
Lattal, K. A.   128, 211
Lawicka, W.   166
Le Cao, K.   191
Lee, A. E.   18
Lejeune, H.   49, 50, 73, 79, 95, 107, 118, 123, 132, 135, 152, 208, 217, 226
Levitsky, D.   188, 190
Levy, M. N.   164
Liberson, W. T.   157
Limpo, A. J.   64, 200, 201, 230
Lincé, R.   28, 188, 189, 191, 192, 196, 197
Lindsley, D. B.   155, 156, 163
Linwick, D. C.   125
Lockart, J. M.   183, 185
Logan, F. A.   9, 30, 60, 67
Lolordo, V. M.   121
Lorens, S. A.   251
Loughead, T. E.   64, 66, 74, 83, 103, 191, 192, 196
Low, M. D.   159
Lowe, C. F.   100, 106, 116, 128, 211
Lubar, J. F.   166
Luce, R. D.   38, 39, 41
Lund, C. A.   210
Lydersen, T.   189
Lyon, D. O.   212

Macar, F.   28, 30, 60, 159, 162, 200, 201, 202, 204, 227, 229
MacGrady, G. J.   189
Mackintosh, N. H.   100
Madigan, R. J.   114
Malagodi, E. F.   108, 109, 129, 212
Malone, J. C.   79, 224, 225
Malott, R. W.   64, 66, 71
Mantanus, H.   36, 37, 40, 73, 108, 118, 208, 229
Marcucella, H.   60, 64, 203, 204
Marley, E.   86, 124, 125
Marr, M. J.   20, 128, 211, 212
Maurissen, J.   64, 86, 87
McAdam, D. W.   159
McCallum, W. C.   159

McCleary, R. A.   145, 166
McFarland, D. J.   93, 198
McGill, W. J.   23
McKearney, J. W.   53, 108, 109, 129
McLeod, A.   63, 64, 210
McMillan, D. E.   29, 191, 207
McSweeney, F. K.   121
Mechner, F.   21, 28, 120, 188
Mednikova, T. U. S.   160
Meltzer, D.   114, 115
Melvin, K. B.   87, 97, 98
Mesker, D. C.   86
Migler, B.   62
Millenson, J. R.   33, 64
Miller, L.   125
Milner, D.   111, 153
Milstein, V.   157
Mintz, D. E.   171, 174, 212
Moffitt, M.   61
Mogenson, G.   112
Molliver, M. E.   29, 60
Morell, F.   157
Morgan, M. J.   211
Morse, W. H.   17, 22, 54, 57, 64, 86, 108, 109, 110, 111, 117, 119, 124, 125, 126, 188, 208
Mostofsky, D. I.   44, 71
Mott, F. W.   176
Mowrer, O. H.   166
Mulvaney, D. E.   207
Münsterberg, H.   165
Myers, L.   114
Myers, R. D.   86

Narver, R.   161
Nauta, W. J. H.   145
Neil, E.   164
Nelson, T. D.   63
Neuringer, A. J.   21, 50, 53, 128, 210, 211
Nevin, J. A.   28, 189
Newlin, R. J.   59
Nilson, V.   44
Nonneman, A. J.   152
Notterman, J. M.   171, 174
Numan, R.   152, 166

O'Hanlon, J. F.   183, 185
Olds, J.   111, 153
Ornstein, K.   121
Orsini, F.   127
Owens, J. A.   111

Palmer, J. D.   3, 138, 182
Patton, R. A.   29
Pavlov, I. P.   7, 11, 12, 131, 222, 224, 225, 226, 230, 232
Pear, J. J.   68
Pellegrino, L.   153
Pepler, R. D.   183
Perikel, J. J.   36, 37, 73
Perkins, D.   189

Persinger, M. A. 187
Peterson, J. L. 189
Peterson, W. W. 42
Pfaff, D. 183
Piaget, J. 7, 124
Pickens, R. 111, 116, 117
Pieron, H. 181
Pinotti, O. 164
Pittendrigh, C.S. 181
Platt, J. R. 29, 60, 61, 62, 65, 66, 74, 83, 191, 217
Pliskoff, S. S. 63, 64, 111, 112, 117, 154
Poole, E. W. 165
Popov, N. A. 8, 157
Popp, R. J. 218
Poulton, E. C. 169
Pouthas, V. 132, 188, 192, 196
Powell, E. W. 151, 152
Powell, R. W. 22, 87, 98, 103, 104, 106, 117, 118, 210
Provins, K. A. 183
Pubols, L. M. 145

Quesada, D. C. 171

Rabin, A. 161
Rachlin, H. 93
Randich, A. 59, 226
Randolph, J. J. 189
Ray, R. C. 23
Razran, G. 106
Reberg, D. 126, 194
Rebert, C. S. 159
Reese, E. P. 87, 97, 98
Reese, T. W. 87, 97, 98
Reinberg, A. 3
Requin, J. 161, 162, 163, 165
Rescorla, R. A. 30
Revusky, S. H. 189, 191
Reynolds, G. S. 36, 37, 38, 60, 63, 64, 71, 114, 192, 200, 201, 210, 230
Ricci, J. A. 200
Richards, R. W. 189, 230
Richardson, W. K. 23, 59, 60, 61, 64, 65, 66, 74, 83, 103, 123, 191, 192, 196, 210
Richelle, M. 7, 37, 93, 105, 107, 126, 168, 175, 178, 179, 180, 196, 222, 229
Richter, C. P. 4, 5, 106
Rilling, M. 23, 57, 60, 64, 71, 189, 191, 230
Riusech, R. 188
Rixon, R. H. 183
Roberts, S. 84
Robinson, S. 184
Rodriguqez, M. 121
Roper, T. J. 188, 191, 195
Rosenkilde, C. E. 9, 152, 166, 189
Ross, G. S. 23, 59, 157
Rowland, V. 157, 159
Rozin, P. 87, 99, 185
Rubin, H. B. 86, 93
Ruch, F. L. 9, 10
Ruchkin, D. S. 159
Ruzak, B. 6

Sachs, B. D. 188
Sams, C. F. 9
Sanger, D. 177
Saslow, C.A. 30, 192, 198
Schaub, R. E. 207
Schmaltz, L. W. 125, 146, 148, 150, 151, 152
Schmidt, R. A. 170, 171, 175
Schneider, B. A. 15, 16, 18, 19, 21, 49, 50, 53, 57, 82, 114, 128, 210
Schnelle, J. F. 166
Schoenfeld, W. N. 16, 19, 33, 57, 58, 71, 128, 200, 202, 206, 210
Schuster, C. R. 23, 59, 111, 116
Schwartz, B. 30, 60, 100, 121, 189, 195
Schwartzbaum, J. S. 145, 146, 166
Scobie, S. R. 87, 99
Scull, J. 121
Sears, G. W. 30
Segal, E. F. 177, 188, 190, 193, 200, 201, 204, 206, 219
Segal-Rechtschaffen, E. 191
Seitz, C. P. 187
Seldeen, B. L. 168, 186
Shagass, C. 157
Shanab, M. E. 189
Shepard, R. N. 41
Sherman, J. G. 57
Sherrington, C. S. 166, 176
Shettleworth, S. J. 93, 135, 193, 195
Shimp, C. P. 61, 65, 69, 204
Shull, R. L. 21, 22, 50, 128, 201, 210, 211
Sidman, M. 30, 31, 32, 33, 34, 44, 60, 71, 83, 154, 210
Siegel, R. K. 44
Silberberg, A. 123
Silby, R. 93, 198
Silverman, P. J. 201, 212, 213
Simmelhag, V. L. 93, 126, 193
Simpson, A. J. 183
Singer, G. 188
Singh, D. 191, 192, 197, 198, 218, 224
Skinner, B. F. 7, 14, 15, 22, 28, 54, 55, 57, 58, 64, 100, 114, 117, 119, 126, 166, 188, 193, 200, 202, 222, 225, 226
Skuban, W. E. 191
Slonaker, R. L. 189, 191
Smith, J. B. 110, 191
Smith, R. F. 121, 151
Snapper, A. G. 19, 20, 36, 37, 77, 188
Sokolov, E. V. 155, 161
Spence, K. W. 54, 166
Squires, N. J. 200, 201
Staddon, J. E. R. 22, 52, 64, 66, 67, 68, 69, 71, 93, 107, 114, 116, 121, 126, 128, 191, 192, 193, 195, 196, 197, 198, 210, 211, 212, 221, 231, 233, 234
Stamm, J. S. 145, 146, 148, 150, 152
Stebbins, W. C. 114, 115
Stein, N. 189, 190
Stevenson, J. A. F. 183
Stevenson, J. G. 29
Stokes, L. W. 138, 139
Stott, L. H. 10
Stretch, R. 109

Stubbs, A.   212
Stubbs, D. A.   36, 37, 39, 42, 43, 54, 73, 74, 76, 77, 83, 192, 213
Swets, J. A.   42

Tang, A. H.   111
Taub, E.   176
Tecce, J. J.   159
Terman, J. S.   111, 119, 154
Terrace, H. S.   230
Thomas, J. R.   36, 37, 200, 201
Thompson, T.   111, 116, 200, 201, 202
Thor, D. H.   135
Thorne, B. M.   151
Tierney, T. J.   63
Tinbergen, N.   260
Todd, G. E.   86, 94
Todorov, J. C.   123
Tolman, E. C.   9
Topping, J. S.   121, 168, 172, 189
Trapold, M. A.   128
Treisman, M.   41, 42, 43, 182
Trowill, J. A.   112
Trumble, G.   118

Uramoto, I.   200, 201, 202, 219
Urbain, C. A.   86, 105
Urban, F. M.   39

Valliant, P. M.   187
Van Heosen, G. W.   148
Velluti, J. C.   157
Vigier, D.   164
Villareal, J.   189, 190
Vitton, N.   159
Voronin, L. L.   160

Wagman, A. M. I.   152
Wall, A. M.   53, 55, 82, 198
Wallace, M.   161
Waller, M. B.   86
Walter, W. G.   158, 159
Wasserman, E. A.   69

Watanabe, M.   160, 161
Webb, W. B.   224
Weber   41, 42, 65, 73, 74, 83
Weinberg, H.   155, 158
Weisman, R. G.   230
Weiss, B.   25, 26, 64, 71, 189, 200, 201
Wenger, G. R.   177
Wertheim, G. A.   32
Wertheim G. W.   187
Westbrook, R. F.   121
Wetherington, C. L.   188
Wever, R.   222
Wheeler, L.   191
Wickens, D. D.   218, 224
Wilkie, D. M.   68, 223
Williams, B. A.   44, 60, 72
Williams, H.   120
Williams, D. R.   30, 120, 189
Willis, F.   114
Wilson, M. P.   23, 59, 64, 71, 188, 189, 190, 193
Winograd, E.   44
Witoslawski, J. J.   86
Witte, R. S.   166
Witter, J. A.   94
Wolach, A. H.   225
Wolf, M.   87, 99
Woodrow, H.   10
Wright, A. A.   40
Wyckoff, L. B. Jr   206

Xhignesse, L. V.   170

Yagi, B.   9

Zeier, H.   152
Zeiler, M. D.   20, 22, 55, 57, 69, 80, 126, 128, 208, 211, 212, 229
Zieske, H.   164
Zimmerman, J.   23, 30, 59, 60, 200
Ziriax, J. M.   123
Zucker, I.   6
Zuili, N.   124
Zuriff, G. E.   191, 192

Subject Index

Abrupt training 125
Acceleration 168, 186
Acceleration index (Acc) 18
Activation 146, 147, 159, 216
Active inhibition 79, 80, 81, 107, 198, 222
Activity cycles 139
Activity wheel 138, 188, 190, 194, 196–197
Adjunctive behaviour 126, 130, 194, 198, 230, 234
Adjusting method 37
Adventitious reinforcement 126, 193, 194, 195
Alpha blocking 156, 157
Alpha rhythm 156, 157
Amygdala 152, 153, 160
Anticipatory timing 169, 176
Apes 97
Arabinol cytosine 3
Archistriatum 151
Arousal 161
Attack 189, 194, 229
Attention 35–36, 107, 150, 153, 159, 162, 168, 169
Autocontingencies 210
Autocorrelation 26, 27
Automaintenance 120
Autoshaping 101, 120, 195, 200
Average response rate (ARR) 18
Aversive stimulation 9
Aversive stimulus 14, 30, 62, 129
Avoidance 14, 30, 59, 108, 129, 146, 207
    fixed cycle 32
    limited interval 33–34
    non-discriminated (Sidman) 31, 110, 157, 162, 189, 201, 218, 220, 226

Baboon 86, 97
Baroreceptors (carotid sinus) 164, 165
Bat 86
Bee 87, 99, 100, 107, 138
Behavioural clock 6, 138, 140
Behavioural contrast 230, 231
Beneficial anticipation 170, 171
Biochemical (internal) clock 182
Biological function of the reinforcer 109, 110
Birds 97–98, 124
Bisection 37, 76, 77, 78, 82, 83
Blood chemistry 187
Blood pressure 162
Break-point 19, 50, 82
Break-and-run pattern 15, 18, 57, 93, 98, 217, 233, 235
Budgerigar 44, 86, 97

Cardiac cycle 161–164
Cardiac frequency 162
Cat 28, 29, 49, 50, 64, 85, 86, 100, 107, 108, 124, 126, 130, 132, 152, 153, 159, 172, 178, 189, 201, 202, 219, 226, 234
Caudate nucleus 152
Celestial cues 138
Cellular activity 159, 160
Cellular membrane 160
Cerebral-lesions 121, 125, 143–153, 217
Cerebral-stimulation 153–155
Chaining 53, 79, 175, 196, 198
Chicken 86, 124–125
Chimpanzee 86, 97
Chronobiology 3, 4, 6, 26, 57, 82, 83, 106, 124, 135–140, 182, 197, 200, 217
Cingulum 152, 153, 160
Circadian rhythms 3, 135, 137–140, 183
Circatidal rhythms 3
Clock
    behavioural 52, 78–82, 83, 134, 140, 177, 190, 222
    biological 3, 82, 140, 161, 177, 179, 181, 197
CNS 143–161
Collateral behaviour 28, 53, 64, 79, 103, 132, 166, 168, 174–175, 178, 184, 188–199, 210, 233–234
Compensatory mechanisms (see Inhibition) 80–81, 175, 199, 233
Concept (of time) 7, 124
Conditioned inhibition 222
Conditioned reinforcement–secondary reinforcement 206–207, 216–217, 221
Conditioned reflex to time 224
Conditioned suppression (CER) 226
Constant condition 138, 139, 217
Constant stimuli (method of) 37
Contingencies of reinforcement 13, 108–110, 111, 124, 207–214
    negative 14, 30–33
    positive 13–14
Contingent negative variation (CNV) 158, 159, 160
Copulation 188
Cortex 159
Counter model 73, 74
Counting behaviour 175, 196
Covert collateral behaviour 192, 198
Critical fusion frequency 185
Crow 87, 98, 103, 104, 105
Cued DRL schedule 148, 151, 168
Cyclic interval schedule 52, 68–69, 221
Cyclic mixed DRL schedule 67
Cyclochrony 157, 158

Day, time of   135–139
Deafferented limb   176
Decision rules   38
Delayed conditioning   11, 29, 106, 153, 157, 224
Delay (critical length of in DRL)   59, 64, 132, 137, 204, 227
Delay to reinforcement hypothesis   53–54, 55
Deprivation   90, 98, 117–120, 148–149, 210
Developmental factors   124–125, 134
Differential reinforcement of low rates schedule (DRL)   22–30, 57–71, 78–82, 85, 88, 132, 133, 135, 137, 143, 157, 162, 166, 177, 188, 189, 200, 201, 203, 208, 223, 224, 229
Differential reinforcement of response duration schedule (DRRD)   28, 60, 66, 82, 108, 122, 132, 160, 171, 172, 189, 196
Differential reinforcement of response latency schedule (DRRL)   29, 60, 64, 66, 82, 201, 202, 219–220
Differential threshold   9, 10, 36, 73, 74
Discharge   222
Discrete trial DRL schedules   27–30
Discrete trial FI schedules   20, 21, 117, 208
Discrete visual tracking   169, 170, 171
Discrimination (see also Duration)   9, 12, 14, 33–45, 54–56, 63–65, 74, 79, 80, 150, 152
Discriminative stimulus   13, 20, 21, 28, 30, 36, 53, 63, 175, 200–207, 211, 221
Disinhibition   12, 13, 166, 179, 218, 224–226, 232
Displacement activities   126, 198, 234
Dog   85, 86, 106, 131, 155, 158, 159, 226
Double avoidance   9, 10
Drive   145
Drug   111, 116, 117, 118, 153, 154, 157, 168, 176–181, 190
  amphetamine   153, 157, 177
  atropine   177
  benzodiazepines   178, 179
  chlordiazepoxide   168, 178, 179
  chlorpromazine   179
  cocaine   111, 116, 118, 214
  deuterium oxide (D$_2$O)   181
  diazepam   178
  morphine   111
  neuroleptics   179
  pentobarbital   111, 157
  phenothiazines   179
  sulpiride   180
  tranquillizers   178, 179
  tremorine   179
Duration (estimation of)
  in FI   54, 55, 79
  of external events   33–45, 79, 80
  of IRTs   63, 64, 65, 79, 80
  of confinement   9, 152
  of response   79, 80
  of response latency   63, 79, 80
Dwelling time   25, 26, 27

EEG   155–160
Electrocardiogram (ECG)   161, 163–164
Electrophysiological correlates   155–160

Emotion   151, 153
Endogenous maturation   125
Environmental pressure   107
Escape   14, 108, 129
Ethology   94, 107, 126, 198
Evoked potential   158
Evolution   6, 104, 105–107
Excitation   131, 222, 233, 234
Expectancy   83
Expectation density   26
Experimental history (effects of)   110, 125, 126, 127, 128
Experimental neurosis   228, 229
External clock   103, 200–207, 230, 234
Exteroceptive cues   166, 200–221, 234
Extinction   145, 146, 190, 201, 226
Extrapyramidal system   152, 153

Factors influencing temporal regulations   108–140
Fechner's law   78
Feedback clock   200
FI-DRL paradox   78, 79, 80, 81, 108
First-half-ratio F(T/2)   18
First-order effects (or deviations)   57
Fish   87, 98, 100
Fixed interval schedule (FI)   14–22, 49, 52–56, 78–82, 85, 88, 101, 132–133, 143, 185, 188, 189, 195, 198, 200, 202, 206–207, 212–214, 223, 224, 229, 233, 234, 235
Fixed-mean-interval-schedule   19
Fixed-minimax-interval schedule   16
Fixed-minimum-interval schedule   16
Fixed minimum interval schedule (FMI)   28, 189
Fixed ratio schedule (FR)   126, 137, 187, 211–212, 229
Fixed time schedule (FT)   22, 80, 82, 93, 101, 126–130, 138, 188, 189, 193, 194, 195, 211, 234
FKC (curvature) index   18, 19, 89, 94, 116, 123, 133
Food postponement (schedule of)   109, 110
Fornix   151
Frontal lobes   152
Frustration   146, 231

Generalization gradient   37, 38, 40, 41, 44, 54, 71, 72, 73, 82, 83, 222, 223, 230
Genetic factors   134
Gerbil   86
Globus pallidus   153–154
Goldfish   87, 99, 185
Gonads   187
Gourami Fish   87, 99
Gradient of excitation   222, 223
Gradient of inhibition   222, 223
Gravitation   168
Grooming   188
Guinea pig   86, 88–90, 105

Hamster   86, 93, 135, 194
Handshaping   125
Hapalemur   86, 94–95

Heart-rate 155
Heat 183
Heavy water 181
Hippocampal seizure 153
Hippocampus 143–151, 152, 153, 156, 157, 160
Hoffman reflex 161
Horse 86
Human 135, 140, 145, 157, 159, 161, 164–165, 169–171, 181, 189, 207
Hyperactivity 145, 150
Hyperthermia 185
Hypothalamus 151, 155, 159, 160
Hypoxia 187

Individual differences 130–135, 165
Individual histort 110, 125–130, 134, 138, 146–148, 213, 234
Induction 230
Information 207
Inhibition 12, 79–81, 107, 124, 131, 143, 145–150, 152, 153, 156, 164, 175, 198, 212, 222–235
Insect 87, 99
Instrumental procedures 8, 9–10
Interim activities 93, 105, 194, 195, 198, 229, 234
Interindividual differences 130, 131, 132, 133, 134
Interresponse time (IRT) 22–23, 32, 60–65, 79, 88, 90, 92, 95, 103, 111, 119, 143, 146, 151, 153, 168–169, 177, 184, 191, 202, 208, 221, 223
Intracerebral self-stimulation 111, 112, 117, 119, 154, 155, 157
IRTs per opportunity (IRT/OP) 24, 25, 27
Isobias curve 42, 43
Isosensitivity curve 42, 43

Just noticeable difference JND 41, 73–74

Kinesthesis 166

Latent learning 197
Law of effect 193
Lesions 143–153
Limited hold (LH)
   in DRL 22, 28, 103, 203
   in FI 16
Limits (method of) 36
Lion 107
Locomotion 192
Lymbic system 144, 151, 153, 154

Major tranquillizers 179
Mamillary bodies 151
Mammals 86, 87–97
Mathematical indices 101, 133
   discrimination 38, 43
   DRL 23–27, 65
   FI 16–19
Maturation 125
Maze 9

Medial forebrain bundle 153, 154
Mediating behaviour 53, 79, 147, 175, 193, 195–198
Medullar reflexes 161
Metabolism 185–186
Microelectrode 160
Minor tranquillizers 178–181
Mixed DRL 67
Mnemonic deficit 145
Monkeys 29, 30, 36, 53, 57, 64, 85, 86, 100, 106, 108, 109, 110, 111, 117, 118, 119, 124, 129, 130, 146, 151, 152, 157, 159, 160, 162, 176, 202, 214
Motivation 83, 145, 151, 169
Motor control 80, 107, 143, 169
Motor programme hypothesis 176
Motor skills 169–171, 176
Mouse 64, 78, 86, 87–88, 100, 107, 124, 234
Movement (voluntary) 175–176
Moving average 26
Multiple schedules 58, 111, 128, 129, 230
Muscle relaxation 168

Natural behavioural repertoire 121, 124, 234
No-cue FI schedule 20, 215–217, 227–228
"Novel" stimulus 224

Observing response 36, 200, 206, 207
Occipital area 155, 160
Olfactory bulbs 151
Omission (reinforcement) 116, 120, 147, 212–214, 231–232
Ontogeny 124–125
Operant
   behaviour 193
   conditioning 7, 193
   procedures 12–48
   technical advantages 7
Optional clock 200
Oscillator 222
Ovariectomy 187

Pacemaker 222
Pause 15, 16–17, 49–52, 82, 98, 109, 116, 117, 126, 128, 143, 201, 210
Pavlovian conditioning 10–12, 29, 126, 153, 156, 160
Pentobarbital 157
Perceptual anticipatory timing 169–171
Periodic reinforcement 14
Periodicity—periodic (structure of schedules) 80, 200, 207–214, 214–217, 234
Peripheral mechanisms 161–176
Perseveration 146
Pharmacology 120, 176–181 (see Drugs)
Photic driving 157–158
Physical restraint 191, 192, 197, 198, 210, 226
Plasticity 106
Pleasure centres see Intracerebral self-stimulation
Point of subjective equality (PSE) 41
Polydipsia 105, 188, 189, 194

Post-synaptic potential   160
Potto   86, 95–97, 100, 135
Power function   66, 83
Prairie dog   86, 94
Prefrontal cortex   152
Principle of parsimony   222
Progressive interval schedule (PI)   51–52
Progressive training   125, 145, 148
Prolonged exposure
  to DRL   71, 98, 110
  to FI   57
Prosimian   86, 94–97, 100, 135
Proprioception   148, 150, 153, 165–176, 186, 218
Proprioceptive input hypothesis   170, 171, 175
Proprioceptive trace hypothesis   170, 171
Psychogenetics   134, 135
Psychology of time   6
Psychometric function   39–42, 73
Psychophysics   6, 36, 66, 101, 124
Psychotropic drugs   176–181
Pulse rate   161
Punishment   14, 129

Quail   87, 97, 100
Quarter life (Q)   17

Rabbit   86, 93, 156, 158, 160
Raccoon   86
Rate of responding   14, 23, 57, 58–64, 109, 114, 126, 129, 145, 146, 177, 202, 206, 210, 211, 231
Rate-dependency effect   177
Raven   87, 98
Reaction time   30, 158, 161, 163
Reinforcement   13, 58–64, 65, 100, 108, 110, 114, 195, 207, 212, 217, 232
Reinforcer   111–120
  drug   111, 116, 117, 118, 214
  intracerebral self-stimulation   111–112, 117, 119, 153, 154, 157
  water   111
Relative time
  and FI   49–52, 82
  and DRL   66–68, 82
  and DRLL   66, 82
  and DRRD   66, 82
  and duration discrimination   83
  and FT   60, 82
  and sequences of behaviour   69, 70
Relaxation   222
Respiratory cycle   161–165
Respondent behaviour   126
Response   37, 38, 120–124
  arbitrariness   78, 101, 120, 121, 195
  biting   130
  consumatory   121
  drinking   130, 194
  equivalence in various species   100
  force   167–168, 171–174
  head poking   121
  immobility   29
  latency   29–30, 53, 55

  operant   225
  pavlovian   225
  pecking   121, 123, 194–195, 234
  running   123
  salivary   225
  treadle pressing   121–123
  tube licking   121
Response bursts   23, 28, 32, 64, 71, 83, 103, 123, 146, 147
Response feedback   217–221
Reticular formation   159
Rhythms
  as a dimension of stimulus   44
  biological   3, 57, 106, 135–140, 200, 210, 217
ROC curve   42–43
Running rate (RR)   18
Runway   9

Satiation   117–120
Scaling   74, 75, 76, 77, 78
Scallop   15, 16, 18, 57, 90, 98, 99, 129, 155, 198, 202, 204, 210, 217, 233, 235
Schedules   see Avoidance
  cued DRL
  cyclic interval
  cyclic mixed DRL, mixed DRL
  differential reinforcement of low rate (DRL)
  differential reinforcement of response duration (DRRD)
  differential reinforcement of response latency (DRRL)
  discrete trial DRL
  discrete trial FI
  Fixed Interval (FI)
  Fixed-mean-interval
  fixed-minimax-interval
  fixed-minimum-interval
  Fixed minimum interval (FMI)
  fixed ratio (FR)
  fixed time (FT)
  food postponement
  no-cue FI
  progressive Interval (PI)
  Second order
  Titration
  variable interval (VI)
  waiting
  yoked and multiple
Second-order effects (or deviations)   56
Second-order schedules   109, 111, 212–214, 217
Secondary reinforcement–conditioned reinforcement   121, 206–207, 216–217, 221
Sensitization   157
Septal rage   145
Septum   143–151, 153, 155, 157, 166
Sequential analysis (of IRTs)   25, 26, 28, 70, 71, 169
Sex   135, 136, 187
Shock   108–110, 112, 129, 155, 211, 214
Shock delay procedure   31
Shock deletion procedure   32, 210, 220
Sigmoid gyrus   155
Signal detection theory   42–43, 74, 119

Simultaneous delays   68, 106
Single unit   160
Sleep   157, 196
Species-specific (characters)   60, 64, 78, 85–107, 110, 118, 121, 128, 135, 234
Species-specific defence reactions   121
DRL control (specificity of)   60–64
Spectral analysis   26, 172
State-dependent learning   137
Stereotyped behaviour   64, 155, 188, 193, 195
Stimulus   36, 212, 214–217
Stimulus control   148
Stimulus (negative)   222
Striate cortex   155
Subcallosal cortex   145
Subjective equality   41
Superstitious behaviour   22, 126, 193, 194, 195
Suprasegmental properties
    DRL   70–71
    FI   56–57
Synchronizer   3, 138, 139, 210, 217, 234

$t(1)$ 16
$t(4)$ 17
Tail nibbling   188
Temperature (body)   181–186
Temporal conditioning (Pavlovian)   11, 22, 79, 126, 157, 158, 160, 224
Temporal discrimination   see discrimination duration
Temporal information   200–221
Temporal regulation
    of the subject's own behaviour   14–33, 49–71, 108
    required   14, 22–30, 57–71, 108, 217, 233
    spontaneous   14–22, 32, 49–57, 108, 217, 233
Tendinous reflex   161

Tension   222, 232, 235
Terminal activities   194, 195
Thalamus   148, 152, 153, 159
Thêta rhythm   156, 157
Thirst   145
Threshold (differential)   36, 73–74
Tilapia   87, 98, 100
Time-out   20, 229
Time (of day)   135–140
Titration schedule   37
Trace conditioning   11, 29, 153, 157, 224
Two response procedures   20–21, 27–28, 29, 37, 60, 76, 120, 122, 162, 172, 190, 201, 227
Typology   131, 133, 226

Variable interval   58–60, 148, 153, 177, 207, 226
    interval percentile schedule   61, 64
    paced VI   61
    percentile reinforcement schedule   61
    with DRL   58, 189
Velocity (movement)   44, 45
Visceral pacemakers   161–165
Visual tracking   169
Vultur   86

Waiting schedule   28, 29, 30
Water intake   145
Weber fraction   41, 65, 73, 74, 83
Weber's law   41, 42, 65, 73, 74, 82, 83
Weber model   73, 74

Yoked schedules   58–60

Zona incerta   152